COMPARATIVE VERTEBRATE ANATOMY

A LABORATORY DISSECTION GUIDE

COMPARATIVE VERTEBRATE ANATOMY

A LABORATORY DISSECTION GUIDE

KENNETH V. KARDONG, Ph. D.
Washington State University

EDWARD J. ZALISKO, Ph. D.
Blackburn College

With considerable artistic contributions by
FAYE C. PREVEDELL

WCB
McGraw-Hill

Boston, Massachusetts Burr Ridge, Illinois Dubuque, Iowa
Madison, Wisconsin New York, New York San Francisco, California St. Louis, Missouri

WCB/McGraw-Hill

A Division of The **McGraw-Hill** *Companies*

TITLE: COMPARATIVE VERTEBRATE ANATOMY

This book is printed on recycled, acid-free paper containing 10% postconsumer waste.

1 2 3 4 5 6 7 8 9 0 QPD/QPD 9 0 9 8 7

Cover image: © John Cancalosi/Tom Stack & Associates

ISBN 0-697-37879-9

Editorial director: *Kevin T. Kane*
Publisher: *Michael D. Lange*
Sponsoring editor: *Margaret J. Kemp*
Developmental editor: *Kathleen R. Loewenberg*
Marketing manager: *Thomas C. Lyon*
Senior project manager: *Kay J. Brimeyer*
Production supervisor: *Laura Fuller*
Designer: *Katherine Farmer*
Art editor: *Jodi K. Banowetz*
Compositor: *Carlisle Communications, Ltd.*
Typeface: *10/12 Times Roman*
Printer: *Quebecor Printing Book Group/Dubuque, IA*

Zalisko, Edward J.
Comparative vertebrate anatomy: a laboratory dissection guide / Edward J. Zalisko, Kenneth V. Kardong ; with considerable artistic contributions by Faye C. Prevedell. -- 1st ed.
p. cm.
Includes index.
ISBN 0–697–37879–9
1. Vertebrates--Dissection--Laboratory manuals. 2. Anatomy, Comparative--Laboratory manuals. I. Kardong, Kenneth V. II. Title.
[DNLM: 1. Vertebrates--anatomy & histology--laboratory manuals. 2. Dissection--laboratory manuals. 3. Anatomy, Comparative--laboratory manuals. 4. Vertebrates--anatomy & histology--atlases. 5. Anatomy, Comparative--atlases. 6. Dissection--atlases. QL 812.5 Z22c 1998]
QL812.5.Z35 1998
571.3'16--dc21
DNLM/DLC
for Library of Congress 97-27278
CIP

www.mhhe.com

"For Willemina, Kyle, and Jason,
and for Amy, Ben, and Sarah"

Contents

Preface

One of the most effective ways to master a subject, and make it one's own, is to meet it directly by engaging the material hands-on. A laboratory, accompanying a course in comparative vertebrate anatomy, provides such direct, practical experience. Students are encouraged to become active participants in their own mastery and understanding of vertebrate design. Certainly, there is no single "vertebrate morphology," and no single way to run a laboratory. But a course on vertebrate anatomy includes modern expectations in a discipline that should be a centerpiece of a university curriculum in organismal biology. First and foremost, the basic animal architecture, the anatomy, should be understood in a clear and concise way. Second, the significance of the discovered anatomy must be considered, beyond the business of putting names on parts.

Students need to understand that the anatomy raises questions about how the systems work and holds clues to the remarkable story of how vertebrates evolved. An understanding of these form-and-function relationships and the evolutionary history gives meaning and significance to the structure that is studied.

With this philosophy in mind, we have prepared a laboratory manual that sets the foundations and tone for instructors who wish to provide a sound laboratory course that suits their needs within a modern university curriculum. First, we have written a text that carefully guides students through dissections, and is richly illustrated. This brings together under one cover both a dissection guide and a pictorial atlas. Students must work through the written text to discover the anatomical parts for themselves, but the integrated figures allow them to do so in an efficient and economical manner. We expect that, in addition to reading the text and studying the figures, students engaged in active dissections should make a few simple sketches for their own use.

Second, we include at selected points in the text, short boxes that note patterns and connections, and form and function relationships. As students complete a section of the text, these boxes invite them to pause and consider the larger implications of the anatomy just explored. These boxes are not intended to be complete in themselves. Instead, they help set the tone for the laboratory. While the student is actively engaged in dissection, they are encouraged to consider the significance of what they find. What is the functional significance of the positions, patterns, shapes, and interrelationships of the anatomy under dissection? What might this suggest about the evolution of characters that define derived groups of vertebrates? These questions help correlate the laboratory exercises and the lectures. The instructor can use the lectures to more fully develop these larger conceptual issues related to the function and evolution of vertebrates.

Further, we have included features of design and organization that make the text more accessible for students, and less confusing about what they are to do.

To help students understand the context in which the dissections are undertaken, each chapter and some chapter sections, begin with a general introduction to the anatomy. With this before them, the text then carefully guides students through the material, using precise anatomical terminology to verify identifications. To best distinguish between the background text and specific dissection directions:

1) The dissection directions are set off from the textual narrative by a beginning arrow and indentation,
2) Terminology and nomenclature are differentiated so that the student may easily refer to them,
3) Common synonyms are indicated in parentheses,
4) Chapters are written to permit flexibility in the course sequence.

Each chapter can be studied as a distinct unit, which gives instructors the flexibility to depart from our sequence and build a laboratory that suits their needs. However, we believe the sequence of chapters in this manual is a reasonable compromise of practical and conceptual goals. For example, we begin with a traditional classification, thereby introducing the representative animals. We follow this classification with a chapter on the protochordates, thus introducing the historical background to the vertebrates. We have placed the examination of external anatomy just before the muscular systems. This is a matter of convenience. Most laboratories begin with "dry" material, then move later to "wet" preserved specimens. Dragging out a wet, preserved cat, shark, *Necturus,* or other vertebrate during the first days of a course is much less convenient. Instead, bringing these preserved animals out just prior to the muscle dissections is an economical use of the instructor's time and makes a nice transition from dry to wet dissections for students.

5) A flexible depth of coverage is used in the text.

In most chapters, instructors will find a depth of coverage that is sufficient for a traditional course, yet structured to allow less detailed coverage where desired.

6) Form and Function Boxes emphasize functional relationships.

This manual was written to support and aid a course in vertebrate anatomy that focuses upon system function and evolution. These boxes promote this perspective by allowing students to reflect upon the functional significance of the anatomy just examined.

7) Patterns and Connections Boxes indicate broad conceptual connections.

These short boxes invite students to reflect upon patterns of vertebrate design and reinforce concepts likely presented in lecture.

Acknowledgments

In addition to the support we have received from family, friends, and colleagues, we would like to express our appreciation for the constructive advice offered by reviewers William A. Schutt, Jr., Bloomfield College; Alexander J. Werth, Hampden-Sydney College; and Michael H. Kessner, Indiana University of Pennsylvania. We also appreciate the enduring support of many associates at Wm. C. Brown, including Marge Kemp, Kathy Loewenberg, Tom Lyon, and the production staff. Finally, we would especially like to thank Faye Prevedell, Julie Brown, and Hazen Audel for their artistic contributions. Special recognition goes to T. L. Kiene, whose personal and professional support have been so important and appreciated during the preparation of this laboratory manual.

Introduction

1

Classification and Comparison

Comparative vertebrate anatomy examines the structure of vertebrates classified in the phylum Chordata, subphylum Vertebrata. To best understand the history of vertebrate evolution, we need to examine representative forms, such as those indicated in figure 1.1. Unfortunately, detailed anatomy, especially of key fossils, is not always known or easily studied. Instead, we must focus on convenient representative animals that exhibit primitive and derived traits. For example, this laboratory manual provides for a detailed dissection of a shark, a salamander, and a cat. These three animals generally represent a cartilaginous fish, an amphibian, and a mammal, respectively. However, these modern animals have evolved, sometimes considerably, since their ancestors first appeared hundreds of millions of years ago. Therefore, as we examine these living representatives, we must remember that they are also specialized for current lifestyles.

Although only three animals will be dissected in detail, many more animal groups will be examined in the protochordate, integument, and skeletal system chapters (chapters 2–5). Thus, you should learn a chordate classification scheme (table 1.1) and be able to use it. We suggest that for each animal you study, you know its traditional classification from the phylum down to class (when available). In addition, you should be able to completely classify the three animals that are dissected in detail (table 1.2).

Defining the Chordates

Chordates are neither the most diverse nor the largest of the animal phyla, although in terms of the number of species, they come in a respectable fourth behind arthropods, nematodes, and molluscs. Chordates have a fluid-filled internal body cavity termed a **coelom.** Chordates, along with other animals possessing such a coelom, are grouped together as the **coelomates.** Among these coelomate animals, two apparently distinct and independent evolutionary lines are present. One line of coelomates, called the **protostomes,** includes molluscs, annelids, and arthropods together with assorted smaller phyla (figure 1.2). The other line, the **deuterostomes,** includes several other small phyla, plus echinoderms, hemichordates, and chordates. The distinction between protostomes and deuterostomes rests upon certain embryological characteristics (table 1.3; figure 1.3). On the basis of these criteria, chordates are classified as deuterostomes: the mouth forms opposite to the blastopore; generally, cleavage is radial; the coelom is an enterocoelom; and the skeleton arises from mesodermal layers of the embryo.

We should be clear from the beginning, however, about the character of the chordate phylum itself. It is easy to forget that two of the three chordate subphyla are technically *invertebrates*. The phylum Chordata traditionally divides into three subphyla of unequal size: the Cephalochordata, commonly represented by amphioxus; the Urochordata, commonly known as sea squirts or tunicates; and the largest group, the Vertebrata, or vertebrates. Strictly speaking, the "invertebrates" include all animals except members of this last subphylum, the "vertebrates."

The earliest chordate fossils appear in the Cambrian period. Although later chordates would evolve hard bones and teeth that preserved well, ancestors to the first chordates likely had soft bodies. Early chordates left little fossil trace of the evolutionary pathway taken from prechordate to chordate. Thus, to decipher chordate origins we use evidence from anatomical clues carried in the bodies of living forms. To be able to evaluate the success of these attempts at tracing chordate origins, we first need to identify the features that define the chordates. We will then attempt to discover the animal group that is the most likely evolutionary source for chordates.

Chordate Characteristics

At first glance, the differences between the three chordate subphyla are more apparent than the similarities that unite them. Most vertebrates have an endoskeleton, a system of rigid internal elements, of bone or cartilage, beneath the skin. The endoskeleton participates in locomotion, support, and protection of delicate organs. Some chordates are terrestrial, and most use jaws to feed. But cephalochordates and urochordates are all marine animals, none are terrestrial, and all lack a bony or cartilaginous skeleton, although their support system may involve rods of collagenous material. Cephalochordates and urochordates are suspension feeders, using a sticky sheet of mucus to collect food from streams of water passing over a filtering apparatus. All three subphyla, despite these superficial differences, share a common body design similar in five fundamental features: a notochord, pharyngeal slits, endostyle or thyroid gland, a dorsal, hollow nerve tube, and a postanal tail (figure 1.4). These five features unite all chordates and, taken together, distinguish them from all other phyla.

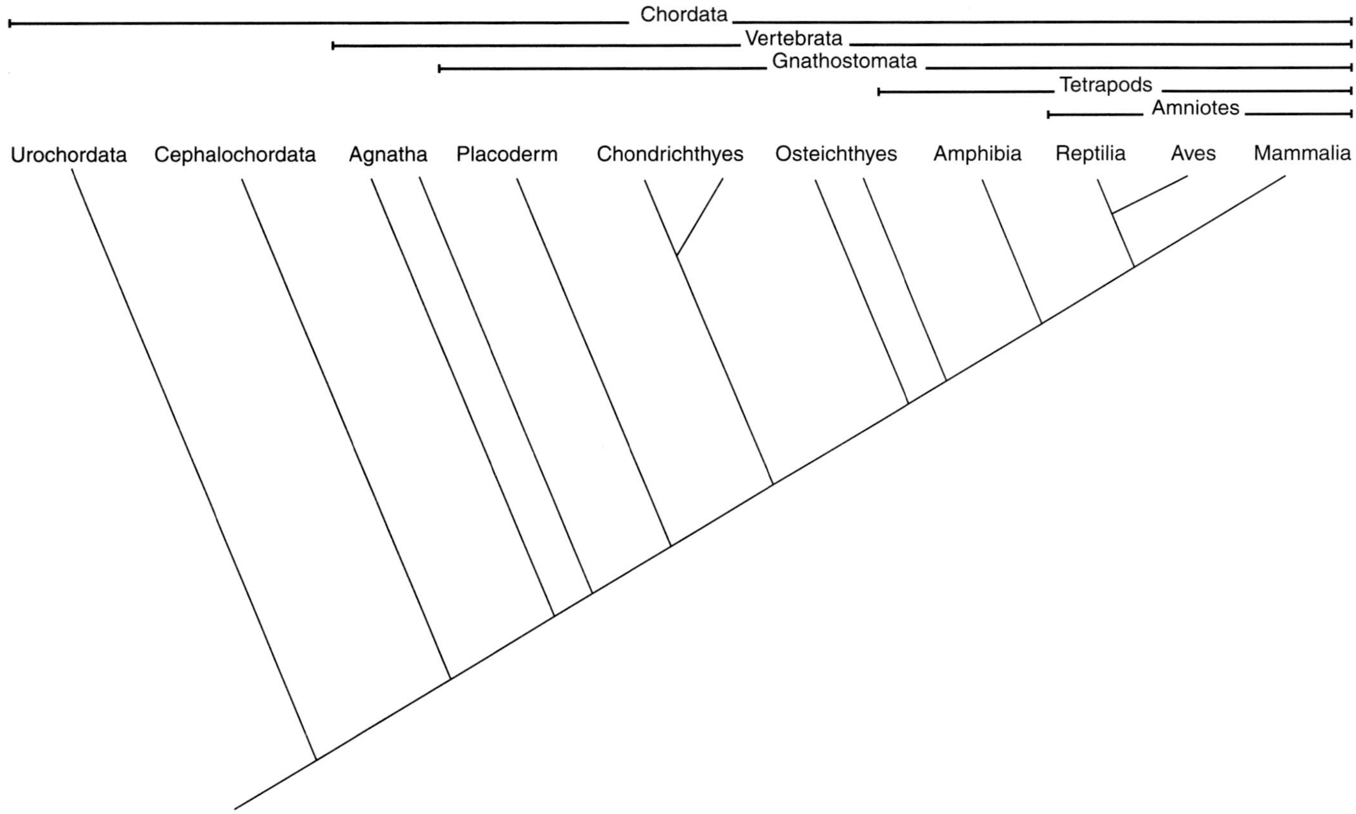

Figure 1.1
Chordates. One possible evolutionary sequence is shown using traditional names. An understanding of these groups reveals how vertebrates changed through time. But because many of these early groups do not survive or are not easily accessible, we study selected living vertebrates which we hope provide a reasonable introduction into these evolutionary events. Students must remember that these modern examples may have changed significantly since their groups first evolved.

Table 1.1

Traditional Classification

Phylum Chordata
- Subphylum Urochordata (tunicates)
- Subphylum Cephalochordata (amphioxus/*Branchiostoma*)
- Subphylum Vertebrata (extant groups)
  - Class Myxini (hagfish)
  - Class Cephalaspidomorpha (lampreys)
  - Class Placodermi
  - Class Chondrichthyes
    - Subclass Elasmobranchii (sharks and rays)
    - Subclass Holocephali (chimaeras)
  - Class Osteichthyes
    - Subclass Actinopterygii (paddlefish, sturgeon, gar, bowfin, bass, tuna etc.)
    - Subclass Sarcopterygii (lungfish and lobe-finned fishes)
  - Class Amphibia (frogs, salamanders, and caecilians)
  - Class Reptilia
    - Subclass Anapsida (turtles)
    - Subclass Diapsida (snakes, lizards, tuatara, and crocodilians)
  - Class Aves
  - Class Mammalia
    - Subclass Prototheria (monotremes: duckbill platypus and echidna)
    - Subclass Theria
      - Infraclass Metatheria (marsupials: kangaroo, opossum etc.)
      - Infraclass Eutheria (placentals: rodents, apes, cats etc.)

Table 1.2

Common Name	Dogfish Shark	Mudpuppy	Domestic Cat
Subphylum	Vertebrata	Vertebrata	Vertebrata
Class	Chondrichthyes	Amphibia	Mammalia
Subclass	Elasmobranchii	Lissamphibia	Theria
Order	Selachimorpha	Urodela	Carnivora
Family	Squalidae	Proteidae	Felidae
Genus	*Squalus*	*Necturus*	*Felis*
Species	*acanthias*	*maculosus*	*domestica*

Notochord

The notochord is a slender rod that lies dorsal to the coelom but beneath and parallel with the central nervous system. The phylum takes its name, Chordata, from this structure. The notochord is typically composed of a core of cells and fluid that are encased in a tough sheath of fibrous tissue. The notochord is a hydrostatic organ with elastic properties that resist axial compression. It lies along the body axis to allow lateral flexion but prevents collapse of the body during locomotion.

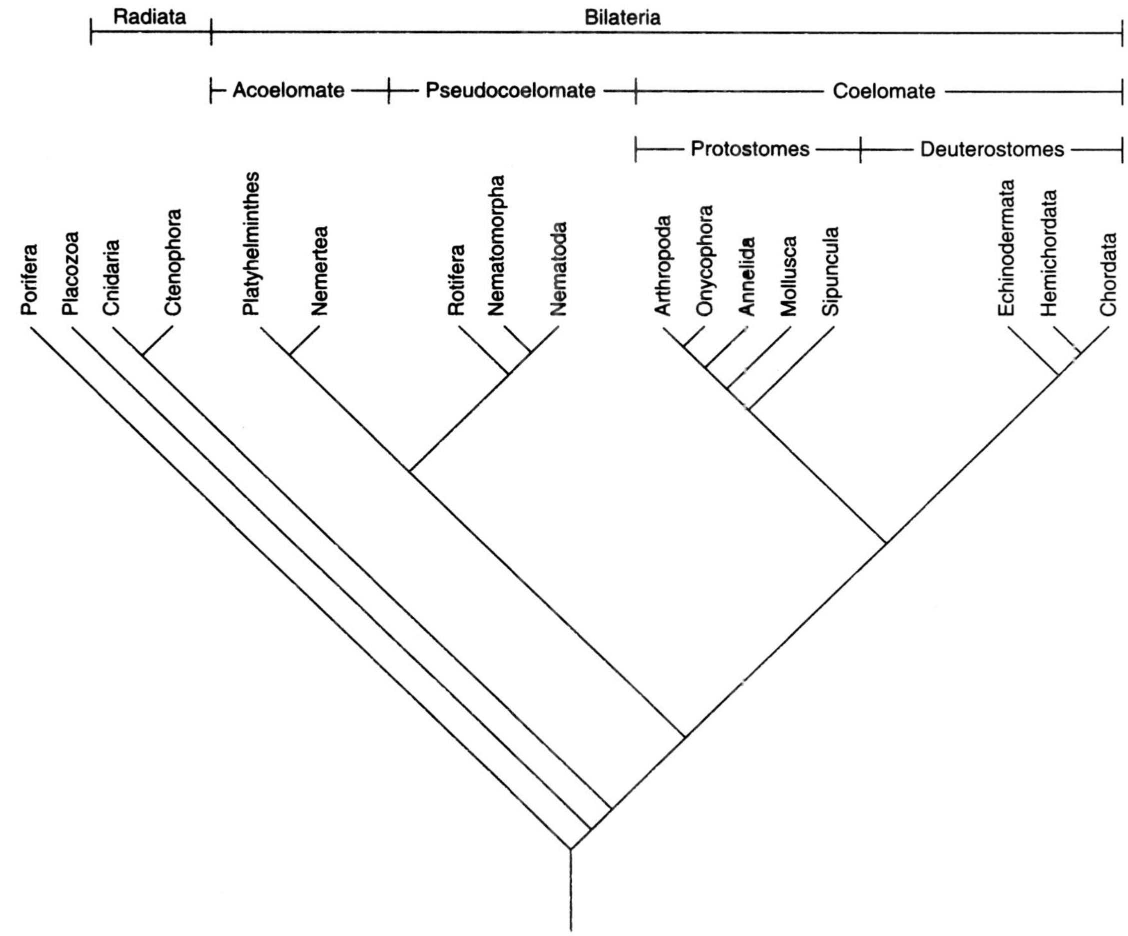

Figure 1.2
Diagram of the relationships of the major metazoan groups. The coelomate body plan and deuterostome form of development separates the echinoderms, hemichordates, and chordates from all other phyla.

Table 1.3

Protostomes	**Deuterostomes**
Blastopore (mouth)	Blastopore (anus)
Spiral cleavage	Radial cleavage
Schizocoelic coelom	Enterocoelic coelom
Ectodermal skeleton	Mesodermal skeleton

Pharyngeal Slits

The **pharynx** is a part of the digestive tract located immediately posterior to the mouth. During some point in the lifetime of all chordates, the walls of the pharynx are pierced, or nearly pierced, by a longitudinal series of openings, the **pharyngeal slits.** The term "gill slits" is often used in place of pharyngeal slits for each of these openings in the wall of the pharynx, but a "gill" proper is a specialized structure composed of tiny plates or folds that harbor capillary beds for respiration in water. In many primitive chordates these openings serve primarily in feeding, but in embryos they play no respiratory role. Therefore, the term "gill slits" is misleading.

Pharyngeal slits may appear early in embryonic development and persist into the adult stage; or pharyngeal slits may be overgrown and disappear before the young chordate is born or hatched. Whatever their eventual embryonic or adult fate, all chordates show evidence of pharyngeal slits at some time in their lives.

When slits first arose, they likely aided feeding. As openings in the pharynx, they allowed the one-way flow of a water current: in at the mouth and out through the pharyngeal slits (figure 1.4). Secondarily, when the walls defining the slits became associated with gills, the passing stream of water also participated in respiratory exchange with the blood circulating in the capillary beds of these gills. The water entering the mouth could bring suspended food and oxygen to the animal. As it exited through the slits and across the gills, carbon dioxide was given up to the departing water and carried away. The current of water can thus simultaneously support feeding and respiratory activities.

In primitive chordates, the pharynx itself is often expanded into a pharyngeal or branchial basket, and the slits on its walls are multiplied in number, increasing the surface area exposed to the passing current of water. Sticky mucus lining the

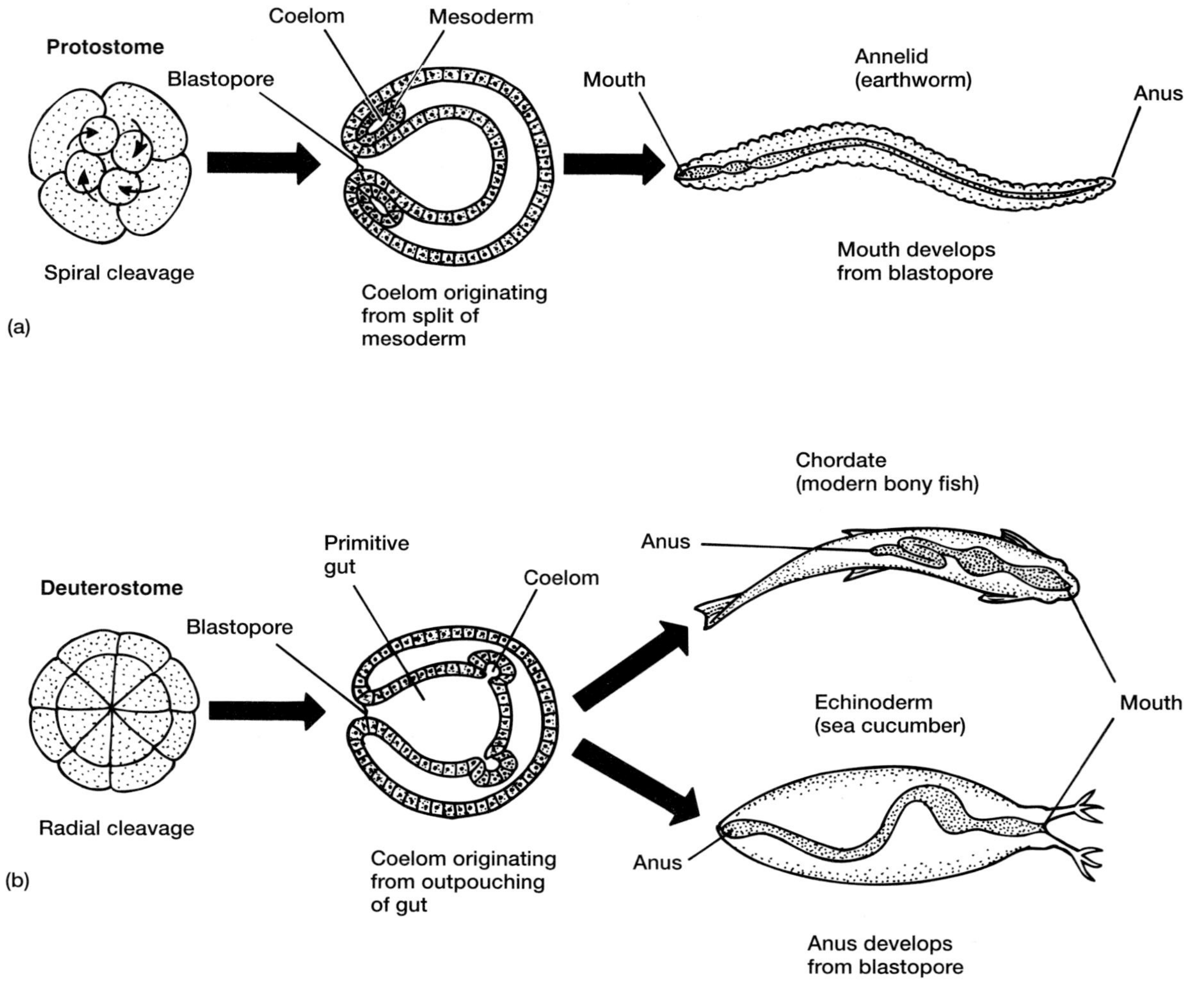

Figure 1.3
Protostomes and Deuterostomes. The coelomates are divided into two major groups on the basis of embryonic characteristics. (a) Protostomes show spiral cleavage, coelom formation by splitting of mesoderm, and derivation of the mouth from the blastopore. (b) Deuterostomes show radial cleavage, an enterocoelom, and derivation of the anus from or in the vicinity of the blastopore.

pharynx adheres to food particles from suspension. Sets of cilia, also lining the pharynx, produce the water current. Other cilia gather the food-laden mucus and pass it into the esophagus. This mucus-and-cilia system is especially efficient in small organisms that are **suspension feeders,** those that extract food floating in water. Such a ciliary mucus feeding system is prevalent in primitive chordates and in groups that preceded them.

Endostyle or Thyroid Gland

The thyroid is involved in iodine metabolism, and so is the endostyle, leading some to suggest that one (endostyle) is the predecessor of the other (thyroid). If so accepted, then all chordates (but not hemichordates) have endostyles (urochordates, cephalochordates, larval lamprey) or thyroids (adult lamprey, all other vertebrates).

Dorsal and Tubular Nerve Cord

The nerve cord in chordates lies above the gut and is hollow along its entire length (figure 1.4a,c). The major nerve cord in most invertebrates is ventral in position, below the gut, and solid (figure 1.4b). The advantage, if any, of a tubular rather than a solid nerve cord is not known, but this distinctive feature is found only among chordates.

Postanal Tail

Chordates possess a postanal tail that represents a posterior elongation of the body beyond the anus. The tail is primarily an extension of the chordate locomotor apparatus composed of segmental musculature and a notochord. The tail provides an increased surface area important for thrust and maneuverability.

Chordate Body Plan

What is common, then, to all chordates are these primary features: notochord, pharyngeal slits, thyroid gland or related structure, dorsal and tubular nerve cord, and postanal tail. These may be present only briefly during embryonic development, or they may persist into the adult stage. But, all

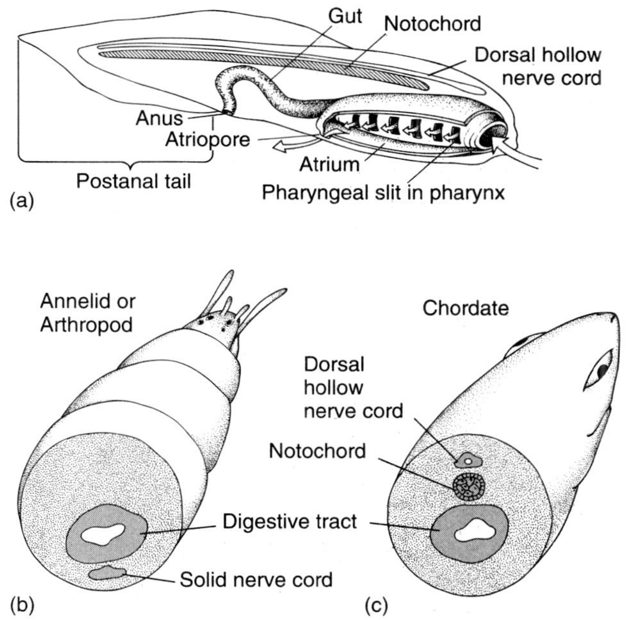

Figure 1.4
Basic body plans of chordates and nonchordates. (a) Generalized chordate. A single stream of water enters the mouth into the pharynx, then exits through several pharyngeal slits. In many lower chordates, water exiting through the slits enters a common enclosing chamber, the atrium, before returning to the environment via the single atriopore. (b) Basic body plan of an annelid or arthropod. In such animals, a definitive nerve cord, when present, is ventral in position, below the gut. (c) Basic chordate body plan. The nerve cord of chordates lies in a dorsal position above the gut and notochord. Its core is "hollow," or more correctly, a fluid-filled central canal.

chordates exhibit these distinctive characteristics at some poin during their lifetimes. Taken together, they are a suite of characters found only among chordates. Chordates are also bilaterally symmetrical and show segmentation. Blocks of muscle, or **myomeres,** are arranged sequentially along the body and tail as part of the outer body wall.

Now that we have described the basic and secondary characteristics of chordates, we turn our attention to the ancestry of this group. Biologists interested in such questions often consult an assortment of primitive animals that in some ways seem in their structure and design to have preceded chordates phylogenetically and in other ways seem to represent chordates at their earliest stages. These animals are the protochordates, and they will be addressed next in chapter 2.

Studying Advice

Laboratory Strategy

Association, correspondence between parts, and facility with descriptive terminology are as important to learn as the names of structures. In reading through the written descriptions, you are encouraged to learn the parts more by their association with one another than just as a picture atlas. For example, the certainty in identifying a blood vessel comes from confirming the region it serves, a muscle from noting its attachments, a bone from its shape and position, and so on. These features are less obvious or less likely to be noted if you rely on pictures rather than working through, and hence actually thinking through, a specimen described in words.

Uncertainty, impatience, and a lust for speed in completing an assignment will tempt you to rely more on picturebooks than upon your own thinking resources. Yes, the material can be learned by brute force, but such knowledge vanishes quickly and your skill in becoming conversant and comfortable with descriptive anatomy will remain rudimentary. We thus urge you to work with and hence think through the written descriptions.

Approaches

Two approaches to descriptive laboratory anatomy are presently in vogue. One is "Regional Anatomy," wherein a region of the body, for example the forelimb, is dissected and the bones, muscles, blood vessels, nerves, and other structures simultaneously examined. The advantages are that you are introduced to everything on one dissecting pass, and philosophically the integrated nature of structures is stressed. The second approach is a "Systems Anatomy." Each system (skeletal, muscular, digestive etc.) is examined separately. Our text takes this approach. Systems are treated as entities rather than as integrated with one another. That is a disadvantage. Special responsibility thus lies with each student to add systems functionally together as each new one is learned. Knowledge and understanding should accumulate, not displace each other. The advantage to a systems approach is that it facilitates and focuses comparisons between representative animals. Making comparisons and understanding the adaptive reasons for changes so revealed are the major goals of the course.

In the laboratory manual, "left," "right," "dorsal," "ventral," and other directional words refer *to the animal's* left, right, and so on, not to yours as you peer down on the specimen. The planes of the body are illustrated in figure 1.5.

Study Tips

Read the assigned section of the laboratory manual before each laboratory. This will give you an overview, help organize your work, and introduce you to the structures to be located.

Make a few, simple drawings after completing a section. The drawing should be uncomplicated. Delicate shading and subtle coloring should be avoided. The drawing should be an outline of the structure: forelimb bones, jaw muscles, branches of the dorsal aorta, and so on. These drawings will not be checked or graded, but are a personal reference for you and a personal check to see if you have identified structures in a way that makes sense to you. To make a simple, but accurate drawing, you must have done a good dissecting job and intellectually have some idea of the topography discovered. Don't overdo it. Don't make lots of drawings; just a few simple, quick, but accurate ones for your use and benefit.

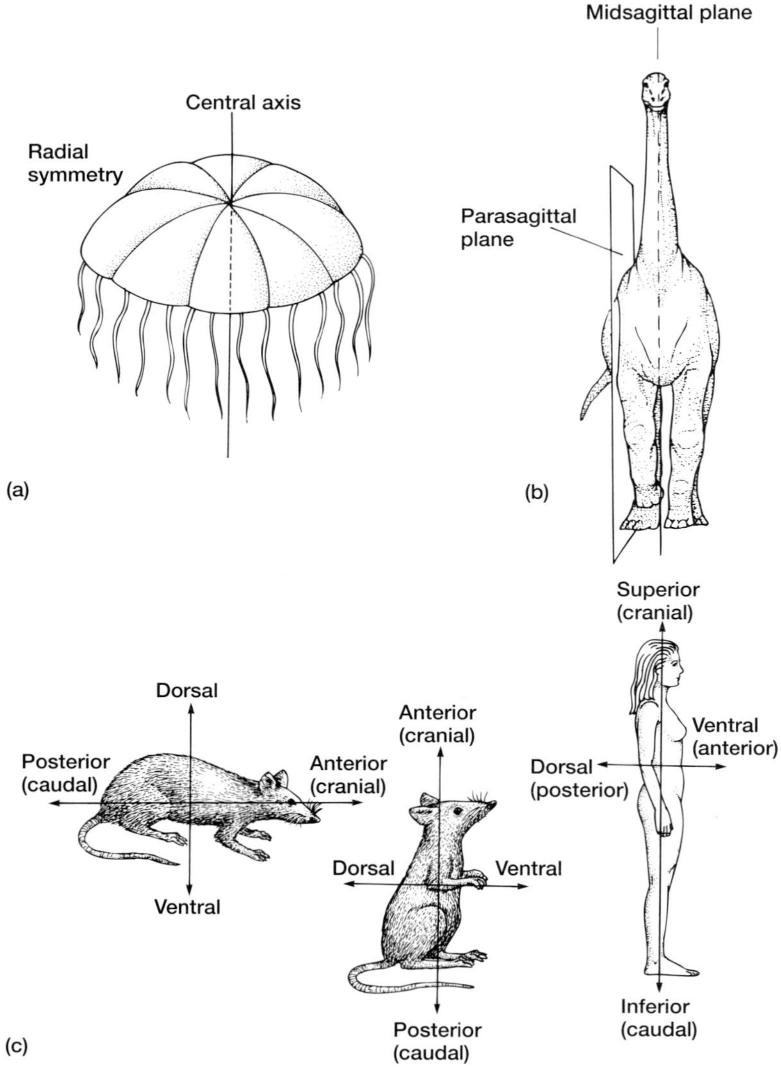

Figure 1.5
Body symmetry. Radial and bilateral are the two most common body symmetries. (a) Radially symmetrical bodies are laid out regularly around a central axis. (b) Bilaterally symmetrical bodies can be divided into mirror images only through the midsagittal plane. (c) Dorsal and ventral refer to back and belly, respectively, anterior and posterior to cranial and caudal ends, respectively. In animals that move in an upright position (e.g., humans and birds), superior and inferior apply to head (cranial) and posterior (caudal) ends.

The most effective studying occurs frequently and for relatively short periods of time (an hour or less). Just as athletes practice their sports on a regular schedule, and not just the day before a game, reviewing material and mastering terminology comes not with a sudden burst of effort just prior to an exam. Instead, regular and steady progress, with frequent breaks to rest the mind, yields the greatest depth of retention and recall.

Designing for Students

As a student, the style of your participation in a laboratory differs from how you approach the lecture. But keep in mind that the subject is the same. Lab and lecture are simply different methods to the same goal, namely to becoming conversant with the descriptive terminology and to understanding vertebrate form, function, and evolution, To help set the conceptual context in which you undertake dissections, each chapter and some chapter sections begin with a general introduction. The actual examination of specimens and identification of parts then follows. To distinguish between background material and specific dissection directions, arrows and indented text are used. This is intended to make it clear as to what you are to read as background and what you are to do as part of the actual dissection. The structures feasible to locate on the specimens are underlined; important terms are boldface. Take a moment to scan a few of the later chapters and notice how this design feature of the laboratory manual is used.

Protochordates

2

Protochordates

Between the vertebrates and other invertebrates lie the "protochordates," which usually includes three groups: the hemichordates, urochordates, and cephalochordates. Although they are an informal assemblage of animals and are not a proper taxonomic group, protochordates share some or all five features of the fundamental chordate body plan. Living protochordates are themselves products of a long evolutionary history, independent from other phyla. Their anatomy is relatively simple, and their phylogenetic position ancient. The hope is that within living members of the protochordates are traces of the evolutionary steps from prechordate to chordate. A closer look at these protochordates will indicate why they provide tantalizing clues to the origin of the chordates.

All protochordates are marine and feed by means of cilia and mucus. But, as young larvae they often live quite different lives than they do as adults. As larvae, they may be **pelagic,** living in open water between the surface and the bottom. Although unattached, most free-floating larvae have limited locomotor capability and are therefore **planktonic,** being moved from place to place primarily by currents and tides rather than by their own efforts. But as adults, they are usually **benthic,** living on or within a bottom marine substrate. Some **burrow** into the substrate or are attached to it—that is, they are **sessile.** Some adults are **solitary,** living alone. Others are **colonial** and live together in associated groups. Some are **dioecious** (literally, "two houses"), with male and female gonads in separate individuals; others are **monoecious** ("one house") with both male and female gonads in one individual.

Phylum: Hemichordata

Members of this phylum are marine "worms" with some links to chordates and other links to echinoderms. They share with chordates unmistakable pharyngeal slits (figure 2.1a) and similar embryonic origin of the nerve cord. Although typically solid, parts of the nerve cord (figure 2.1b) may be tubular in some species. However, hemichordates lack a notochord and postanal tail, hence the name "hemi-" or half-chordates. As larvae, some pass through a small, planktonic stage, the **tornaria larva** (figure 2.1c). This planktonic larva is equipped with ciliated bands on its surface and a simple gut. The ciliated structure, simple digestive system, and planktonic lifestyle of the tornaria larva resembles the larva of echinoderms. In fact, some scientists argue that the similarities testify to a phylogenetic link between hemichordates and echinoderms.

Further, hemichordates, like both echinoderms and chordates, are deuterostomes. The mouth forms opposite to the embryonic blastopore, and hemichordates exhibit the characteristic deuterostome patterns of embryonic cleavage and coelom formation. The similarities of hemichordates to the larval design of echinoderms on the one hand, and adult chordates on the other, are tantalizing. Perhaps hemichordates stand close to the evolutionary route taken by early chordates and still hold clues to the evolutionary origin of the chordate body plan. But, it must be remembered that living hemichordates are themselves millions of years departed from the actual ancestors they might share with early prechordates. Their own evolution has dealt them specialized structures serving sedentary habits.

Within the hemichordates are two taxonomic groups: the **enteropneusts**, burrowing forms examined below, and the **pterobranchs** (figure 2.2a–b), usually sessile forms not discussed further.

Class: Enteropneusta—Examination of an "Acorn Worm"

→ Examine the anatomy of an adult acorn worm, usually embedded in a plastic block. The enteropneusts, or acorn worms, are marine animals of shallow waters. Some species reach over a meter in length, but most are shorter than this. They live in burrows lined by mucus and have a body with three regions—proboscis, collar, and trunk—each with its own coelom (figure 2.1b). Identify these three regions of the body.

The proboscis, used in both locomotion and feeding, includes a muscular outer wall that encloses a fluid-filled coelomic space. Muscular control over the shape of the proboscis gives the animal a useful probe to shape a tunnel or inflate against the walls of the burrow to anchor the body in place. Tucked away in their burrows, many species ingest loosened sediment and extract the organic material. After passing through their simple gut, the remaining material is expelled out the anus as a casting, deposited on the surface of the substrate. Acorn worms are fragile, so when collected they commonly break. If this specimen is complete, the *anus* (figure 2.1b) will be located at the terminus of the trunk.

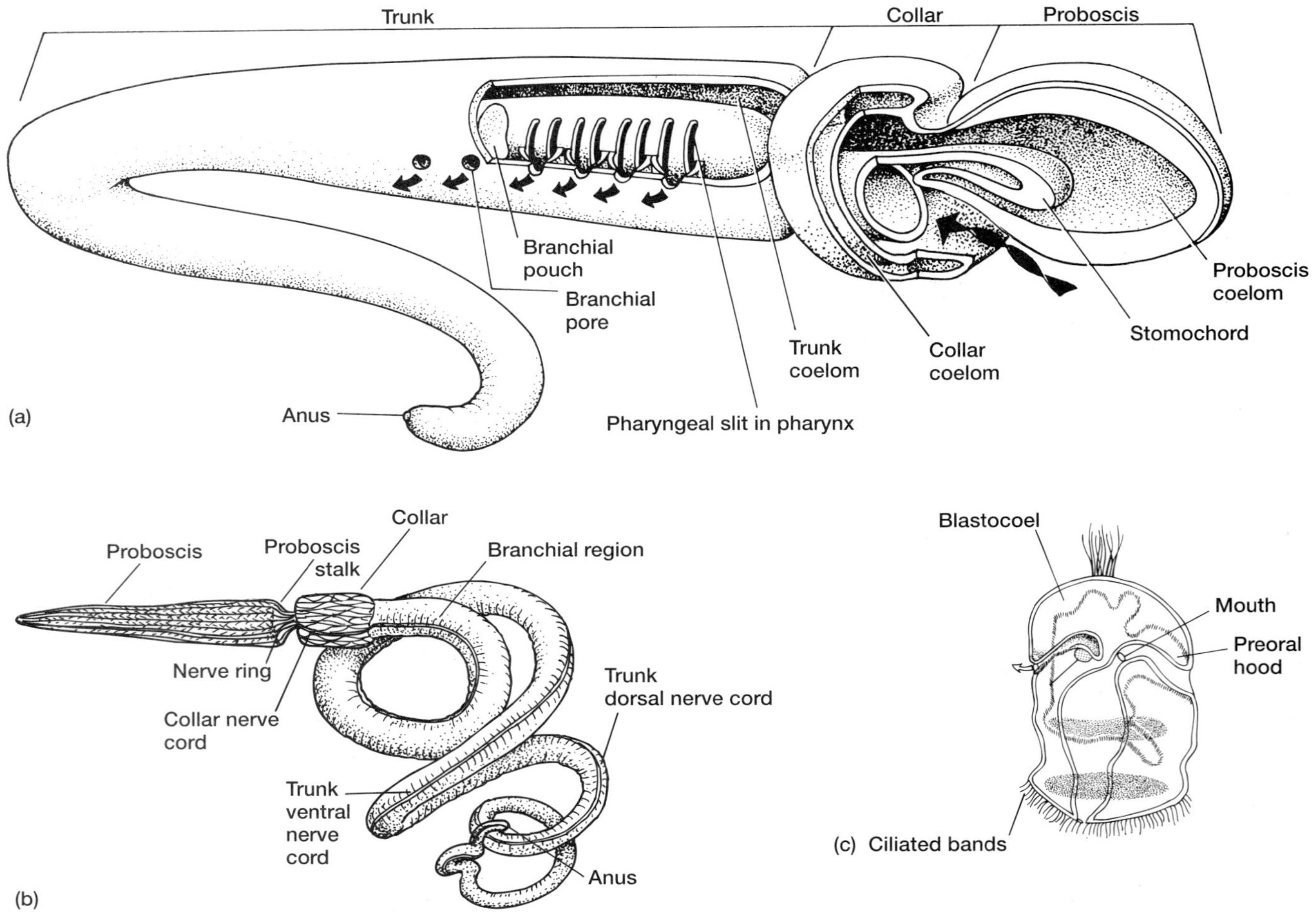

Figure 2.1
Hemichordata, Class Enteropneusta. (a) Generalized acorn worm. Proboscis, collar, and trunk regions are shown in partial cutaway view, revealing the coelom in each and the associated internal anatomy. Within the proboscis is the stomochord, an extension of the gut. The food-laden cord of mucus (spiral arrow) enters the mouth together with water. The food is directed through the pharynx into the gut. Excess water exits via the pharyngeal slits. Several slits open into a common compartment, the pharyngeal pouch, that in turn opens to the environment by a pharyngeal pore. (b) General structure of the acorn worm, *Saccoglossus*. (c) Hemichordate, generalized tornaria larva.
(b) Source: G. Stiasny, 1910, "Zur kenntnis der lebenweise von Balanoglossus clavigerus," Zoolisches Anzeiger 35, Gustav Fischer Verlag.

Other species are suspension feeders, extracting tiny bits of organic material and plankton directly from the water. In these forms, the synchronous beating of cilia on the outer surface of the proboscis sets up water currents that flow across its mucous surface (figure 2.3). Suspended materials adhering to the mucus on the proboscis are swept along ciliary tracks to the mouth. The muscular lip of the collar can be drawn over the mouth to reject or sort larger food particles. The excess water that enters the mouth exits through the pharyngeal slits (figure 2.3).

The **stomochord** (figure 2.1a) arises embryologically as an outpocketing from the roof of the embryonic gut anterior to the pharynx. In the adult, it retains a narrow connection to what becomes the buccal cavity but usually enlarges as it projects forward as a preoral diverticulum into the cavity of the proboscis. The surface of the stomochord is associated with components of the vascular and excretory systems. Its walls consist of epithelial cells, like those of the buccal cavity, as well as ciliated and glandular cells. Its interior is hollow and communicates with the space of the buccal cavity. Excretion probably occurs partly through the skin, but acorn worms also possess a dense network of blood vessels termed the **glomerulus**, within the proboscis, the presumed excretory organ.

→ A dissection scope may be necessary to see the following detail. Numerous slits are located along the lateral walls of the pharynx. Sets of adjacent pharyngeal slits open into a common chamber, the branchial pouch, which in turn pierces the outer body wall to form an undivided exit opening, the branchial pore (figure 2.1a). Excess water departing from the pharynx thus passes first through a slit, then through one of several branchial pouches, and finally through a branchial pore to reach the outside.

Other Systems The circulatory system is represented by two principal vessels: a **dorsal** and a **ventral blood vessel** not easily distinguished in whole specimens. The blood, which contains few cells and lacks pigment, is propelled by muscular pulsations in the major blood vessels. The nervous system

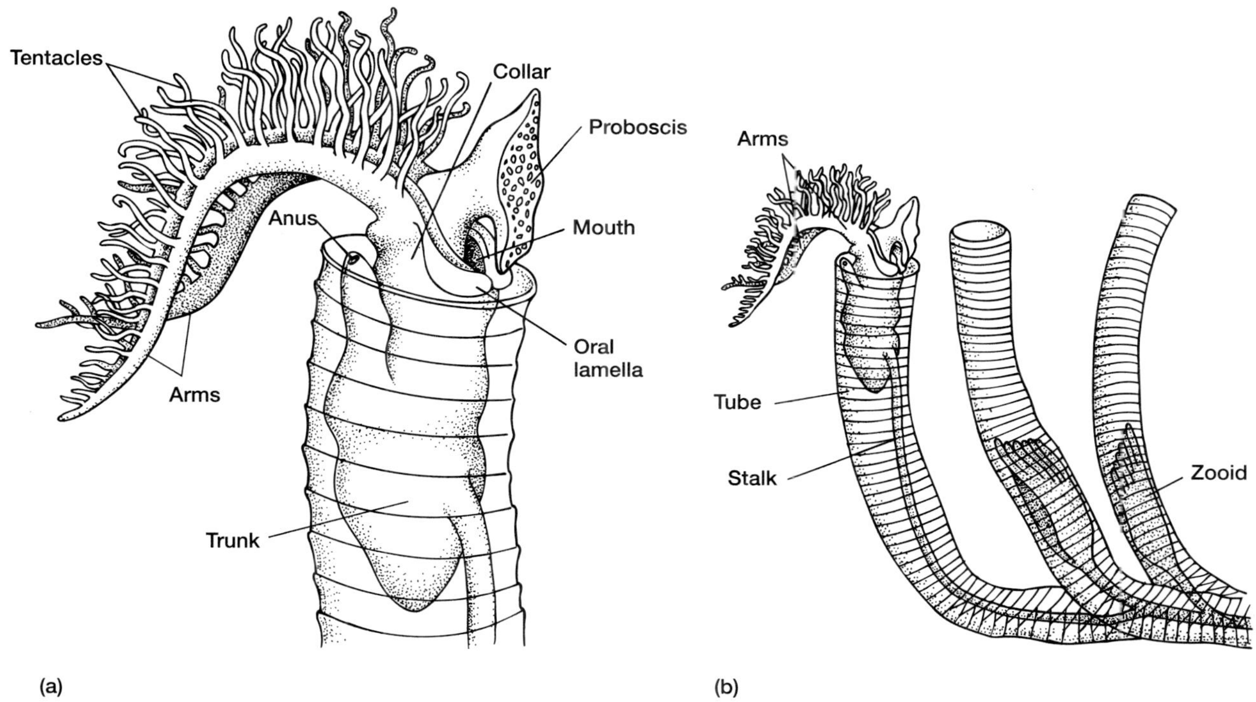

Figure 2.2
Hemichordata, Class Pterobranchia. (a) The sessile pterobranch, *Rhabdopleura*. Note that it is built upon the same body plan as acorn worms—proboscis, collar, trunk—but these are modified, and the whole animal lives in a tube. (b) Pterobranchs in tubes. When disturbed, the stalk shortens to pull them inside to safety. When merged into a colony, each contributing individual is often termed a zooid.
After Dawydoff.

consists mainly of a diffuse network of nerve fibers at the base of the epidermis (figure 2.1b). Dorsally and ventrally, the **nerve network** is consolidated into longitudinal nerve cords joined by nervous interconnections. Often this **collar nerve cord** is partially tubular. The method of embryonic origin from dorsal ectoderm and retention of a tubular structure in a few species suggests homology to the dorsal and tubular nerve cord of chordates. The gonads are housed in the trunk; the sexes are separate, thus dioecious; fertilization is external.

Subphylum: Urochordata

At some point in their life histories, urochordates generally show all five defining chordate characteristics: **notochord, pharyngeal slits, tubular nerve cord, endostyle** or **thyroid gland,** and **postanal tail.** Consequently, they are proper chordates traditionally classified as a subphylum within the phylum Chordata. Urochordates are specialists at feeding on suspended matter, especially very tiny particulate plankton. In some, the pharynx is expanded into a complex straining apparatus, the **branchial basket.** In other species, the filtering apparatus includes mucus that is secreted by the epidermis and surrounds the animal. There are about 2000 species, all of which are marine.

Urochordates are divided into three taxonomic classes. Larvae of the class Ascidiacea are free swimming until they become attached to firm rock surfaces, piles, or floats and metamorphose into sessile adults. Members of the classes Larvacea and Thaliacea are permanently pelagic. Urochordate literally means "tail back-string," a reference to the notochord. The more common name, **tunicates,** is inspired by the flexible outer body cover, the **tunic,** secreted primarily by the underlying epidermis. This tunic (sometimes called a *test*), characterizes members of the largest class, Ascidiacea, the only class we examine further.

Class: Ascidiacea—"Sea Squirts"

Ascidians, or sea squirts, are marine animals that are often brightly colored. Some species are solitary, others colonial. Larvae are planktonic, but adults are sessile.

Larva The larva, sometimes called the **ascidian tadpole,** does not feed during its short sojourn of a few days as a free-living member of the plankton. Instead, the larva disperses to and selects the site at which it will undergo a dramatic metamorphosis and take up a permanent residence as an adult. Only the larval stage exhibits all four chordate characteristics simultaneously (figure 2.4a).

→ Under a compound microscope, examine the mounted larval specimen to identify four of the main chordate characteristics. Note the swollen body and thin tail (1). The tubular nerve cord (2), and notochord (3) which supports it, both extend from the body to the tail (figure 2.4a). The branchial basket, an enlargement of the pharynx, is recognized by the presence of tiny, regular

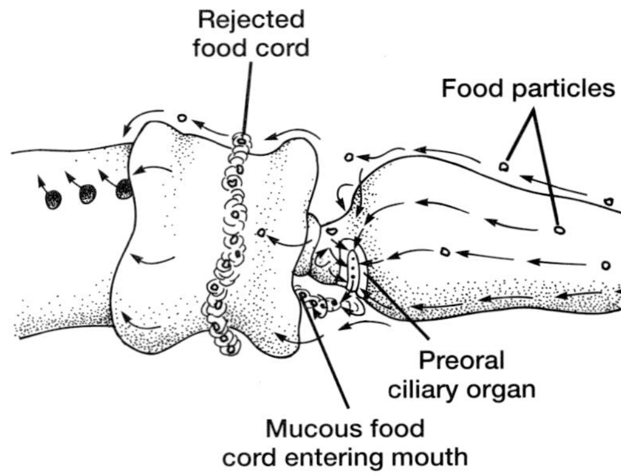

Figure 2.3
Ciliary mucus feeding. The direction of movement of food and mucus is indicated by arrows. Food material, carried along in the current produced by surface cilia, travels across the proboscis and is directed to the mouth where it is captured in mucus and swallowed. Rejected food material collects in a band around the collar and is shed.
After Burdon-Jones.

rows of perforations, the pharyngeal slits (4) (figure 2.4a). Water is drawn in through an incurrent siphon and expelled through the excurrent siphon. The **anus** empties internally into the excurrent siphon and thus does not clearly mark the point beyond which the tail continues.

Other features can be identified. Several adhesive papillae on the anterior end of the body are involved in eventual attachment of the larva when it settles down (figure 2.4a). Usually two very black spots are present: the tiny, marble-shaped spot, the otolith, is a sensor of gravity; the irregular black spot, the ocellus, is a light-sensitive eyespot (figure 2.4a).

Other organs are present but difficult to identify on most slides.

Metamorphosis of larva to the adult Most of the chordate features that debuted in the larva, namely the **notochord, tail,** and **dorsal and tubular nerve cord,** disappear in the forming adult (figure 2.4b). Although the pharynx persists and even expands, it becomes highly modified. The slits in its walls proliferate and each subdivides repeatedly, producing smaller openings, termed **stigmata.** This radically remodeled pharynx forms the barrel-shaped **branchial basket** (expanded pharynx plus numerous stigmata) of the adult sea squirt (figure 2.4c).

Adult The tunic, composed of a unique protein, **tunicin,** a polysaccharide similar to plant cellulose, forms the body wall. The tunic and epidermis attach the base of the animal to a secure substrate and enclose the branchial basket, the viscera, and a large atrial cavity (figure 2.4c). **Incurrent** (branchial) and **excurrent** (atrial) **siphons** form entrance and exit portals for the stream of water that circulates through the body of the tunicate (figure 2.4c). Particulate matter is extracted from the passing stream of water by a netlike sheet of mucus lining the branchial basket. Rows of cilia forming a **ciliated band** deliver the food-laden mucus to the **stomach,** which in turn conveys it to the **intestine** (figure 2.4c). The current of water sieves through the **stigmata** and passes out of the branchial basket to the space between basket and tunic, the **atrium** (figure 2.4c). From here it exits the animal via the excurrent siphon.

On adult specimens that are mounted in plastic and not cleared and stained, it is only practical to see the two *siphons* (one incurrent, the other excurrent) and the tough external covering, the tunic.

The nervous system consists of a **cerebral ganglion** located between the siphons (figure 2.4c). From each end of the ganglion arise nerves that pass to the siphons, gills, and visceral organs. Contractions of bands of smooth muscle running the length of the body and encircling the siphons bring about changes in the shape and size of the adult.

Blood in a tunicate is propelled to the organs and tunic by contractions of a tubular heart. After a few minutes, the flow reverses to return blood along the same vessels to the heart. No specialized excretory organ has been found in tunicates.

Tunicates reproduce sexually and asexually. Both sexes occur in the same individual (they are thus monoecious), although self-fertilization occurs only in some species. Asexual reproduction, known only in colonial species, involves **budding**. The rootlike **stolons** of the base may fragment into pieces that produce more individuals, or buds may arise along blood vessels or viscera.

Subphylum: Cephalochordata

Cephalochordates are clearly chordates, built upon the characteristic pattern that includes pharyngeal slits, dorsal and tubular nerve cord, notochord, and postanal tail (figure 2.5a). Yet they are anatomically simple, with an approach to food gathering we have seen in other protochordates, namely suspension feeding based upon a pharyngeal filtering apparatus involving an atrium. Amphioxus[1], or the lancelet, is the most commonly studied cephalochordate, with a thin and elongate body extending only a few centimeters long. It prefers coarse sediments in coastal waters and lagoons, well aerated by tides but not churned by heavy wave action. Adults can swim but usually live buried with the front of their head protruding into the water.

Cephalochordate Systems

The general structure of cephalochordate body systems will first be described. Then you will be directed to identify specific structures in whole mounts and representative cross sections.

[1]The term *Branchiostoma* wins claim to the official scientific generic name for most members of this subphylum, because this name was bestowed, alas, two years before the now better known and more widely used term *Amphioxus.* We thus use *Branchiostoma* as the Latin name of the genus, but in a spirit of charity will use "amphioxus" (without italics, or capital) as the common name.

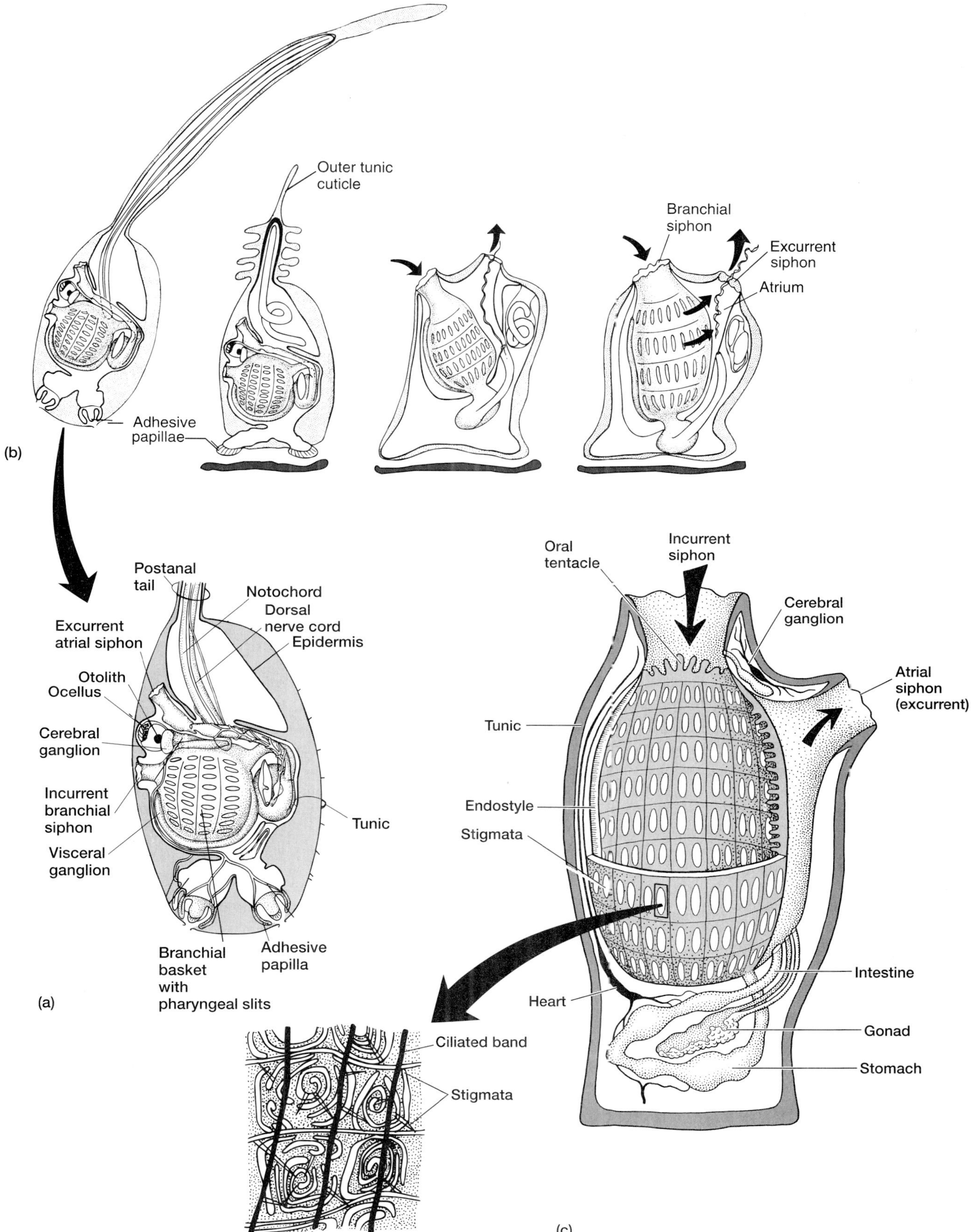

Figure 2.4

Urochordate, Class Ascidiacea. (a) Larval head of the ascidian *Distaplia occidentalis*. (b) Metamorphosis of ascidian larva *Distaplia*. Left to right, the planktonic, nonfeeding larva settles and attaches to a substrate. Adhesive papillae hold the arva, the tail contracts, and the outer cuticle of the tunic is shed. (c) Generalized adult urochordate. The flow of water is shown as it enters the incurrent siphon and passes through modified pharyngeal slits (stigmata) into the atrium and out the excurrent siphon. At the right, several of the highly subdivided pharyngeal slits are shown.

(a) Based on the research of R. A. Cloney.

Filter Feeding The **oral hood** encloses the anterior entrance to the pharynx and supports an assortment of food-processing equipment (figure 2.5b,c). Projecting from the free edge of the oral hood are **buccal cirri** (figure 2.5b) that prevent entrance of large particles. The inside walls of the oral hood hold ciliated tracts that sweep food particles into the mouth. The coordinated motion of these cilia gives the impression of rotation and inspired the name **wheel organ** for these tracts (figure 2.5b,c). The posterior wall of the oral hood is defined by the **velum,** a partial diaphragm, that supports short sensory **velar tentacles** (figure 2.5b).

Major ciliated food corridors line the pharynx. The ventral channel is the **endostyle** (figure 2.5b, 2.6), the dorsal channel is the **epibranchial groove** (figure 2.6), and the inside edges of the primary and secondary pharyngeal bars carry **ciliary tracts** (figure 2.6). One of these dorsal tracts, usually located below the right side of the notochord, bears a ciliated invagination known as **Hatschek's pit** or **Hatschek's groove** (figure 2.5b,c) that secretes mucus to help collect food particles. It is an invagination in the roof of the oral cavity, a similarity shared with the vertebrate **pituitary gland**, part of which forms similarly by invagination from the roof of the buccal cavity. This has led some to propose that Hatschek's groove may have an endocrine function like the pituitary gland.

Mucus, secreted by the endostyle and secretory cells of the pharyngeal bars, is driven up the walls of the pharynx by cilia. Food particles adhere to the mucus, which is collectively swept dorsally to form a thread in the epibranchial groove and then conveyed to the gut. Slits in the walls of the expanded pharynx allow exit of a one-way feeding current driven by cilia. Excess water passes out these **pharyngeal slits** (figure 2.5b) to the atrium and finally departs posteriorly via the single **atriopore** (figure 2.5a).

Digestive System Parts of the cephalochordate digestive system may be precursors to vertebrate organs. For instance, the **endostyle** of amphioxus collects iodine, as does the pharyngeal endocrine gland of vertebrates, the thyroid gland. The midgut cecum, a forward extension of the gut, is thought by some to be a forerunner of both the liver (because of its position and blood supply) and of the pancreas (because cells in its walls secrete digestive enzymes). Whatever their phylogenetic fate, these and other parts of amphioxus are structurally distinct from vertebrate organs, a reflection of the specialized demands of suspension feeding.

Nervous System and Muscles The tubular nerve cord of amphioxus does not enlarge anteriorly into a differentiated brain. Muscles along the body used for swimming and burrowing make contact with the spinal cord in a unique fashion. Most vertebrates have motor nerves that reach out peripherally from the spinal cord to the muscles. But in amphioxus, thin processes of the muscles reach centrally to the surface of the spinal cord.

Unique Notochord Unlike notochords of other chordates, the notochord of amphioxus consists of a transversely arranged series of striated muscle cells. The developmental origin of the cephalochordate notochord is from the roof of the embryonic gastrocoel, as in most other chordates. But the muscle cells of the cephalochordate notochord are unique and set it apart from all other protochordates and vertebrates. When these muscle cells of the notochord contract, the tough notochordal sheath prevents ballooning, so internal pressure rises, and the notochord stiffens. Stiffening may strengthen burrowing, or it could increase the intrinsic vibration rate of the animal to aid fast swimming.

Circulatory System Blood flow and circulation in amphioxus occurs in the same general pattern as in vertebrates. However, in two major respects, circulation in amphioxus departs from that of vertebrates. First, there is no heart. A swelling at the confluence of the returning veins is, because of its location in the circulation, generally termed a **sinus venosus** (a vertebrate heart chamber), but it lacks any pulsations. Instead, the **hepatic vein, ventral aorta** (endostylar artery), and others pump blood using specialized **myoepithelial cells** in their walls. The second difference from the vertebrate circulatory system occurs in the pharynx. Here, several parallel vessels travel in each pharyngeal arch from the ventral to the dorsal aorta, rather than the single aorta typical of vertebrates (figure 2.6). Furthermore, each primary pharyngeal bar contains a narrow extension of the body cavity, the coelomic space (figure 2.6).

Examination of Amphioxus Structure

→ **External Anatomy (Whole Mounts) (Figure 2.5a)** In a whole mounted specimen, note the head with oral hood ending in projecting buccal cirri, the tail with expanded caudal fin, and the body with a single dorsal fin running lengthwise along its dorsal surface. Also notice repeated v-shaped blocks of muscle, or myomeres, that are arranged sequentially along the body and tail as part of the outer body wall. Myomeres are also connected to the notochord.

→ **Internal Anatomy (Whole Mounts) (Figure 2.5 a,b)** Also on this whole mount, note the long notochord beginning well into the head and continuing to the base of the caudal fin. Above the notochord and parallel with it is the dorsal and tubular nerve cord. Within the oral hood can be seen fingerlike projections of the wheel organ actually applied to the inner surface of the oral hood. The velum, bearing short velar tentacles, is an incomplete partition at the back of the oral hood, but from this lateral view is seen as a line defining the posterior boundary of the oral hood. Behind the velum begins an extensive region, the pharynx, pierced by pharyngeal slits. Midventrally the pharynx bears an endostyle and middorsally an epibranchial groove, both seen to best advantage in cross section.

Posteriorly the pharynx constricts into an esophagus that continues into the gut itself divisible into a forward projecting midgut cecum, and caudally into the

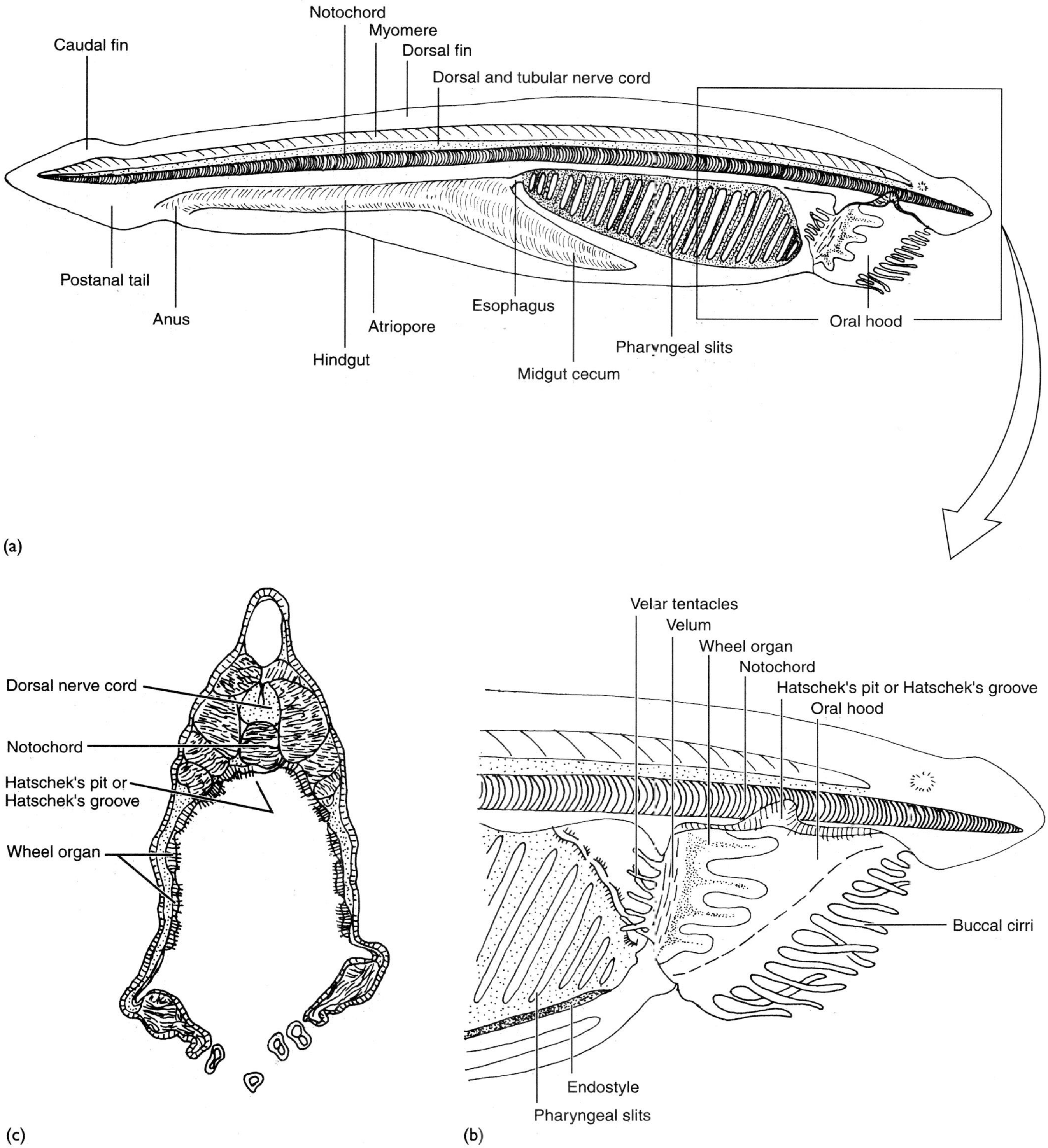

Figure 2.5
Cephalochordate, amphioxus. (a) Lateral view. (b) Enlargement of oral and pharyngeal regions. (c) Cross section through oral hood.

hindgut that empties through the anus near the base of the caudal fin. About halfway between pharynx and anus is the atriopore that allows escape of water that streams over branchial arches, into the atrium, and exits via the atriopore.

Specimens used to prepare whole mounts are usually small and sexually immature, making the identification of the gonads difficult. The gonads of sexually mature specimens are found along the ventral edge of the body (figure 2.6).

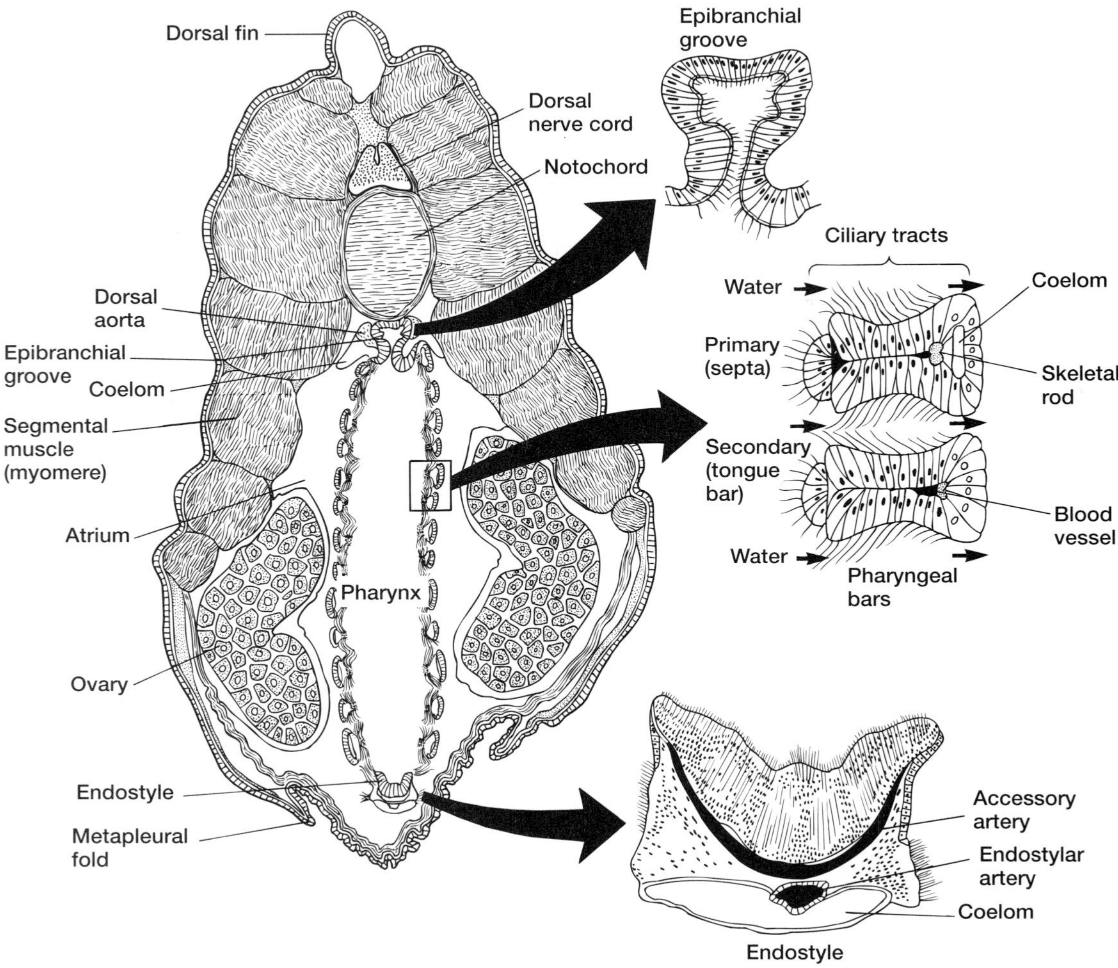

Figure 2.6
Amphioxus in cross section. The slanted pharyngeal bars encircle the pharynx. Enlarged at right are individual pharyngeal bars cut transversely at right angles to their long axis. The coelom continues into some of the branchial bars.
After Smith; Moller and Philpott; Baskin and Detmers.

→ **Representative cross sections (Figures 2.5c and 2.6)** The smallest cross section is through the anterior end of the body (oral hood region, figure 2.5c). Other cross sections are from selected sites more posteriorly. Examine at low magnification the cross section which passes through the pharynx (figure 2.6). The notochord lies in about the center of this section; note its structure, outer sheath of fibrous tissue enclosing irregular material (cells), and spaces (fluid filled in life). Again notice the notochord below the nerve cord. Below the notochord lies the pharynx; slanting branchial arches are cut in cross section and so define the pharynx with an oval series of represented arches. In cross section, the epibranchial groove forms the middorsal, and the endostyle the midventral channel of the pharynx.

Examine two adjacent branchial arches at higher magnification. In each, the core of pink tissue is the pharyngeal bar, the space between arches the pharyngeal slit. Slightly darken the field by closing the substage diaphragm. This should help show the surface cilia on each arch that project into the gill slit. (A magnification near 4003 may be required, but check with your instructor before using oil immersion, probably the highest magnification on your microscope.)

Open the substage diaphragm, return to low magnification, and move to the last cross section taken from farther posteriorly in the amphioxus body. Reidentify structures already mentioned: notochord, dorsal and tubular nerve cord, and pharynx (branchial arches, epibranchial groove, endostyle). Next to the pharynx lies the tubular midgut cecum, and next to it, but still within the body cavity, the gonad (testis or ovary depending upon the sex of the individual).

Finally, identify in cross section the myomeres, metapleural folds, and the dorsal fin (figure 2.6).

Overview

Although unsettled and controversial in its specifics, the origin of chordates lies somewhere among the invertebrates, a transition occurring in remote Proterozoic times. Within the

Patterns & Connections

Larval and adult protochordates typically have different structures and lifestyles. For example, although protochordates possess some (hemichordates) or all (urochordates, cephalochordates) five characteristics that define a chordate—notochord, pharyngeal slits, dorsal and tubular nerve cord, endostyle or thyroid gland, and postanal tail—these features often occur in only the larva or adult. In addition, adults are usually benthic, living on the ocean bottom, while their larval stages are planktonic, floating unattached in water. Yet despite these differences in structure and lifestyle, most protochordates share a common mode of feeding: extracting suspended particles from a stream of water moved by cilia. These food particles are collected on sheets of mucus and directed to the gut. Excess water drawn in with the food is diverted to the outside of the body through lateral pharyngeal slits.

The structures associated with filter feeding often occupy a significant portion of the body, up to about half the total body surface in cephalochordates. The composition of these filtering devices is typically complex and delicate. It is therefore no surprise that the larger protochordates (adult tunicates and adult cephalochordates) typically have a protective sheath or wall which guards the filtering region from damage by the environment.

So why do these various animals share a common feeding system? One way to look at this would be from a functional point of view. Given the relatively small size of the larvae and adults, suspended food particles (composed of microscopic organisms and pieces of larger organisms) may be the most abundant food source. Another way would be to consider their evolutionary history. Each of these protochordates has inherited a system of mucus production and cilia that are well suited for moving currents of water and extracting suspended food. Finally, other feeding systems that chew food typically require hard skeletal systems or tough plates that have not evolved in these groups.

chordates arose the vertebrates, a subphylum of vast diversity that includes some of the most remarkable species of animals ever to grace the air, waters, and surface of Earth. Within the early chordates the basic body plan was established (figure 2.7). Feeding depended upon the separation of suspended food particles from the water and involved a specialized area of the gut with walls lined by cilia to conduct the flow of food-bearing water. Pharyngeal slits allowed a one-way flow of water. Locomotor equipment included a notochord and segmentally arranged muscles extending from the body into a postanal tail.

Feeding and locomotion were activities that favored these novel and specialized structures in chordates. Subsequent evolutionary modifications would center around feeding and locomotion and continue to characterize the wealth of adaptations found within the later vertebrates, the subject of chapters 4–10.

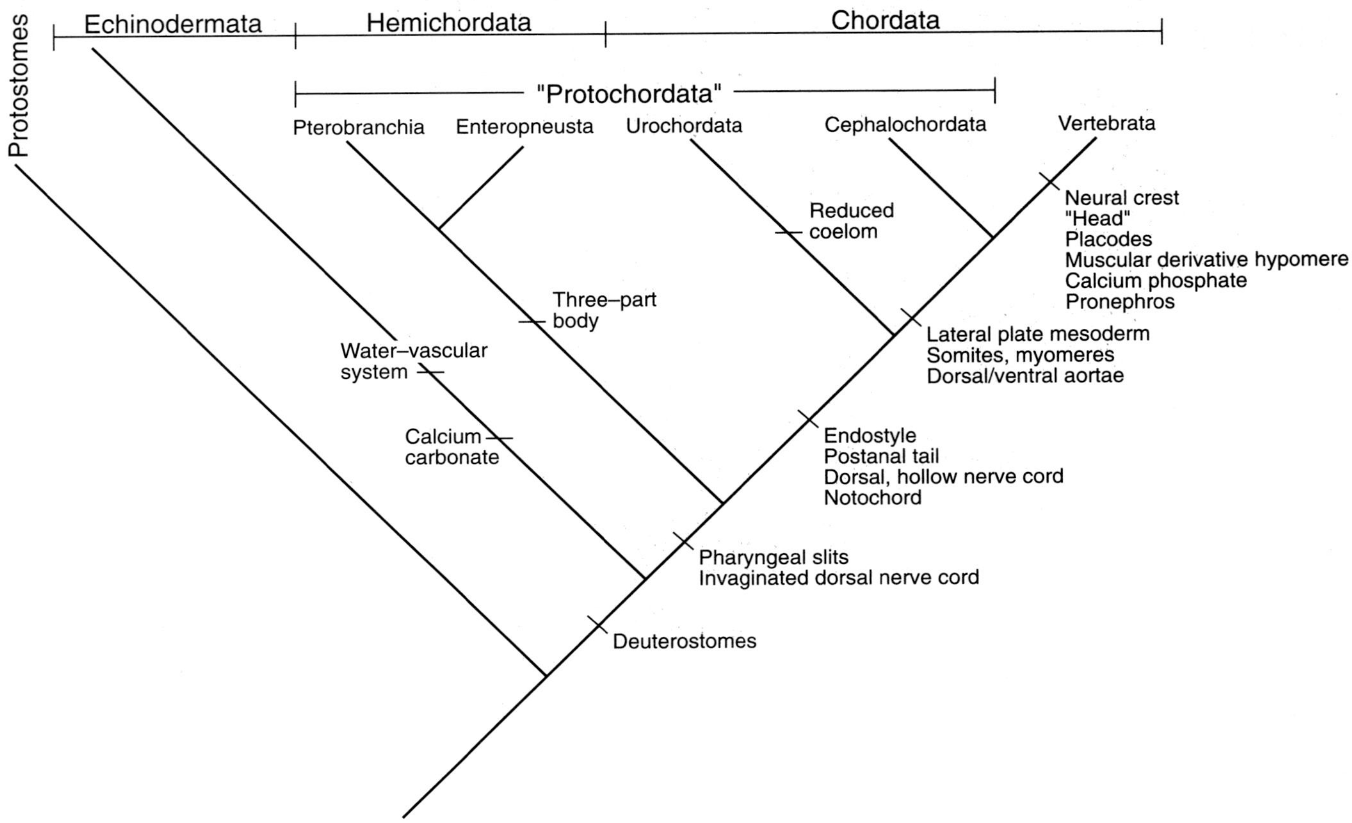

Figure 2.7
The evolution of chordate characteristics.

Agnathans—Subphylum Vertebrata Examination of a Primitive Vertebrate: The Lamprey

3

The remaining chapters focus on members of the subphylum Vertebrata (= Craniata). As a first sampler of vertebrates, to see some primitive structures, and to examine a possible early phylogenetic link to protochordates, a lamprey and its larva are examined first.

Agnathans—Three Vertebrate Classes in the Phylum Chordata

Table 3.1

Two Ways to Classify Agnathans

Informal Categories	
Ostracoderms—All extinct	
Cyclostomes—Lampreys and hagfish, extant	
Formal Classes	
Class Pteraspidomorphi—Only ostracoderms, all extinct, with two nasal openings	
Class Cephalaspidomorphi—lampreys, extant, and ostracoderms (extinct) not in the Class Pteraspidomorphi	
Class Myxini—Only hagfish, living forms	

The vertebrate story begins with agnathans, the "jawless" fishes. These animals are considered to be jawless because they lack a biting apparatus derived from branchial arches. Agnathans are subdivided into the informal categories of ostracoderm and cyclostome, which denote extinct and living jawless fishes, respectively. Ostracoderm (meaning "bone" and "skin") applies loosely to the now extinct but earliest agnathans, typically minnow-sized fish not more than a few centimeters in length. **Ostracoderms** were characterized by an extensive bony **exoskeleton** that encased the body in bony armor just beneath the epidermis. Although ostracoderms were the first vertebrates to develop any sort of bone, their **endoskeleton** was not well developed and, when present, was usually made of cartilage. **Cyclostome** (meaning "round and mouth") applies to all living agnathans: the lampreys and hagfishes. These fishes lack bone altogether and share a defining feature of a single median nostril.

As convenient as the terms ostracoderm and cyclostome are, neither seems to be a natural evolutionary category. Ostracoderms are not a single taxonomic group, but instead comprise at least two formal vertebrate classes. Lampreys are most likely related to one of these ostracoderm classes, so they should be classified with their extinct cousins. Hagfishes are distinct from all other jawless fishes and may be the most primitive of all agnathans. They are placed in their own vertebrate class separate from lampreys and any ostracoderms. These relationships are perhaps best recognized formally by dividing the living agnathans into three classes (as noted in table 3.1, and described below).

Class: Myxini—Hagfishes

The hagfishes are a very primitive vertebrate with no true vertebrae, no teeth, and reduced eyes with limited vision. They are eel-like scavengers that feed on the insides of dead or dying animals, for example invertebrates and other fishes (figure 3.1a–b). They use teethlike processes on their muscular tongues to rasp flesh from prey. Slime glands beneath the skin release mucus through surface pores. This mucus, or "slime" as it is called, may serve to release them from the grip of a predator or clog its gills.

Ovaries and testes occur in the same individual, but only one is functional, so hagfishes are not practicing hermaphrodites. Eggs but no larval stages have been found, so development is thought to be direct, without metamorphosis.

Like lampreys, hagfishes lack bone or surface scales, which was once taken as evidence of close affinity between them. However, lack of bone is now thought to be a secondary loss, occurring independently in both hagfishes and lampreys.

Class: Cephalaspidomorpha—Lampreys and Some Ostracoderms

The body shape of cephalaspidomorphs is quite varied (figure 3.2a–c), suggesting diverse lifestyles. However a unique and distinctive opening on top of the head unites all members of this diverse group.

A lamprey uses its oval mouth to grasp a stone and hold its position in a water current. In parasitic forms, the mouth clings to live prey so that the rough tongue can rasp away flesh or skin, allowing the fish to open blood vessels below and drink of the body fluids. The marine forms often migrate long distances to reach spawning grounds where fertilized eggs are deposited in a prepared nest of loose pebbles. A lamprey larva, termed the **ammocoete larva,** hatches from an egg. Unlike its parents, the ammocoete is a suspension feeder that lies buried in loose sediment with only its mouth protruding up from the bottom. Upon metamorphosis, the

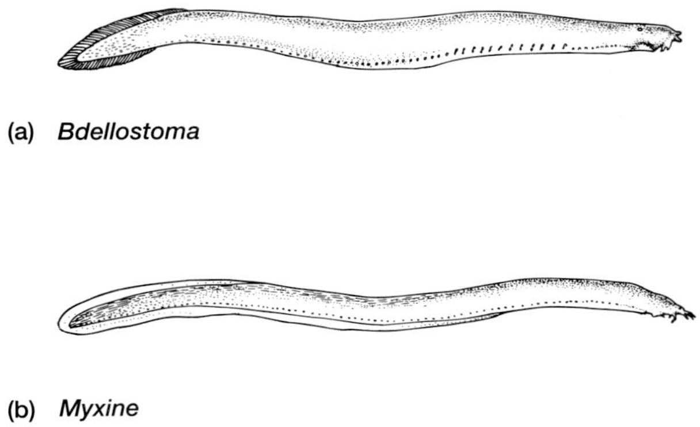

Figure 3.1
Myxini, the hagfishes. (a) The slime hag *Bdellostoma.* (b) The hagfish *Myxine.*
After Romer and Parsons, 1986, The Vertebrate Body, Saunders College Publishing, from Dean.

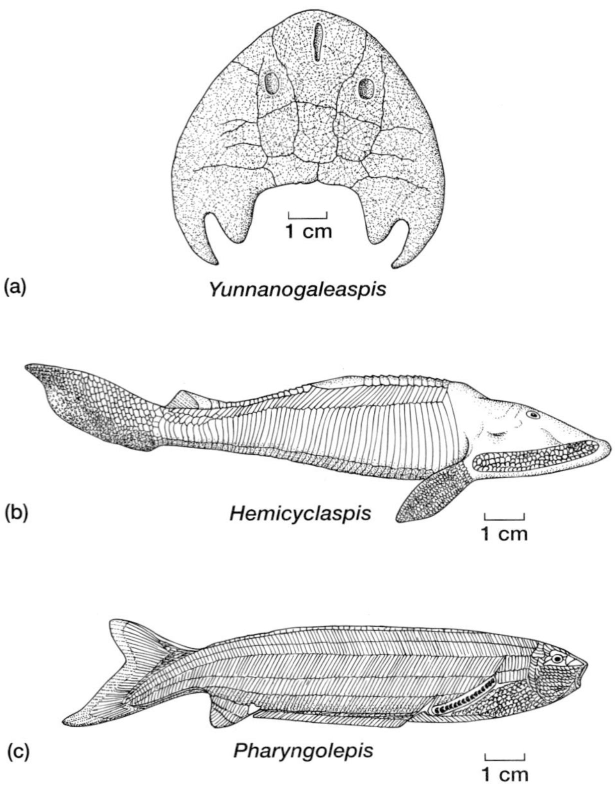

Figure 3.2
Cephalaspidomorphs. (a) The galeaspid *Yunnanogaleaspis,* for which only the head shield is known. (b) The osteostracean *Hemicyclaspis.* (c) The anaspid *Pharyngolepis.*
(a) After J. Pan and S. Wang, 1980: 1, "New finding of Galeaspirotmes in South China," Acta Palaeotologica Sinica, Beijing.
(b) Redrawn from J.A. Moy-Thomas and R.S. Miles, 1971, Paleozoic Fishes, W.B. Saunders Company, (b) After Stensio and and A. Heintz;
(c) after Ritchie.

ammocoete transforms into a parasitic adult. In some species, the larval stage may last up to seven years, at which time metamorphosis yields a nonfeeding adult that reproduces and soon dies.

Lampreys have single medial fins but lack paired fins or lobes. Individual blocks of cartilage ride atop a lamprey's prominent notochord, forming a relatively simple vertebral column (figure 3.3). Like their ostracoderm relatives, lampreys are jawless. But unlike ostracoderms, they lack the bony exoskeleton. In fact, lampreys lack bone entirely. Despite these differences, the single medial nasal opening and strikingly similar brain and cranial nerves are sufficient to classify lampreys with similar ostracoderms.

Class: Pteraspidomorphi—Extinct Ostracoderms from the Paleozoic

The pteraspidomorphs are now viewed as the oldest vertebrates, as they have the earliest documented fossil record (however, some extinct groups, proposed as possible vertebrates, are known from earlier periods). Pteraspidomorphs appeared in the late Cambrian, although they are represented at first by only splinters of bone lacking true bone cells. The class extends into the late Devonian, where more complete fossils have been found. Although some species are incomplete, the presence of paired nasal openings seems to distinguish pteraspidomorphs from all other agnathans.

Most pteraspidomorphs had head shields formed by fusion of several large bony plates (Figure 3.4a–c). Behind the head shield, the exoskeleton was composed of small plates and scales. Occasionally, lateral and dorsal spines projected from this shield. These spines may have offered protection from attack or anchored their heads in mud or sediment to prevent them from being dislodged.

The Origin and Evolution of Vertebrates: Shifts in Feeding Mechanisms

In chapter 2, protochordates were examined and discussed as links between invertebrates and vertebrates. But what became of the first prevertebrates? Some evolved into modern cephalochordates and urochordates, while others evolved into agnathans and other vertebrates (figure 3.5). It is this agnathan story that we explore next.

The evolution of early vertebrates was characterized by increasingly active lifestyles, hypothesized to proceed in three major steps. These steps primarily involve changes in feeding strategies and the related evolution of new pharyngeal structures.

Step 1 Prevertebrates

As noted in chapter 2, hemichordates, urochordates, and cephalochordates rely upon suspension feeding using ciliary pumps to propel water past a mucous-covered surface. This system was similarly shared by the first prevertebrate. But as the prevertebrates evolved, two changes occurred in the structure of the pharynx that produced a muscular instead of ciliary pump. First, the pharynx developed an encircling band of muscle, which when contracted, forced water out of the pharyngeal slits. And second, cartilage replaced the collagen of

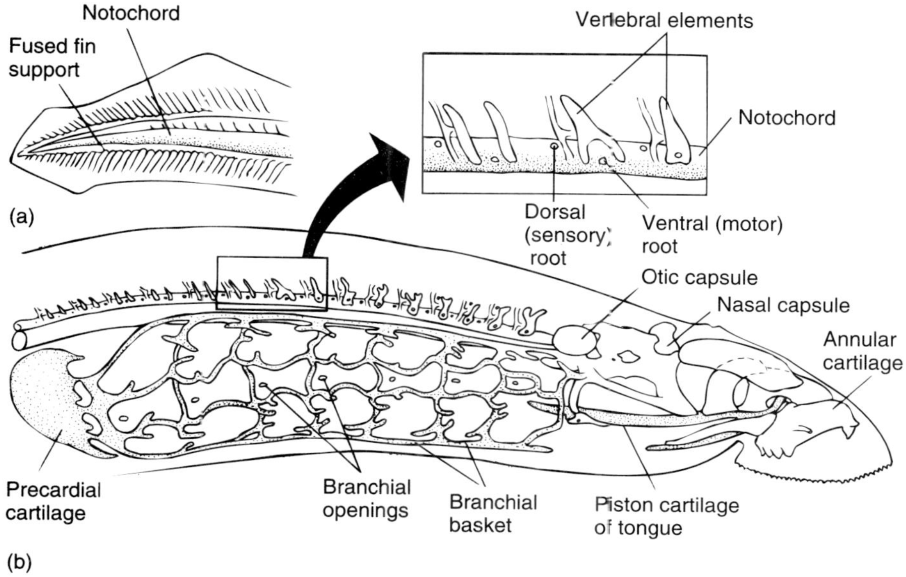

Figure 3.3
Lamprey skeleton. (a) Enlargement of caudal body section of the lamprey. (b) Anterior end of the lamprey with enlargement of the axial skeleton illustrating the prominent notochord. Note that only a few cartilaginous vertebral elements are present.
After M.J. Wake (ed.), 1979, Hyman's Comparative Vertebrate Anatomy, University of Chicago Press, p. 203

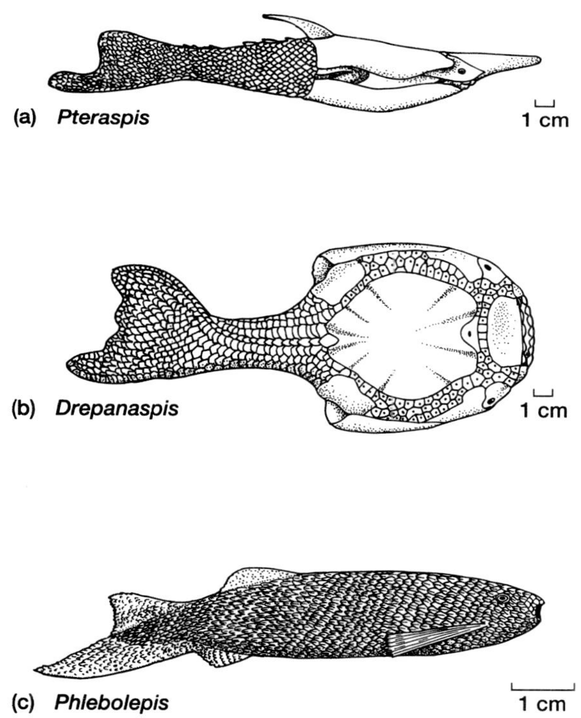

Figure 3.4
Pteraspidomorphs. All are extinct fishes of the early Paleozoic, with plates of bony armor that developed in the head. (a) The heterostracan *Pteraspis*. (b) The heterostracan *Drepanaspis*. (c) The thelodontid *Phlebolepis*.
Redrawn from J.A. Moy-Thomas and R.S. Miles, 1971, Paleozoic Fishes, W.B. Saunders Company, (a) after Gross, (b) after White, and (c) after Ritchie.

pharyngeal bars. Thus, when the muscular band contracted, these cartilage bars were bent, and water was forced through the pharynx. But when the muscular band relaxed, the flexed cartilaginous supports sprang back to expand the pharynx, restore its original shape, and draw in new water. Initially, this new muscular pump probably supplemented the prevertebrate ciliary pumps in moving water through the pharynx. But as animals increased in size, the muscular pumps became more significant as the surface ciliary pumps became less effective. The appearance of the new muscular pump thus removed the size limits imposed by the ciliary pump, and the first agnathans, much larger than the prevertebrates, appeared.

Step 2 Agnathans

The evolution of the new muscular pump and variations on this design resulted in the ensuing diversification of these jawless fishes. At this point, most agnathans were deposit feeders, mud grubbers that pushed their mouths into loose organic or silty mud and drew in sediment, rich in organic particles and microorganisms. Although cilia and mucus of the branchial basket still served to collect these passing particles from suspension and transport them into the esophagus, the new muscularized pharynx, not cilia, forced the stream of rich organic material through the mouth.

Step 3 Gnathostomes: The Jawed Vertebrates

Prevertebrates, with their ciliary pump, and probably early jawless vertebrates, with their muscular pump, fed on suspended particles. Food-bearing currents carried enough organic material past mucus, where some collided with it and was gathered up. But the transition from agnathan to gnathostome involved a switch away from this feeding method.

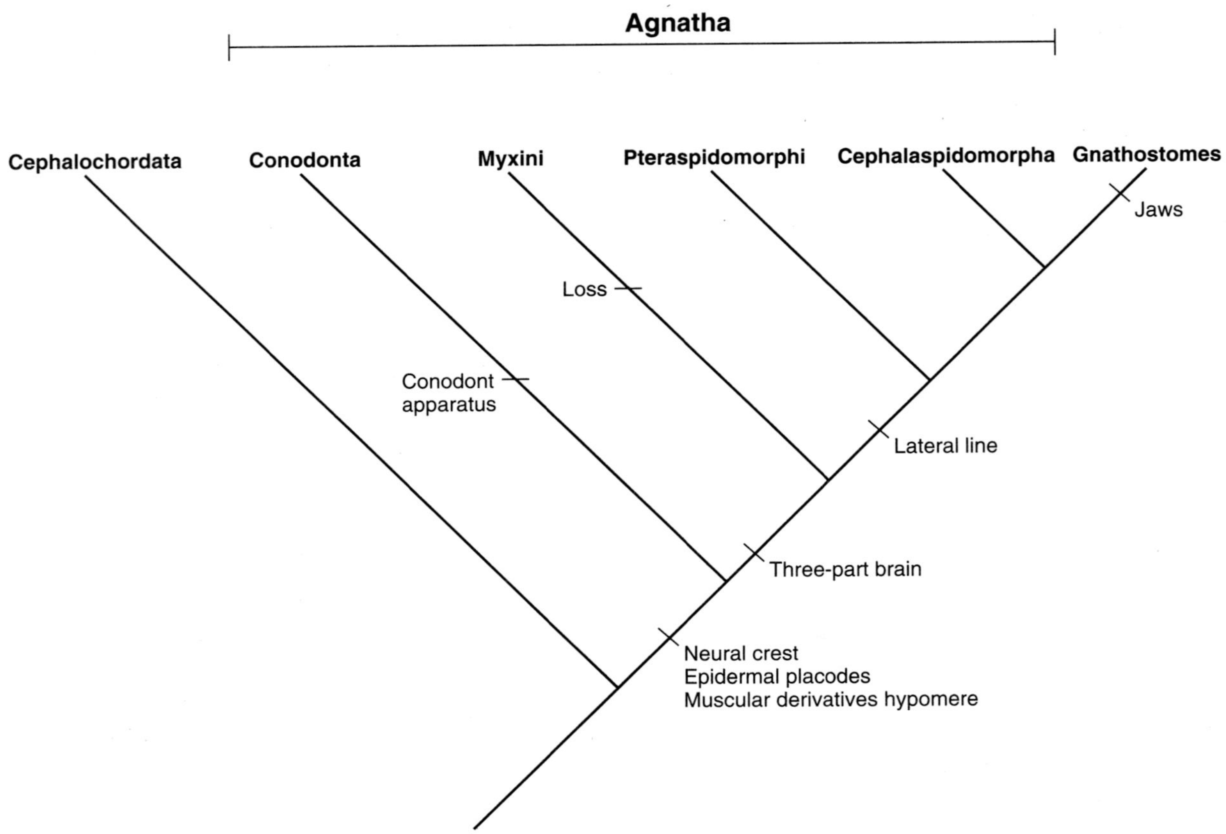

Figure 3.5
The relationships between cephalochordates, agnathans, and gnathostomes. Conodonts are a specialized group of extinct vertebrates. Loss of eyes and mineralized dental tissues occurred in the Myxini.

Transitional species plucked individual food particles selectively from suspension or off of surfaces. These food items may have been escaping zooplankton or heavier pieces of food that required a more powerful effort to be ingested. Thus new designs evolved that allowed for the forceful expansion of the pharyngeal pump followed by secure mouth closure to prevent escape of captured food. These new designs led to the advent of jaws powered by quick muscle action and the evolution of active predation.

Lampreys and hagfish, the only agnathans alive today, represent the second step in the evolution of jawed vertebrates. Both lack bone (and jaws) entirely and are specialized for parasitic or scavenging lives that depend on a rasping tongue to scrape up tissue for a meal. We therefore examine in detail a lamprey as an intermediate between prevertebrates and the gnathostomes.

Adult Lamprey Anatomy

Specific directions for finding structure in sections are indicated at the end of each system. The underlined terms are structures to find in the sections. Boldface terms are key structures in the text descriptions that are not to be identified at that point in the laboratory manual.

External Anatomy

→ Examine a whole preserved lamprey (figure 3.6a) and note the basic parts: the head includes a circular mouth, properly termed a buccal funnel, that is lined by soft papillae and hard horny teeth (figure 3.6b). Near the center of the buccal funnel is the tongue with hard teeth at the tip. Just above the tongue is the mouth opening, which leads to the internal buccal cavity (best seen in the sagittal section). On the sides of the head, behind the mouth, are a pair of eyes. Dorsally, between the eyes on the top of the head lies a single median nostril. Just behind the nostril is a lightly pigmented spot, below which is the pineal gland, or third eye. Behind the eyes on each side of the body run a series of seven paired round openings, the external gill slits.

Follow the body of the lamprey posteriorly until coming upon the midventral opening of the caudal end of the digestive tract; the slightly raised urogenital papilla lies within this opening. The part of the body between the last gill slit and this papilla is the trunk. The region posterior to this papilla is the tail. Two unpaired dorsal fins run along the middorsal side of the body; the tip of the tail is fringed by the caudal fin. Hold the lamprey up to

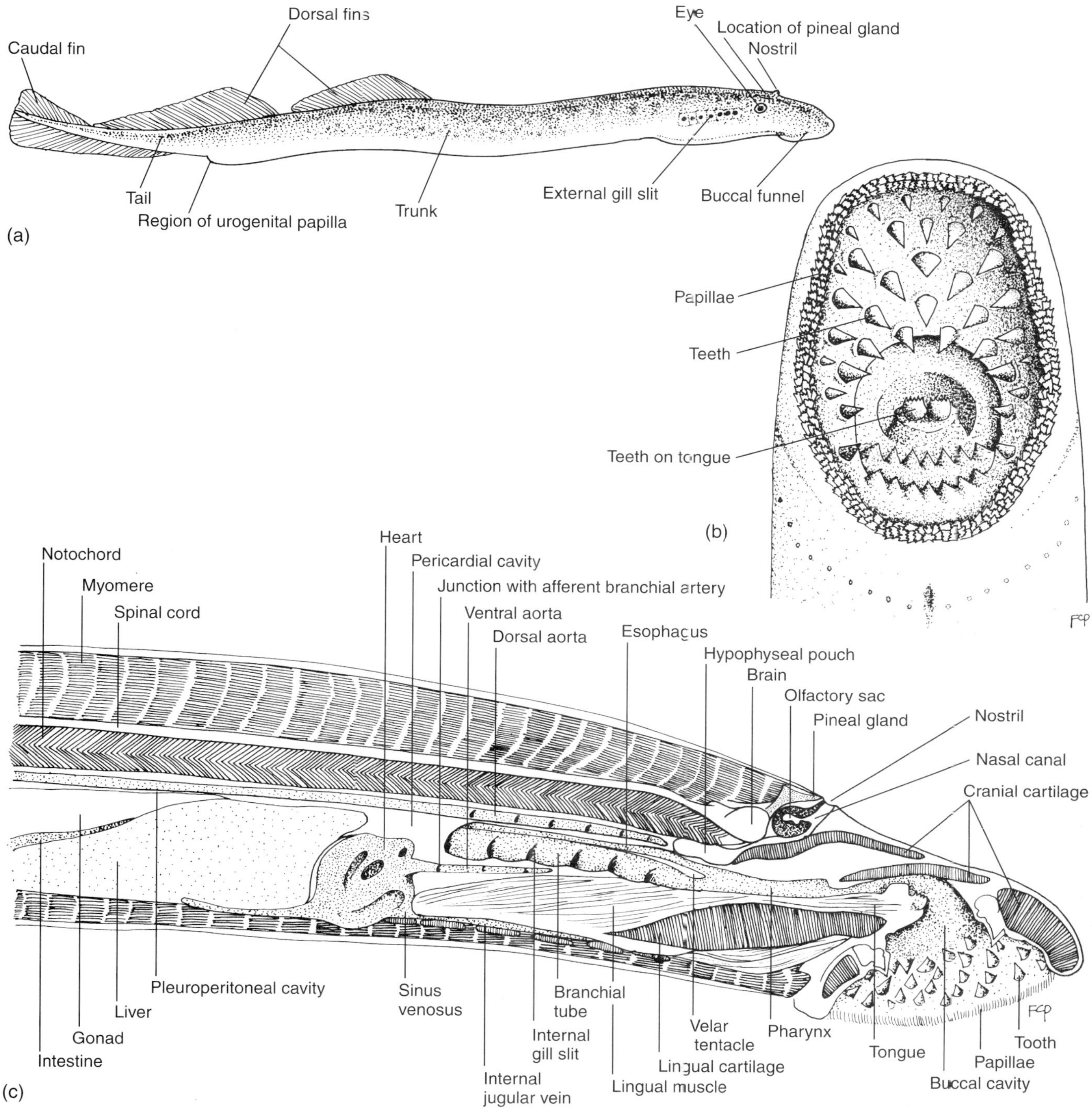

Figure 3.6
The adult lamprey, *Petromyzon*. (a) External morphology of the adult lamprey. (b) Ventral view of the structure of the mouth. (c) Midsagittal section of the anterior half of the body.

the light to back illuminate the numerous fin rays (figure 3.8d), slender cartilages that internally support both dorsal and caudal fins. Return the lamprey to the dissection tray and note the impressions in the skin of underlying myomeres, muscle segments arranged along the body. Myomeres are best seen in the tail region.

Sagittal and Cross Sections

→ Examine a lamprey in which a middorsal sagittal plane extends from the front of the head to a few inches behind the last gill opening (figure 3.6c). Also examine a lamprey cut in cross section at the following levels through:

(1) the pineal gland and eyes, (2) the 3rd or 4th gill slit about midlength along the branchial region, (3) the liver and intestine, and (4) the tail (figure 3.8).

Respiratory System

Vertebrate gills are designed for gas exchange with water. Specifically, gills are dense capillary beds supported by skeletal elements in the branchial region. Ventilation in lampreys involves the muscular pump of the **buccal cavity** actively driving water across the internal gills.

In lamprey species with a prolonged adult stage, the adult lamprey feeds by attaching its circular mouth to the sides of living prey. In such species, the mouth grips the prey, making it unavailable for entry of water through the **pharyngeal slits.** Instead, water exits *and* enters through the external pharyngeal slits (figure 3.7). Muscle compression and relaxation of the branchial apparatus drive this water, which moves tidally in and out of the branchial pouch via the associated slits. A partition divides the **pharynx** into a dorsal **esophagus** and a ventral **branchial tube** (respiratory tube) that serves as a water channel (figure 3.6c). The **velar tentacles** at the tip of the branchial tube selectively close this water channel during feeding to ensure that food moves from the buccal cavity to the esophagus.

→ **Sagittal Section** Identify the path of water that moves from the buccal cavity, through the pharynx, past the velar tentacles, through the branchial tube, and out the internal gill slits to the gills.

Cross Section Examine a cross section through a gill slit in the branchial region (figure 3.8b). Trace the path of water drawn in through the mouth as it next flows laterally from the central branchial tube, through the gill lamellae, into the gill pouch, and then out of the body through the external gill slits.

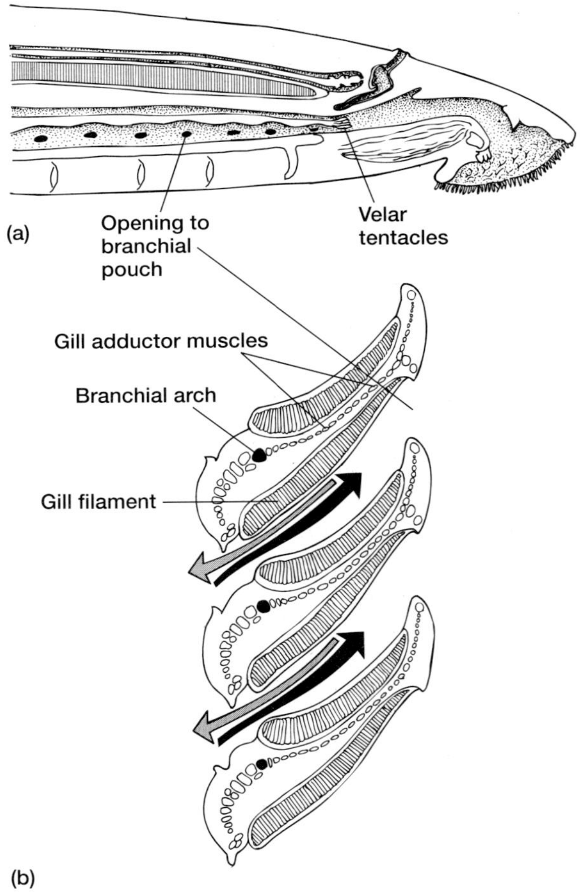

Figure 3.7
Ventilation in the adult lamprey. (a) Longitudinal section. Because the adult lamprey's mouth often is attached to prey, water must alternatively enter as well as exit via pharyngeal slits. Thus, unlike most fishes, gill ventilation in the lamprey is tidal. (b) Frontal section of the left three gill arches. Double open arrows indicate tidal flow of water; black inflow, gray outflow.
After Mallatt.

Digestive System

In lampreys, the alimentary canal is a straight tube leading from mouth to anus without coils, folds, or major bends. The ciliated **esophagus** runs directly from the pharynx to the **intestine.** No distinct stomach is present, and lampreys lack a spleen. Diet includes small particulate matter, blood and tissue rasped from prey, and detritus. Storage in an expanded stomach before entering the intestine would be of little value, so food passes directly from the esophagus into the intestine.

The adult intestine is lined with an epithelium that contains numerous gland cells dispersed along internal folds that form a modest **spiral valve** (figure 3.9). Digestive enzymes are released into the anterior intestine, and mucus is secreted into the posterior intestine. In parasitic lampreys, the anterior intestine is especially important in absorption of fats. In addition, this region of the intestine of marine forms holds swallowed salt water and is important in osmoregulation. The posterior section of the intestine is important in protein absorption and elimination of biliverdin, a pigment of bile produced by the **liver** and released to the intestines. The liver, located just caudal to the heart, and the **intestine** lie within the **pleuroperitoneal cavity,** a portion of the **coelom.**

→ **Sagittal Section** Return to the velar tentacles located at the junction of the esophagus and branchial tube. Trace the esophagus caudally to its position above the heart. The gut continues beyond the heart as the intestine extending ventrally between the gonad and liver. The liver, gonad, and intestine reside within the pleuroperitoneal cavity.

Cross Section Examine a cross section through the liver and intestine (figure 3.8c). Identify the liver on the ventral side and above it, the intestine, both surrounded by the pleuroperitoneal cavity. The walls of the intestine are folded, forming part of the spiral valve.

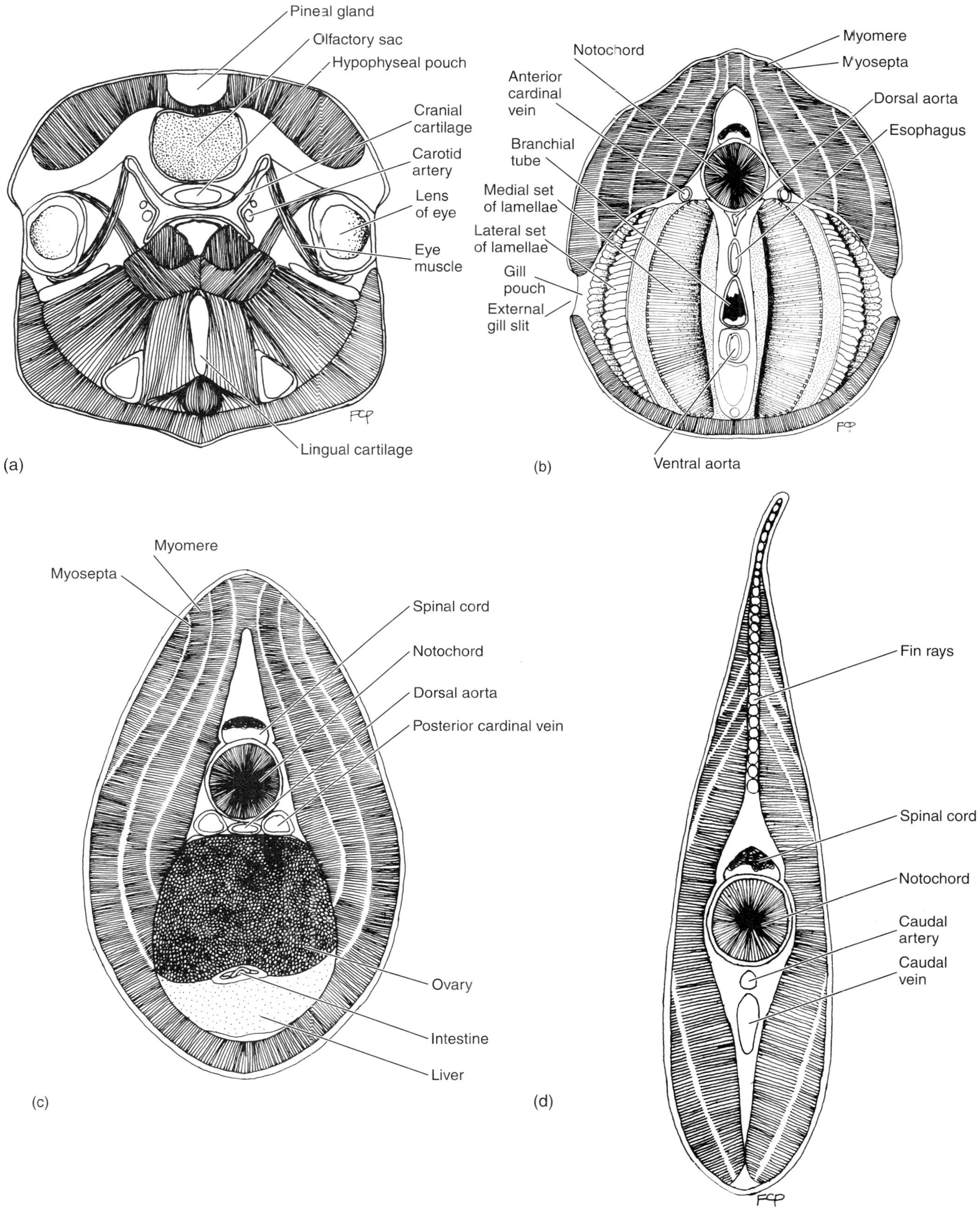

Figure 3.8
Cross sections of the adult lamprey. (a) Section through the pineal gland and eyes. (b) Section through the 3rd or 4th branchial slit. Two sets of gill lamellae are seen. The section passes through the most lateral lamellae in the region of an external gill slit. The medial lamellae are slanted laterally and caudally to communicate with a more posterior gill slit. (c) Section through the liver and intestine. (d) Section through the tail region.

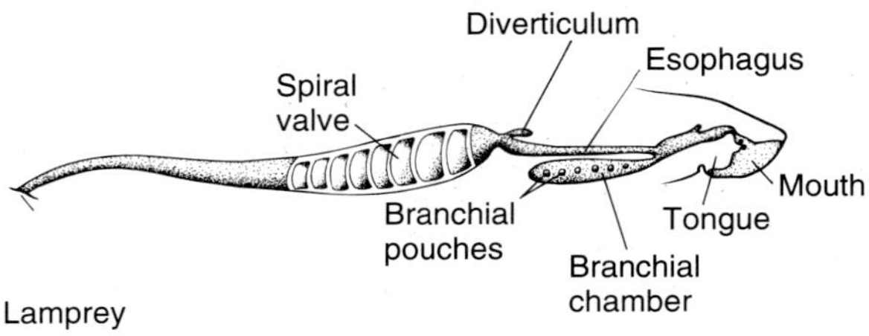

Figure 3.9
Digestive tract of the adult lamprey.
From Dean

Circulatory System

The lamprey heart (branchial heart) is a typical fish heart. It resides in another portion of the coelom, the **pericardial cavity.** Blood flows sequentially through the four heart compartments: the **sinus venosus, atrium, ventricle,** and muscular **conus arteriosus.** One-way valves are present between compartments. The luminal walls of the conus arteriosus are thrown into folds collectively forming the **semilunar valves,** which prevent reverse blood flow. After leaving the conus arteriosus, blood is distributed to the **ventral aorta,** the **afferent branchial arteries,** and finally to the delicate **gill capillaries.** Blood collected from the gills is distributed to the head through the **carotid artery** and to the trunk and tail via the **dorsal aorta** and **caudal artery** (an extension of the dorsal aorta in the tail region).

The **caudal vein** drains the tail into the paired **posterior cardinal veins.** The paired **anterior cardinal veins** drain the head. The left anterior and left posterior cardinal veins fuse with their right counterparts to form the right **common cardinal vein** that enters the **sinus venosus.** Veins draining the gut enter the liver, forming a true **hepatic portal system** common in the vertebrates.

→ **Sagittal Section** A series of blood vessels can be identified cranial to the heart in a midsagittal section (figure 3.6c). (The blood vessels in your specimen may be injected with colored latex making the following identifications easier.) Just ventral to the notochord is the unpaired dorsal aorta. Immediately lateral to the dorsal aorta are the paired anterior cardinal veins. The ventral aorta originates from the cranial side of the heart and extends below the branchial tube. Numerous afferent branchial arteries extend laterally from the ventral aorta. Their points of departure may appear as holes along the length of an uninjected ventral aorta. The internal jugular vein, draining the cranial end of the body into the sinus venosus, runs along the ventral edge of the body. Finally, identify the pericardial cavity surrounding the heart.

Cross Section Examine a cross section through the eyes (figure 3.8a). The paired carotid artery is located below the brain and medial to the eyes. Next examine a cross section through the branchial region (figure 3.8b). The paired anterior cardinal veins are located on the ventral sides of the notochord. The single dorsal aorta, which distributes blood to the trunk and tail regions, is found just ventral to the notochord. Identify the ventral aorta just below the branchial tube. Afferent branchial arteries depart the ventral aorta laterally and extend into the gills, but identification relies upon an unlikely fortuitous section.

Examine a cross section through the liver (figure 3.8c). Just ventral to the notochord, the dorsal aorta is found with the paired posterior cardinal veins on either side. The hepatic portal system extends through the liver but is impractical to distinguish.

Examine a cross section through the tail (figure 3.8d). Just ventral to the notochord is the caudal artery, and just below it the larger caudal vein.

Muscular System

In fishes, the axial musculature arises directly from the embryonic and segmental myotomes. Once fully differentiated into the adult musculature, the blocks of axial musculature retain their segmentation, but they are termed **myomeres** to distinguish them from the formative embryonic myotomes from which they arose. Successive myomeres are separated from each other by connective tissue sheets, the **myosepta.** Myosepta extend inward, become attached to the axial column (primarily the notochord), and join successive myomeres into muscle masses. The horizontal septum of the skeleton is not present in lampreys but is present in all gnathostome fishes, where it divides the myomeres into epaxial (dorsal) and hypaxial (ventral) muscle masses.

The axial musculature of lampreys supplies the major propulsive forces for locomotion, and, not surprisingly, constitutes the bulk of the body's musculature. Viewed from the lateral surface, the myomeres are folded into W-shaped zigzag blocks. A contraction spreading within the axial musculature alternates from side to side, developing characteristic waves of lateral undulation. These powerful bends produced by the axial musculature are responsible for developing the body's lateral thrusts against the water and driving the lamprey forward.

→ **Sagittal Section** Identify myomeres dorsal to the spinal cord (figure 3.6c). The lingual muscles and associated lingual cartilages extend ventrally and caudally from the tongue that they control.

Cross Section Examine a cross section through the liver (figure 3.8c). The lateral body walls are formed by several layers of muscle, each a separate myomere separated from other myomeres by connective tissue sheets called myosepta.

Skeletal System

Most of the skeletal components listed below are not easily seen in intact whole specimens. Limit your identification of these skeletal components to figure 3.3, unless you have been provided with a specially prepared lamprey skeleton.

The skeleton of lampreys is composed of cartilage and tough connective tissues, with no trace of bone (figure 3.3). The cranial endoskeleton consists of several prominent cartilages forming the **olfactory** and **otic capsules** and the **piston cartilage** which helps to manipulate the tongue. In addition, a series of pharyngeal cartilages form an unjointed **branchial basket** that supports the roof of the pharynx, lateral pharyngeal slits, and forms a shield around the heart (**precardial cartilage**). A lamprey's mouth is neither defined nor supported by jaws. Instead, an **annular cartilage** that supports the buccal area is itself supported by the cartilaginous braincase (figure 3.3). The **notochord** provides most of the axial support for the body, extending forward from the tail to a point just ventral to the brain. As muscles apply their forces, the notochord must act as a compression girder, resisting telescoping of the body that might otherwise result.

→ **Sagittal Section** Identify the notochord, lingual cartilage, and portions of the cranial cartilages cut in sagittal section and identified in figure 3.6c. Note how little of the skull is formed by the skeletal elements.

Cross Section Identify the cranial cartilages in a cross section through the eyes (figure 3.8a). The notochord can be seen in the sections through the branchial region, the liver, and the tail (figures 3.8b–d).

Reproductive and Urinary Systems

Adult lampreys are **dioecious,** either male or female. Gametes, either sperm or ova, are produced by an unpaired **gonad** (either **testis** or **ovary**) located above and behind the liver. The gametes are released into the coelom of the body during mating season and collected into short channels at the caudal end of paired kidney ducts. These ducts extend through the **urogenital papilla** to release the gametes out of the body. Although fertilization is external, the male possesses a penislike tube that can be closely applied to the female's cloaca to bring the eggs and sperm into close proximity during mating.

The sex differences in cloacal structure are most pronounced during the mating season. Therefore, unless your specimen was collected at this particular time, distinguishing the sexes based upon cloacal structure will be difficult. You will identify a gonad later in the cross sections.

Lampreys possess **opisthonephric kidneys** generally located just dorsal to the caudal half of the gonad. The paired kidneys are each drained by an **archinephric duct** which unite caudally into a urogenital sinus drained through the tip of the **urogenital papilla.**

→ **Sagittal Section** The cranial end of the gonad, immediately dorsal to the liver, is the only reproductive or urinary structure found in a sagittal section of the anterior end of the body (figure 3.6c). The many follicles of ovaries make them appear to be more granular than testes. Compare the gonad in your lamprey to gonads in other lampreys to speculate on the sex of your specimen.

Cross Section Examine a cross section through the liver (figure 3.8c). The gonad is located just dorsal to the liver and intestine. The many developing follicles make the ovary appear more granular than the texture of the testis. A cross section caudal to the liver is required to see the opisthonephric kidneys, positioned just dorsal to the caudal half of the gonad.

Nervous System and Special Senses

The **spinal cord** is located just dorsal to the notochord, consistent with the basic chordate body plan. The lamprey brain exhibits the basic vertebrate organization, with an immense concentration of neurons in the head. However, the brain is disproportionately smaller than most other vertebrates. The forebrain region associated with smell and the midbrain region associated with vision are well developed. However, the cerebellum and the hindbrain associated with taste and acousticolateral systems are both small. The medulla oblongata is one of the largest brain regions as it is associated with the trigeminal nerve controlling the extensive musculature associated with sucking.

Lampreys possess 10 pairs of cranial nerves which provide sensory information to the brain (e.g., olfactory and optic nerves associated with smell and vision, respectively) and send motor signals out to muscles of the head and pharynx (e.g., oculomotor and trochlear nerves controlling eye muscles). The nerves of lampreys are unique in two ways: (1) the dorsal and ventral roots of the spinal nerves do not join lateral to the spinal cord and (2) all of the nerves are unmyelinated.

The *pineal gland,* identified previously on the middorsal side of the head, functions as a light sensing third eye, and is particularly well developed in lampreys. The gland is actually formed of two sacs of unequal size, in which the right sac is predominant. The gland appears to control fluctuations in body color between night and day. In addition, it may sense light rhythms associated with seasonal activities such as metamorphosis and reproductive cycles.

Lampreys possess only inner ears associated with paired sets of two semicircular canals that provide sensory information about the position of the body. All gnathostome vertebrates possess a pair of three semicircular canals.

The **nasal canal** extends ventrally towards the mouth where it reaches the expanded **olfactory sac.** Near the beginning of the olfactory sac, another duct departs ventrally and expands to form the **hypophyseal pouch.** As the shape of the pharynx changes, this pouch is expanded and compressed, moving water in and out of the nostril. This ventilation of water past the olfactory sac assists the sense of smell by moving more molecules past the sensory surface.

→ **Sagittal Section** Trace the spinal cord cranially until it expands to form the brain, just beyond the cranial end of the notochord (figure 3.6c). Immediately above the brain is the lightly colored pineal gland, located just behind the nostril identified previously when examining the external anatomy. Water is pumped in through the nostril, down the nasal canal, and past the dark olfactory sac to the expanded blind-ended sac called the hypophyseal pouch.

Cross Section Examine a cross section through the eyes (figure 3.8a). Identify the middorsal pineal gland. The hypophyseal pouch is positioned just ventral to the olfactory sac, which is in turn just ventral to the pineal gland.

Anatomy of the Lamprey (Ammocoete) Larva

The larval lamprey looks so much like the cephalochordate amphioxus, that it was originally described in the nineteenth century as an adult cephalochordate and given the name ammocoete. Although we now recognize the ammocoete as a larval lamprey, the name has stuck.

Whole Mount

Like all of the preceding protochordates examined, this larva is a filter feeder. However, the fine organic matter extracted from the stream of water is pumped through the **pharynx,** not by ciliary action, but by the muscular **velum.** Fine food particles are selectively captured in the mucus lining the parabranchial pouches next to the pharyngeal arches using an endostyle like that in cephalochordates. In addition, the endostyle in amphioxus and the larval ammocoetes is similarly involved in thyroid hormone production. In fact, as the ammocoetes larva undergoes metamorphosis, the endostyle is converted into a thyroid gland that releases hormones directly into the circulatory system. Depending upon the species, the ammocoetes larva may persist for several years, before transforming into the sexually mature adult.

→ Obtain a whole mount of a small ammocoete larva and study it using the low power magnification on your microscope. Refer to figure 3.10a for help in finding the underlined structures on your whole mount slide. Examine the gross morphology of the larva, noting the pharynx, oral hood, oral tentacles, dorsal and caudal fin, and segmented myotomes, all similar in structure and function to that seen in amphioxus. The muscles of whole mount specimens have been chemically treated, making them translucent and difficult to identify. Focus carefully up and down and look for these repeated muscle units.

Distinguish between the notochord and the spinal cord just dorsal to it. The spinal cord is confirmed by tracing it forward to the head, where it expands forming the brain. You may identify the three divisions of the brain separated by constrictions defining the prosencephalon (most anterior), mesencephalon (middle), and rhombencephalon (most posterior). Just lateral to the brain note a dark eye spot only slightly sensitive to light.

When feeding, the muscular pharynx expands drawing water through the mouth into the oral cavity surrounded by the velum. Gas exchange between the blood and filtered water occurs as the water passes over the gill filaments in the gill pouch and out the external gill slits. Food particles captured in the mucus pass down the pharynx into the esophagus, then the intestine, and finally out of the anus located near the base of the tail. The liver, heart, and perhaps the gall bladder can be seen on the ventrolateral side of the cranial end of the esophagus. A few tubules of the small pronephric kidney may be seen just dorsal to the heart.

Cross Section Through the Pharynx

→ Use figure 3.10b and a cross section of the pharynx to help you identify the following structures described above: pharynx, gill lamellae, external gill slit (or indentation if the section didn't pass through a gill slit), notochord, spinal cord, and myomeres.

Cross Section Through the Intestine

→ Use figure 3.10c and a cross section through the intestine to help you identify the following structures described above: spinal cord, notochord, kidney tubules, intestine, myomeres, and dorsal and ventral fin fold.

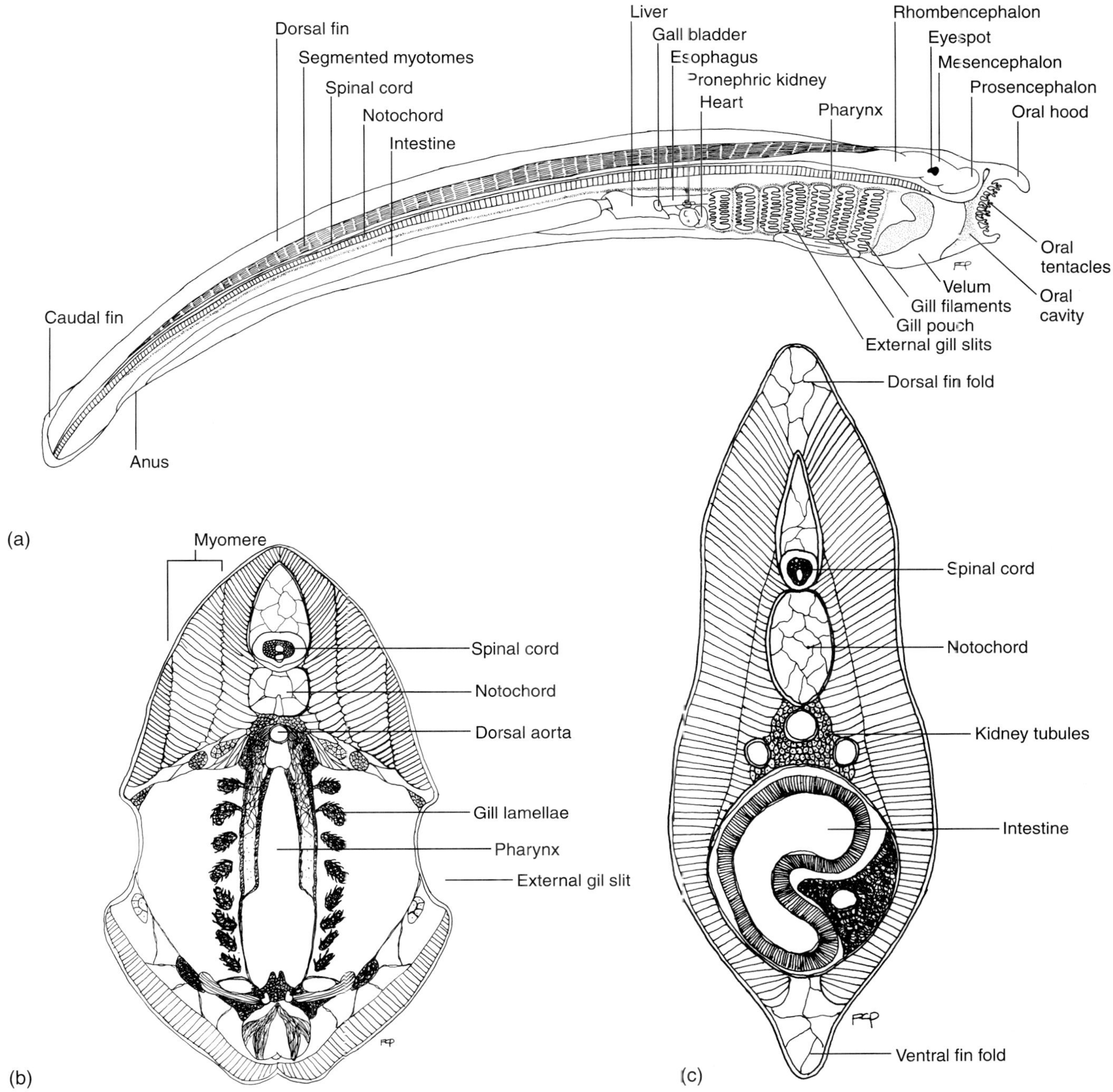

Figure 3.10
Ammocoetes larva. (a) External morphology. (b) Cross section through the pharynx. (c) Cross section through the intestine.

4 Vertebrate Integuments

Introduction

The vertebrate integument is one of the largest organs of the body and forms some of the most varied structures found within vertebrates. For example, the epidermis produces hair, feathers, baleen, claws, nails, horns, beaks, and some types of scale, and the dermis gives rise to dermal bones and the osteoderms of reptiles. Collectively, the epidermis and dermis form teeth, denticles, and the scales of fish. As the critical border between a vertebrate and its environment, the integument or skin has a variety of specialized functions, including:

- forming part of the exoskeleton
- thickening to resist mechanical injury
- establishing a barrier to the entrance of pathogens
- holding the shape of the organism
- participating in regulation of osmotic control
- presiding over the movement of gases and ions to and from the circulation
- absorbing needed heat or radiating the excess
- housing sensory receptors
- contributing feathers for locomotion, hair for insulation, and horns for defense
- containing pigments to block harmful sunlight or to display colors

The integument is a composite organ. The superficial layer, in contact with the outside environment, is the **epidermis,** below it the **dermis,** and between them the **basement membrane** (figure 4.1). The epidermis derives from the ectoderm, and the dermis derives from mesoderm and mesenchyme. Each contributes to a portion of the basement membrane, a layer unique to vertebrates. Below the dermis, between the integument and deep body musculature, is a transitional region made up of very loose connective tissue and adipose termed the **hypodermis** in microscopic examination, or termed the **superficial fascia** in gross anatomical dissection (figure 4.1).

Dermis

The most conspicuous component of the dermis is the fibrous connective tissue composed mostly of collagen fibers. The collagen fibers may be woven into distinct layers, termed **plies,** which are best at resisting forces in specific directions. In addition, the dermis of many vertebrates directly produces plates of bone through intramembranous ossification. Because of their embryonic origin and initial position within the dermis, these bones are termed **dermal bones**. They are prominent in ostracoderm fishes but appear secondarily even in derived groups, such as in some species of mammals.

Epidermis

The epidermis of many fishes and amphibians produces mucus to moisten the surface of the skin. The mucus seems to afford some protection from bacterial infection and helps ensure the laminar flow of water across the surface. In amphibians, the mucus also keeps the skin from drying when the animals are on land.

In terrestrial vertebrates, the epidermis covering the body often forms an outer **keratinized** or **cornified** layer, the **stratum corneum**. Rapidly dividing cells at the base of the epithelium push more superficial cells toward the surface. These superficial cells tend to undergo an orderly self-destruction, accumulating protein products during their demise. Collectively, these various protein products are termed **keratin,** and this orderly process of their formation is **keratinization**. The resulting superficial stratum corneum is a nonliving layer that serves to reduce water loss through the skin in dry terrestrial environments.

Keratinization and formation of a stratum corneum also occur where friction or direct mechanical abrasion damage the epithelium. For example, the epidermis in the oral cavity of aquatic and terrestrial vertebrates often exhibits a keratinized layer, especially if the food eaten is unusually sharp or abrasive. In areas of the body where friction is common, such as the soles of the feet or palms of the hands, the cornified layer may form a thick protective layer, or **callus,** against mechanical damage (figure 4.2).

In addition, scales form within the integument of many aquatic and terrestrial vertebrates. Scales are basically folds in the integument (figure 4.3). Where dermal contributions predominate, especially in the form of ossified dermal bone, the fold is termed a **dermal scale;** where the epidermal contribution predominates, especially in the form of a thickened keratinized layer, the fold is termed an **epidermal scale.**

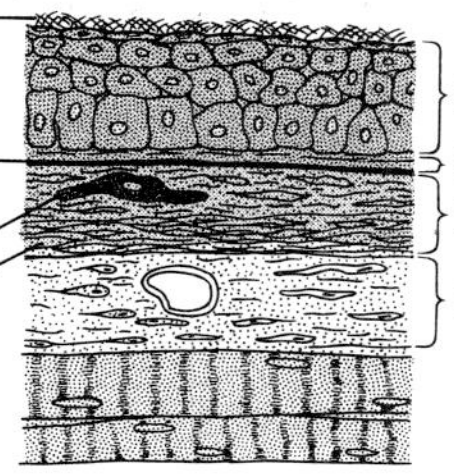

Figure 4.1
Embryonic development of the skin. The epidermis differentiates into a stratified layer often with a mucous coat, or mucous cuticle, on the surface; within the dermis, the collagen often forms distinctive plies, or layers, that taken together constitute the stratum compactum. The basement membrane lies between epidermis and dermis. Beneath the dermis and the deeper layer of musculature is the hypodermis, a collection of loose connective and adipose tissue.

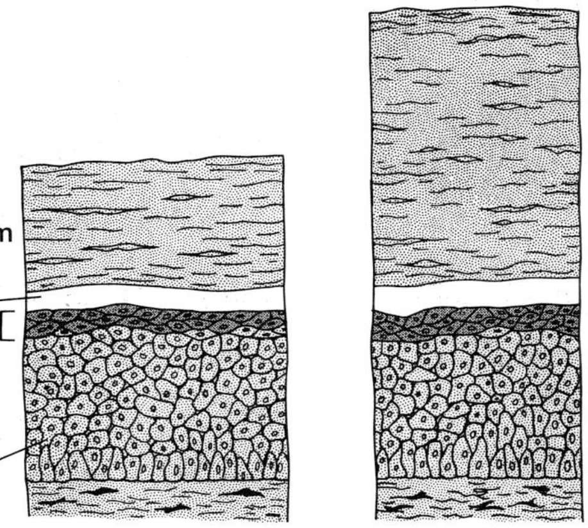

Figure 4.2
Keratinization. Where mechanical friction increases, the integument responds by increased production of a protective keratinized callus, and the stratum corneum becomes thickened as a result.

Examination of Vertebrate Integuments

Fishes

Epidermis

With few exceptions, the epidermis of most living fish is alive and active, with a nonkeratinized surface covered by mucus. The mucous layer is formed from various individual cells in the epidermis with contributions from multicellular glands.

Dermis

Collagen within the dermis is regularly organized into plies that spiral around the body of the fish, allowing the skin to bend without wrinkling. In some fishes, the dermis has elastic properties. When the swimming fish bends its body, the skin on the stretched side of the body stores some of the energy that is then released as the skin recoils to help unbend the body and sweep the tail in the opposite direction.

The fish dermis often gives rise to dermal bone, and dermal bone gives rise to dermal scales of fishes. In addition, the surface of fish scales is sometimes coated with a hard, acellular **enamel,** of epidermal origin, and a deeper layer of **dentin,** of dermal and neural crest origin.

Ostracoderms, Placoderms, and Lampreys

In ostracoderms and placoderms, the integument produced prominent bony plates of dermal armor that encased their bodies in an exoskeleton. The skin of living hagfishes and lampreys departs considerably from that of primitive fossil fishes. Dermal bone is lost, and the surface of the skin is smooth and without scales.

→ **Lamprey Skin** If a piece of lamprey skin is available, it will feel smooth and pliable. Although hagfishes are not eels (teleosts), many "eel-skin" wallets and purses are made from the skin of hagfishes, similar in texture to the skin of lampreys. Examine a cross section of lamprey skin under the microscope. Notice that the epidermis is composed of stacked layers of epidermal cells with unicellular glands interspersed among them. The dermis is highly organized into regular layers of fibrous connective tissue with pigment cells throughout the dermis. The hypodermis includes adipose tissue.

Chondrichthyes

In the cartilaginous fishes, dermal bone is absent, but surface denticles, termed **placoid scales,** persist. These scales are what give the rough feel to the surface of the skin (figure 4.4a). Recent evidence suggests that these tiny placoid scales reduce friction drag as water flows across the skin of a swimming fish. In the epidermis, numerous **secretory cells** are present, in addition to the stratified epidermal cells (figure 4.4b). The placoid scale begins in the dermis but projects through the epidermis to reach the surface. A cap of enamel forms the tip, dentin lies beneath, and a pulp cavity resides within (figure 4.4b). Chromatophores occur in the lower part of the epidermis and upper regions of the dermis.

→ Examine the surface of a shark skin with a hand lens. Try to identify individual placoid scales and their spines. Rub your finger back and forth across the skin noting that the skin feels rougher in certain directions because the spines are curved.

Bony Fishes

The dermis of bony fish is composed of loose and dense connective tissue with chromatophores sprinkled throughout. The most important structural product of the dermis is the scale. In bony fishes, **dermal scales** do not actually pierce the epidermis, but they are so close to the surface they give

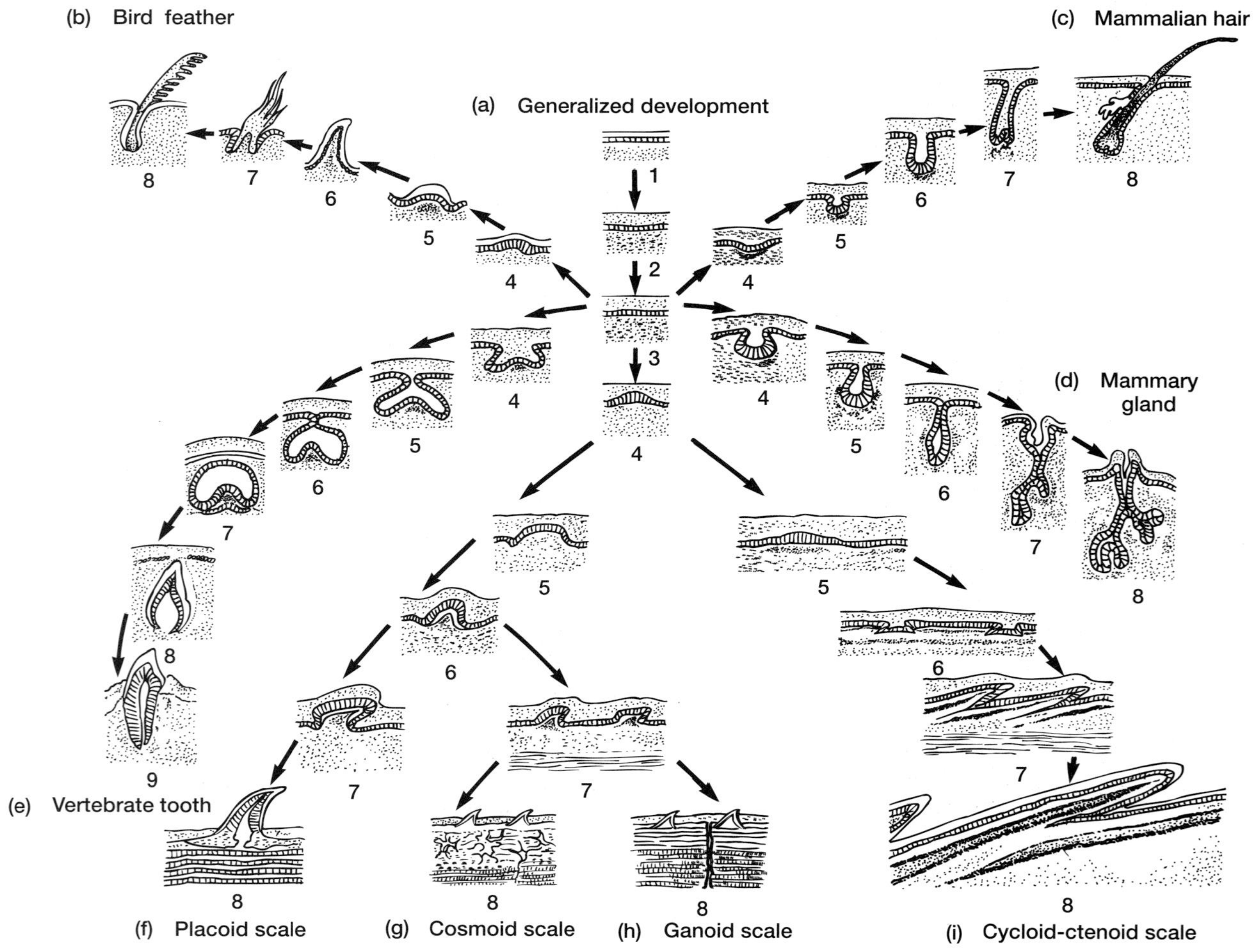

Figure 4.3
Skin derivatives. (a) Out of the simple arrangement of epidermis and dermis, with a basement membrane between them, a great variety of vertebrate integuments develop. The interaction of epidermis and dermis gives rise to feathers in birds (b), hair and mammary glands in mammals (c and d), teeth in vertebrates (e), placoid scales in chondrichthyans (f), and the cosmoid, ganoid, and cycloid-ctenoid scales in bony fishes (g–i).
Based on the research of Richard J. Krejsa.

the impression that the skin is hard (figure 4.5a). The covering epidermis includes a basal layer of cells above which are **stratified epidermal cells** (figure 4.5b). As they move toward the surface, epidermal cells undergo a cytoplasmic transformation, but they do not become keratinized. Within these layered epidermal cells occur single **unicellular glands,** which along with epidermal cells, are the source of the surface "slime."

On the basis of their appearance, several types of scales are recognized among bony fishes (figure 4.3). One type is the **cosmoid scale** present in some living sarcopterygians. This scale type is founded upon a double layer of bone, one layer of which is lamellar and the other, vascular (figure 4.6a). On the outer surface of this bone is a layer that is now generally recognized as dentin, and spread superficially on the dentin is a layer now recognized as enamel. The unusual appearance of these enamel and dentin coats inspired the respective names ganoin and cosmine, on the mistaken belief that ganoin was fundamentally a different mineral from "enamel" and "cosmine" from dentin. Although the chemical nature of these layers is now clear, the earlier names have stuck to give us the terms for distinctive scale types. In the cosmoid scale, there is a thick, well-developed layer of dentin ("cosmine") beneath a thin layer of enamel.

The **ganoid scale,** found in gars and *Polypterus,* is characterized by the prevalence of a thick surface coat of enamel ("ganoin"), without an underlying layer of dentin (figure 4.6b). Dermal bone forms the foundation of the ganoid scale, appearing as a double layer of vascular and lamellar bone (in paleoniscoid fishes) or a single layer of lamellar bone (in other primitive actinopterygians) (figure 4.6b). Ganoid scales are shiny (because of the enamel), overlapping, and interlocking. Those from a gar are rhomboidal in shape. However, in most other lines of bony fishes, the scales are reduced through the loss of the vascular layer of bone and

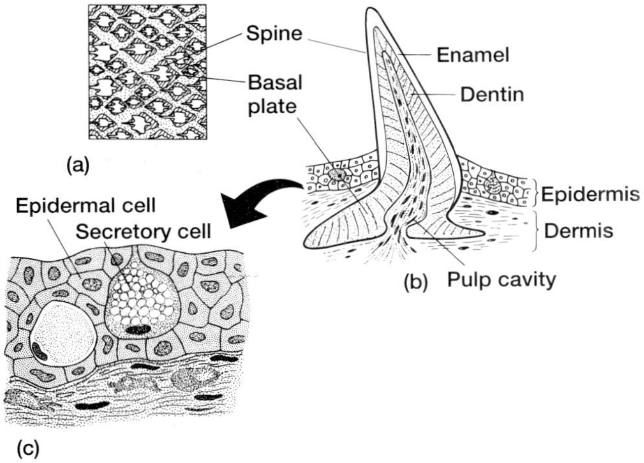

Figure 4.4
Shark skin. (a) Surface view of the skin showing regular arrangement of projecting placoid scales. (b) Section through a placoid scale of a shark. The projecting scale consists of enamel and dentin around a pulp cavity. (c) Close-up of the epidermis, which contains numerous secretory cells.
(a) Based on H.M. Smith, 1960, Evolution of CHordate Structure, Holt, Rinehart and Winson, Inc.
(b) Redrawn from J. Bereiter-Han, et al., Biology of the Integument, Vol. 2, 1986 Springer-Verlag, Berlin

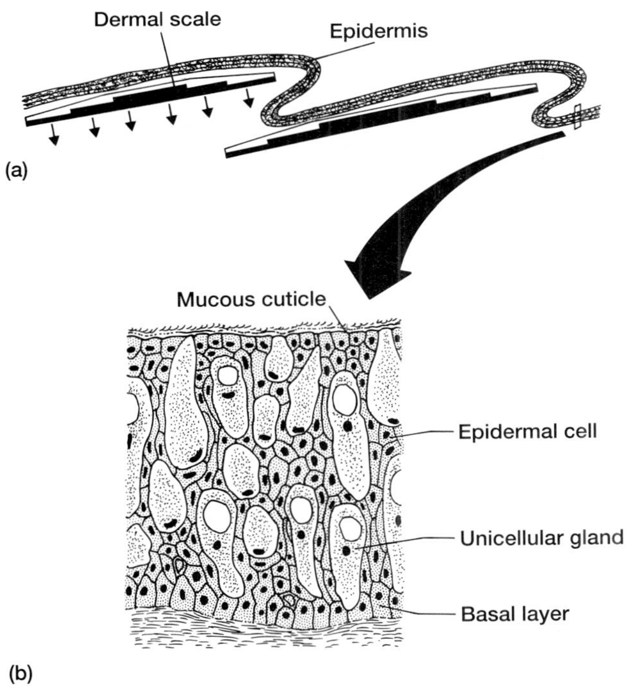

Figure 4.5
Bony fish skin. (a) Arrangement of dermal scales within the skin of a teleost fish (arrows indicate the direction of scale growth).
(b) Enlargement of epidermis. Note epidermal cells and unicellular glands.
(a) After Spearman.

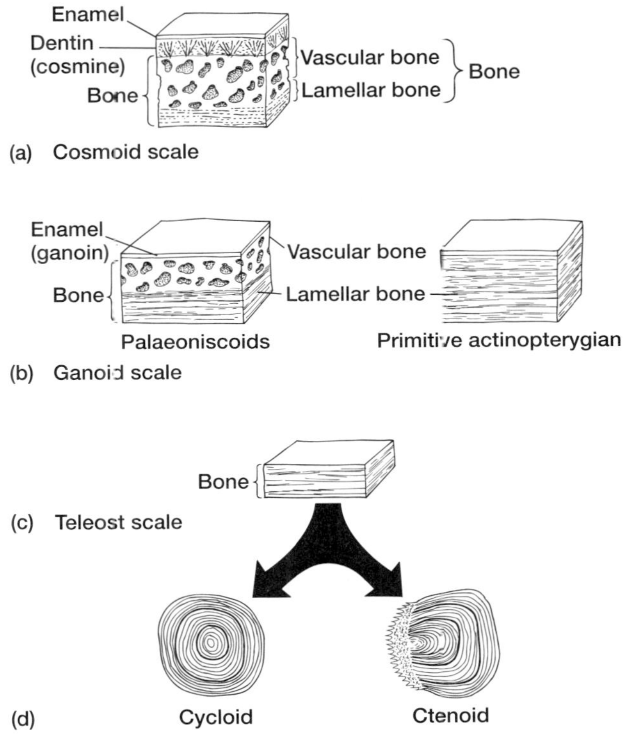

Figure 4.6
Scale types of bony fishes. Cross section of a (a) cosmoid scale, (b) ganoid scale, and (c) teleost scale. (d) Surface views of the two types of the teleost scale, cycloid and ctenoid scales.

loss of the enamel surface. This produces, in teleosts, a rather distinctive scale.

The teleost scale lacks enamel, dentin, and a vascular layer of bone. Only lamellar bone remains, and it is acellular and mostly noncalcified (figure 4.6c). Two kinds of teleost scales are recognized. One is the **cycloid scale,** composed of concentric rings, or **circuli**. The other is the **ctenoid scale,** with a fringe of projections along its posterior margin (figure 4.6d). New circuli are laid down, like rings in a tree, as the fish grows. Annual cycles are evident in the groupings of these circuli and allow general aging of individual fish from this pattern in their scales.

→ Compare the structure of ganoid, cycloid, and ctenoid scales using a hand lens or low power magnification on a microscope. Compare their overall shapes, thickness, and ring patterns. If pieces of dried skin are available, notice how the scales are interconnected.

Tetrapods

Although keratinization occurs in fish, it is among terrestrial vertebrates that it becomes an extensive and a significant feature of the integument. Lipids are often added during the process of keratinization or spread across the surface from specialized glands. These lipids plus the cornified layer increase the resistance of the tetrapod skin to desiccation.

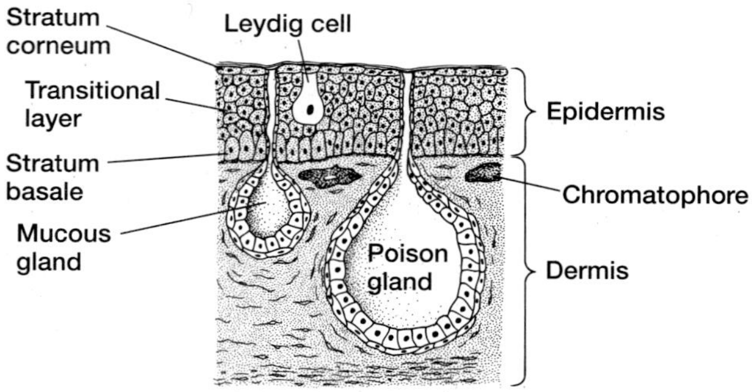

Figure 4.7
Amphibian skin. Diagrammatic view of amphibian skin, showing mucous and poison glands that empty their secretions through short ducts to the surface of the epidermis.

Figure 4.8
Reptile skin. Epidermal scales. The extent of projection and overlap of epidermal scales varies among reptiles and even along the body of the same individual. Snake body scales (top) and tubercular scales of many lizards (bottom) are illustrated. Between scales is a thinned area of epidermis, a "hinge" allowing skin flexibility.
After Maderson.

Multicellular glands are more common in the skin of tetrapods than of fishes. In fishes, the skin is coated by the products of unicellular glands at or near the surface. But in the skin of tetrapods, multicellular glands are usually sunk into the dermis. The products of these glands must pass through ducts that pierce the cornified layer to release their products on the surface.

Amphibians

Amphibians are of special interest, because during their lives they usually metamorphose from an aquatic to a terrestrial form. Phylogenetically, amphibians are also transitional between aquatic and terrestrial vertebrates. In most modern amphibians, the skin is specialized as a respiratory surface for **cutaneous respiration,** a form of gas exchange between the air and capillary beds in the lower epidermis and dermis. In fact, some salamanders lack lungs and depend entirely upon respiration through the skin to meet their metabolic needs.

→ Examine a cross section of frog or salamander skin at low magnification on a microscope. Distinguish between the upper epidermis and the lower dermis. Identify two types of glands that reside in the dermis but whose ducts extend upward through the epidermis (figure 4.7). The mucous glands tend to be smaller; each gland is made up of a little cluster of cells that release their product into a common duct. The poison glands (= granular glands) tend to be larger and often contain stored secretion within the lumen of each gland. You may need to use a magnification of about 400× to distinguish between these glands. The secretions of poison glands tend to be distasteful or even toxic to predators, but are usually of little concern to humans unless they are eaten or injected into the blood. Chromatophores may occasionally be found in the amphibian epidermis, but most reside in the dermis. Capillary beds, restricted to the dermis in most vertebrates, reach into the lower part of the epidermis in amphibians, a feature supporting cutaneous respiration.

Reptiles

The skin of reptiles reflects their greater commitment to a terrestrial existence. Keratinization is much more extensive, and skin glands are fewer than in amphibians. Scales are present, but these are fundamentally different from the dermal scales of fishes, which are built around bone of dermal origin. The reptilian scale usually lacks the bony undersupport or any significant contribution from the dermis. Instead, it is a fold in the surface epidermis; hence, it is an epidermal scale. The junction between adjacent epidermal scales is the flexible **hinge** (figure 4.8).

Reptiles shed pieces of the cornified layer of skin, often in large or complete sections such as in snakes. Since they shed only an upper layer of the epidermis, it is clearly improper to say that reptiles shed their skin. In turtles and crocodiles, sloughing of the epidermis is modest, comparable to birds and mammals, where small flakes fall off at irregular intervals.

The shell of turtles is a composite structure. The **carapace** (figure 5.12) that forms the dorsal half of the shell incorporates dermal bone fused with expanded ribs and vertebrae. Ventrally, the **plastron** represents fused dermal bones along the belly. On the surface of both carapace and plastron, keratinized plates of epidermis cover the underlying bone. Although not usually associated with scales, dermal bone is present in many reptiles. The **gastralia,** a collection of bones in the abdominal area, are examples.

→ Examine a shed "snake skin," noting the thin superficial nature of the cornified layer. Compare it to a piece of skin from a lizard or snake. On the reptile skin, note the shape and arrangement of the individual scales and find the flexible hinges between them. Identify the carapace, plastron, and keratinized epidermal plates of a turtle shell.

On a mounted alligator skeleton, identify the riblike gastralia on the ventral side just caudal to the true ribs. Note that unlike typical ribs, the gastralia do not articulate with the vertebrae.

Birds

Basic Structure The feathers of birds have been called nothing more than elaborate reptilian scales. This oversimplifies the homology but probably not by much. Certainly the

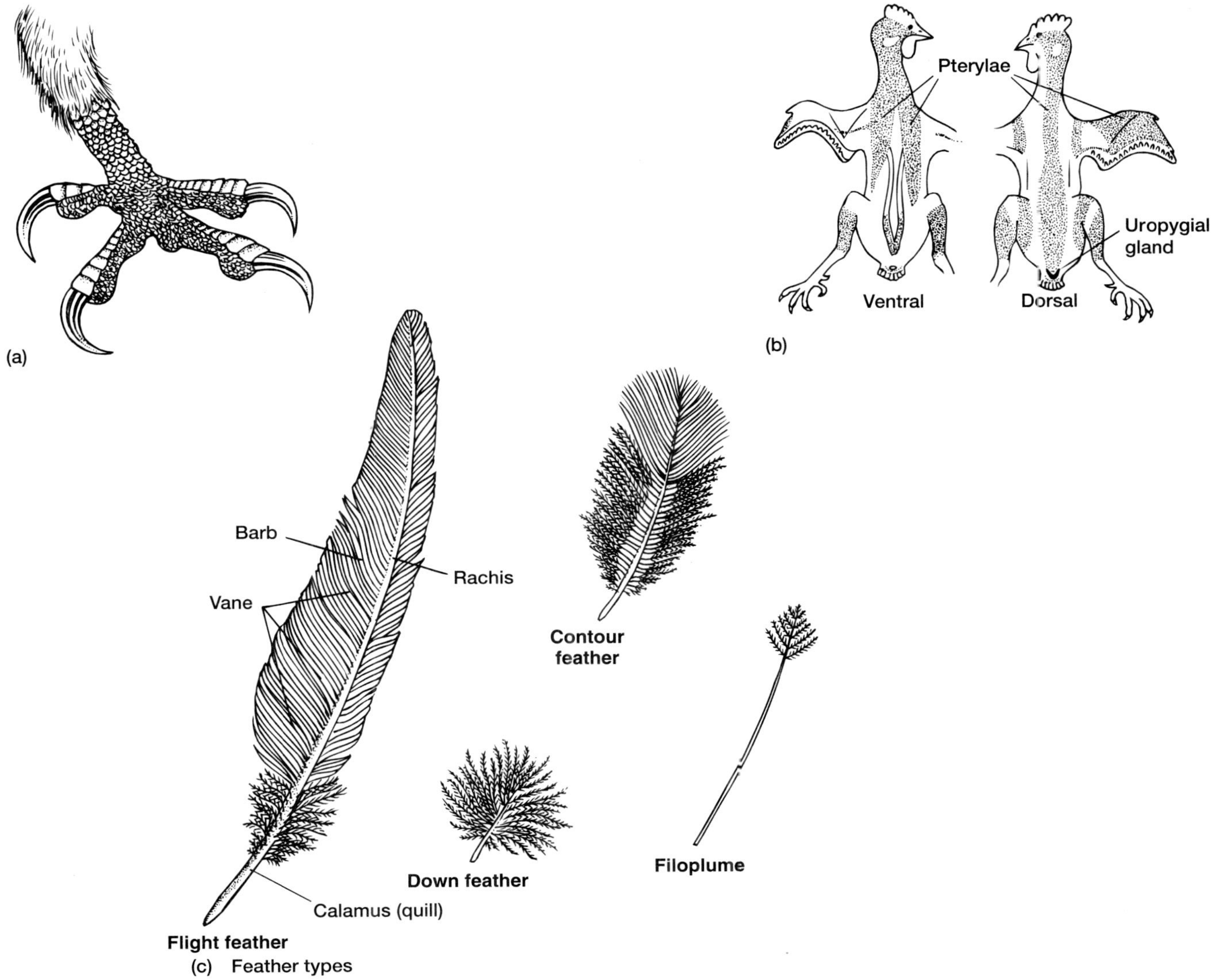

Figure 4.9
Epidermal derivatives in the bird. (a) Epidermal scales are present on the feet and legs of birds. (b) Feathers arise along specific pterylae (or feather) tracts. (c) Feather types. Flight feathers constitute the major locomotor surfaces. Down feathers lie close to the skin as thermal insulation. Contour feathers aerodynamically shape the surface of a bird. Filoplumes are often specialized for display.
After H.M. Smith

presence in birds of **epidermal scales** along the legs and feet testifies to the close phylogenetic relationship between birds and reptiles (figure 4.9a).

→ Examine a piece of skin from the leg of a bird to compare these epidermal scales to those of reptiles.

In the epidermis, two regions are usually recognized: the **stratum basale** and the **stratum corneum**. Between them is the transitional layer of cells transforming into the keratinized surface of the corneum (figure 4.10). Chromatophores and pigment granules are interspersed throughout the epidermis (figure 4.10).

Feathers Feathers distinguish birds from all other vertebrates. Feathers can be structurally elaborate and come in a variety of forms. Yet, feathers are essentially products of the skin, principally of the epidermis and the keratinizing system. They are nonvascular and nonnervous and are laid out along distinctive tracts, termed **pterylae,** on the surface of the body (figure 4.9b). They are replaced during one or more molts each year. Feathers develop embryologically from feather **follicles,** invaginations of the epidermis that dip into the underlying dermis. Here the root, in association with a dermal pulp cavity, begins to form the feather. The feather itself grows outward in a sheathed case (figure 4.11a).

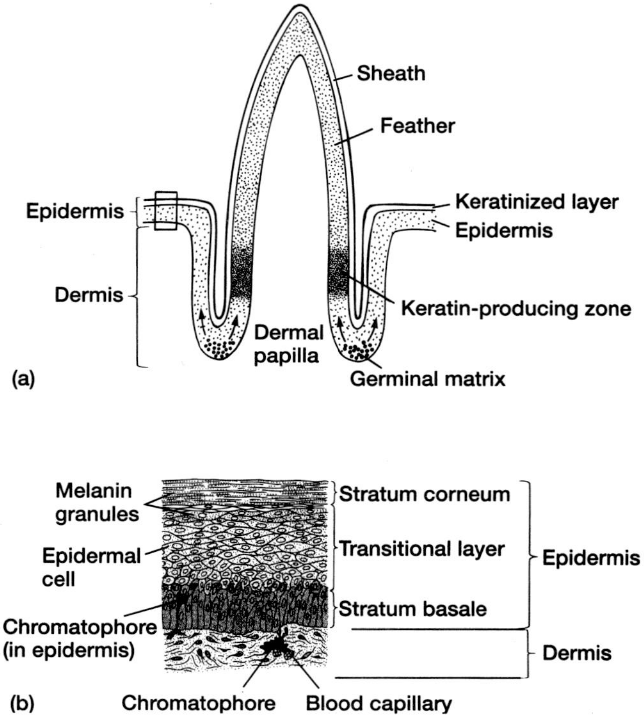

Figure 4.10
Bird skin. (a) Growth of a feather follicle. The feather itself forms within a sheath that, like the feather, is a keratinized derivative of the epidermis. (b) Section of skin showing the stratum basale and the keratinized surface layer, the stratum corneum. Cells moving out of the basal layer spend time first in the transitional layer before reaching the surface. This middle transitional layer is equivalent to the spinosum and granulosum layers of mammals.
(a) After Spearman.

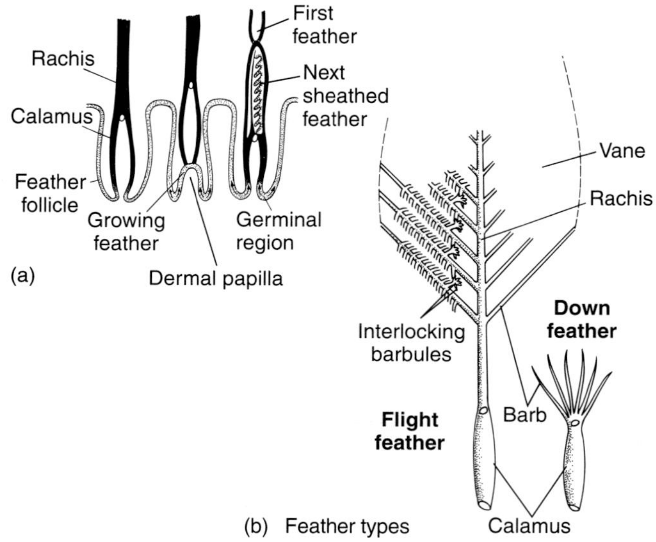

Figure 4.11
Feather growth and morphology. (a) Successive stages in the growth of new feathers. During molting of old feathers, new ones arise in the feather follicles to replace those shed. (b) General morphology of flight and down feathers.
(a) After Spearman.

→ Examine a feather and identify the following components. Within the sheath, the central axis is divided into a distal rachis that bears barbs with interlocking connections, termed barbules, and a proximal calamus that attaches to the body (figures 4.9c and 4.11b). Flight feathers of the wings are characterized by a long rachis and prominent vane, the broad surface of the feather also important in display and insulation.

Mammals

As in other vertebrates, the two main layers of the mammalian skin are epidermis and dermis. Beneath lies the hypodermis, or superficial fascia, composed of connective tissue and fat.

Epidermis The epidermis may be locally specialized into hair, nails, or glands. Epithelial cells of the epidermis are **keratinocytes** and belong to the keratinizing system that forms the dead, superficial cornified layer of the skin. The surface keratinized cells are continually lost and replaced by cells arising primarily from the deepest layer of the epidermis, the **stratum basale**. Cells within the stratum basale divide mitotically, producing cells that restore the stem cell population and epidermal cells that are pushed outward. As the epidermal cells move outward, they pass through keratinizing stages exhibited as distinct, successive layers toward the surface. Progressing toward the surface, several additional layers—**stratum spinosum, stratum granulosum,** often a **stratum lucidum,** and a **stratum corneum**—are encountered (figure 4.12). The process of keratinization is most distinct in regions of the body where the skin is thickest, as on the soles of feet. Elsewhere, these layers, especially the stratum lucidum, may be less distinct.

Chromatophores are another prominent cell in the epidermis. They secrete granules of the pigment **melanin,** which are passed directly to epithelial cells and eventually carried into the stratum corneum or into the shafts of hair. Skin color results from a combination of the yellow stratum corneum, red contributed by underlying blood vessels, and the dark pigment granules secreted by chromatophores.

→ Examine a section of mammalian skin under a microscope beginning with low power and moving to a higher magnification near 400×. Identify the five regions, extending from bottom to top: stratum basale, stratum spinosum, stratum granulosum, stratum lucidum, and stratum corneum of the epidermis. Also identify chromatophores containing dark brown granules. Keep the section of skin on the microscope to examine the components of the dermis described below.

Dermis The mammalian dermis is double layered. The outer **papillary layer** pushes fingerlike projections,

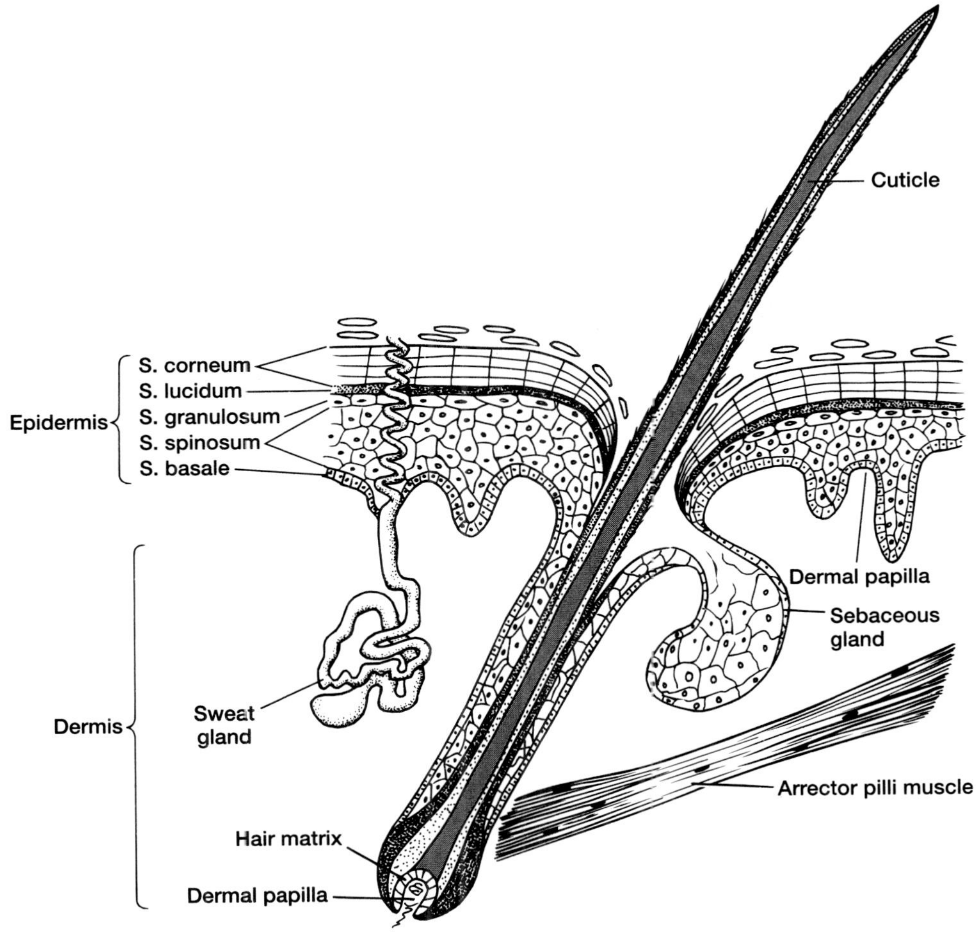

Figure 4.12
Mammalian skin. The epidermis is differentiated into distinct layers. As in other vertebrates, the deepest is the stratum basale, which through mitotic division produces cells that as they age become successively part of the stratum spinosum, stratum granulosum, often the stratum lucidum, and finally the surface stratum corneum. The dermis pokes up dermal papillae that give the overlying epidermis an undulating appearance. Sweat glands, hair follicles, and sensory receptors lie within the dermis. Notice that the sweat ducts pass through the overlying epidermis to release their watery secretions on the surface of the skin.

termed **dermal papillae,** up into the overlying epidermis (figure 4.12). These papillae increase the surface area between the epidermis and dermis and strengthen the connection between these two layers. The deeper **reticular layer** includes irregularly arranged fibrous connective tissue and anchors the dermis to the underlying fascia. **Blood vessels, nerves,** and **smooth muscle** occupy the dermis but do not reach into the epidermis. However, hair originates in the dermis and extends through the epidermis. The base of a hair is the **root**; its remaining length constitutes the **shaft**. The outer surface of the shaft often forms a scaly **cuticle** (figure 4.12). **Arrector pili muscles** in the dermis pull on the base of a hair to cause the hair shaft to thrust upward. The skin immediately surrounding the hair is also pulled up forming a goose bump or goose pimple. This erect hair forms a thicker boundary layer of trapped warm air that better insulates the body when the skin is chilled. In addition, erect hair can also be a defensive maneuver to make a mammal appear larger.

The mammalian dermis still produces **dermal bones,** but these contribute to the skull and pectoral girdle and only rarely form **dermal scales** in the skin. One exception is the armadillo, in which a secondary development of dermal bone arises under the epidermis.

Oil and sweat glands are also typically found in the mammalian dermis. **Sebaceous glands** release an oily material onto the hair shaft to keep the hair flexible. Highly coiled **sweat glands** in the dermis release their product through ducts that extend through the epidermis and out sweat pores at the surface.

→ Return to the section of mammalian skin. In the dermis, identify the papillary layer, dermal papilla, deeper reticular layer with irregularly arranged fibers, blood vessel, hair root, hair shaft, hair cuticle, arrector pili

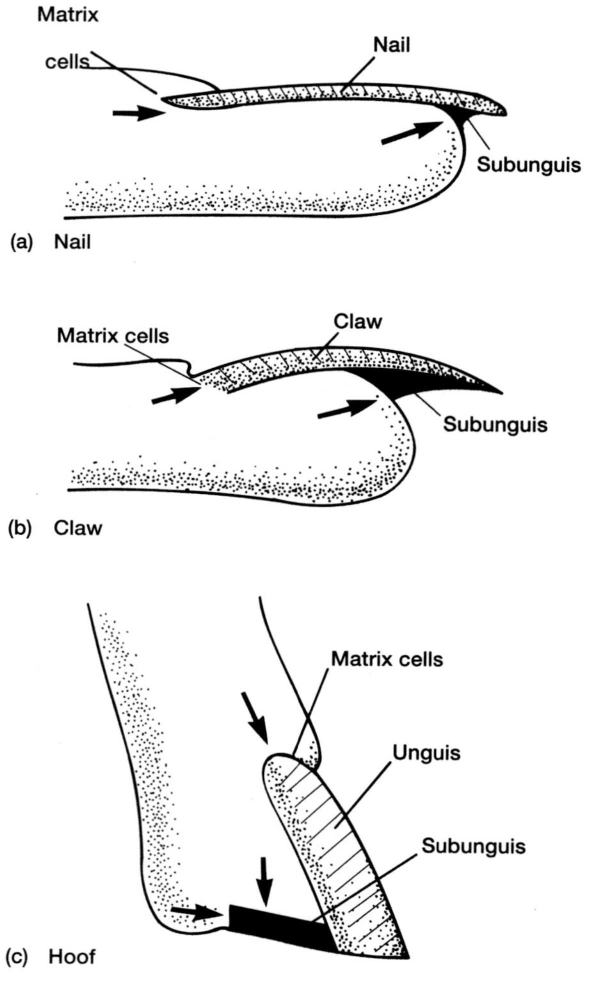

Figure 4.13
Epidermal derivatives. (a) The nail is a plate of cornified epithelium growing outward from proliferating matrix cells at its base and from the subunguis. (b) Claw. (c) Hoof.
After Spearman.

muscle, sebaceous gland, and a sweat gland. If available, examine the unique integument of the armadillo, noting the firm nature of its dermal shell.

Specializations of the Integument

Nails, Claws, and Hooves

Nails are plates of tightly compacted, cornified epithelial cells on the surface of fingers and toes; thus, they are products of the keratinizing system of the skin. The **nail matrix** forms new nail at the nail base by pushing the existing nail forward to replace that worn or broken at the free edge (figure 4.13a). Nails protect the tips of digits from inadvertent mechanical injury. They also help stabilize the skin at the tips of the fingers and toes, so that on the opposite side the skin can establish a secure friction grip on objects grasped.

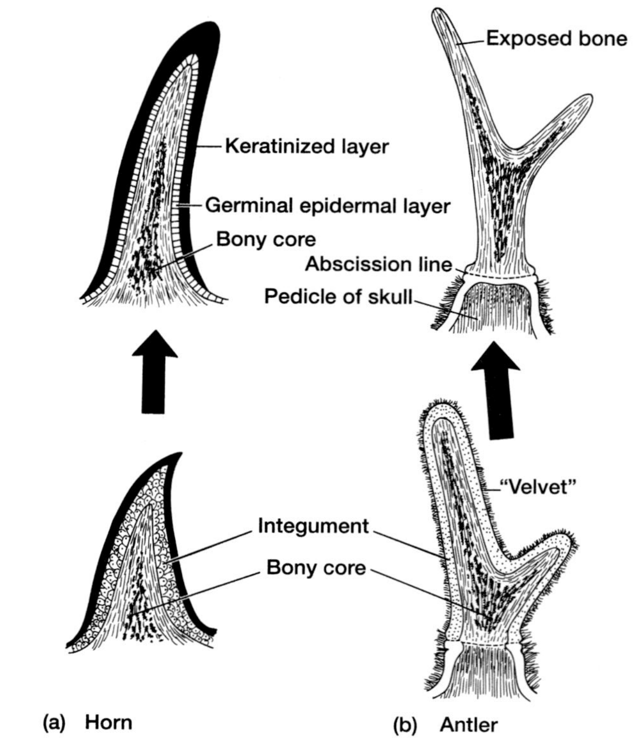

Figure 4.14
Horns and antlers. (a) Horns appear as outgrowths of the skull beneath the integument, which forms a keratinized sheath. Horns occur in bovids of both sexes and are usually retained year-round. (b) Antlers also appear as outgrowths of the skull beneath the overlying integument, which is referred to as "velvet" because of its appearance. Eventually this overlying velvet dries and falls away, leaving the bony antlers. Antlers are restricted to members of the deer family and, except for caribou (reindeer), they are present only in males. Antlers are shed and replaced annually.
After Modell.

Only primates have nails. In other vertebrates, the keratinizing system at the terminus of digits produces claws or hooves. **Claws** (figure 4.13b), or **talons,** are curved, laterally compressed, keratinized projections from the tips of digits. They are found in some amphibians and in most birds, reptiles, and mammals. **Hooves** are enlarged keratinized plates on the tips of the digits of ungulates (figure 4.13c).

→ Compare the relatively simple structure of talons, claws, and hooves that are available for study. Com-pare these structures to your own nails. Note how the shape of each contributes to specific functions that the animal performs.

Horns and Antlers

→ Examine the laboratory demonstrations of horns and antlers. Note the features of each as you read the descriptions that follow.

The skin, together with the underlying bone, contributes to both true horns and antlers. As these structures

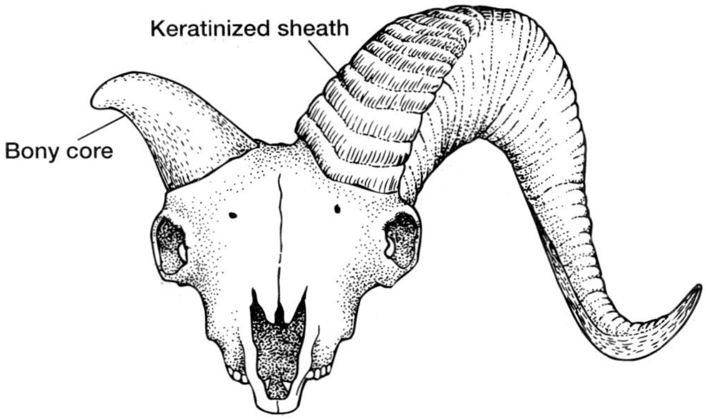

Figure 4.15
True horns of the mountain sheep (Bovidae). The cornified covering of the horn of the mountain sheep is removed on the right side of the skull to reveal the bony core.

take shape, the underlying bone rises up, carrying the overlying integument with it. In **horns,** the associated integument produces a tough, cornified sheath that fits over the bony core (figure 4.14a). In **antlers,** the overlying living skin (called "velvet") apparently shapes and provides vascular supply to the growing bone. Eventually the velvet falls away to unsheath the bare bone, the actual material of the finished antlers (figure 4.14b).

True antlers occur only in members of the family Cervidae (e.g., deer, elk, moose). Typically, only males have antlers. These are branched and shed annually. There are notable exceptions. Among caribou, both sexes have seasonal antlers. In deer, the antler usually consists of a main **beam,** from which branch shorter **tines,** or **points**. In yearling bucks, antlers are usually no more than prongs or spikes that may be forked. The number of tines tends to increase with age, although not exactly. In old age, antlers may even be deformed. In caribou and especially in moose, the main antler beam is compressed and **palmate,** or shovel-like, with a number of points projecting from the rim.

Among mammals, **true horns** are found among members of the family Bovidae (e.g., cattle, antelope, sheep, goats, bison, wildebeest). Commonly, horns occur in both males and females, are retained year-round, and continue to grow throughout the life of the individual. The horn is unbranched and formed of a bony core and a keratinized sheath (figure 4.15). Those of the males are designed to withstand the forces encountered during headbutting combat. In large species, females usually have horns as well, although they are not as large and curved as in males. In small species, females are often hornless.

Unlike the true horns of bovids, horns of the pronghorn, family Antilocapridae, are forked in adult males (figure 4.16a). The old outer cornified sheath, but not the bony core, is shed annually in early winter. The new sheath beneath, already in place, becomes fully grown and forked by summer. Female pronghorns also have horns in which their keratinized sheath is replaced annually, but these are usually much smaller and only slightly forked. The horns of giraffes are different still (figure 4.16b). They develop from separate, cartilaginous processes that ossify, fuse to the top of the skull, and remain covered with living, noncornified skin. The rhinoceros horn does not include a bony core, so it is exclusively a product of the integument. It forms from compacted keratinous fibers (figure 4.16c).

Baleen

The integument within the mouths of mysticete whales forms plates of **baleen** that act as strainers to extract krill

Figure 4.16
Other types of horns. (a) In pronghorns, the bony core of the horns is unbranched, but the cornified sheath is branched. (b) Giraffe horns are small ossified knobs covered by the integument. (c) Rhinoceroses have several horns that rest on a low knob on the skull, but these horns have no inner core of bone. As outgrowths of the epidermis alone, they are mainly composed of compacted keratinized fibers.
After Model.

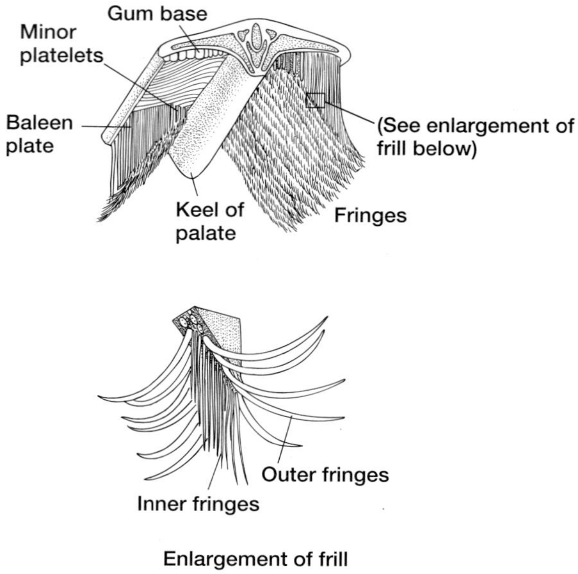

Figure 4.17
Baleen from a whale. The lining to the mouth includes an epithelium with the ability to form keratinized structures. Groups of outgrowing epithelium become keratinized and frilly to form the baleen.
Adapted from A. Pivorunas, 1979, "The feeding mechanisms of baleen whales," American Scientist, July/August.

from water gulped in the distended mouth. Although it is sometimes referred to as "whalebone," baleen contains no bone. It is a series of keratinized plates that arise from the integument. During its formation, groups of dermal papillae extend and lengthen outward, carrying the overlying epidermis. The epidermis forms a cornified layer over the surface of these projecting papillae. Collectively, these papillae and their covering of epidermis constitute the plates of frilled baleen (figure 4.17).

→ Examine a piece of baleen noting the long central keel and lateral fringes.

Skeletal System

5

Tissues of the Skeletal System

Connective tissues include bone, cartilage, fibrous connective tissue, adipose tissue, and blood (figure 5.1). The extracellular *matrix* of connective tissues determines the physical properties of the tissue, and hence its functional role. This matrix is composed of protein fibers and a surrounding ground substance. Cartilage and bone are *specialized connective tissues* (figure 5.1) that constitute the skeletal system proper. *Generalized connective tissues* (figure 5.1) are dispersed widely throughout the body where they hold together other tissues, store nutrients, and perform other functions. *Tendons,* which join bone and muscle, and *ligaments,* which join bones, are examples of general connective tissues associated with cartilage and bone.

The fundamental cell of bone is the **osteocyte,** and of cartilage the **chondrocyte.** In bone, two other cell types can be identified under the microscope: **osteoblasts,** immature osteocytes involved in overall new bone deposition, and **osteoclasts,** involved in overall bone removal through the breakdown of bone matrix. Cartilage and bone also vary in the composition and abundance of the substances deposited within the matrix. These matrix materials give bone and the various types of cartilage unique mechanical properties.

All bone and most types of cartilage are surrounded by a similar appearing coat of fibrous connective tissue. This fibrous sheath is termed the **periosteum** (bone) or **perichondrium** (cartilage) depending upon which tissue it encloses.

Cartilage

Cartilage is made up of chondrocytes and a matrix containing two basic components: a **ground substance** and **protein fibers.** The ground substance is chemically composed of **polysaccharides,** most notably chondroitin sulfates and hyaluronate. Inorganic salts, such as calcium, may also be present. The protein fibers are usually **collagen fibers** or **elastin fibers** in various abundances, depending upon the type of cartilage.

Cartilage tissues do not directly receive a blood supply. Instead, the closest blood vessels are usually in the perichondrium. Thus, nutrients and gases must undergo distant diffusion from blood, through the matrix, to the chondrocytes. Similarly, no nerves directly penetrate cartilage. Cartilage may be heavily invested with calcium salts. Such calcified cartilage is found, for instance, in the shark skeleton and selectively in some stages of bone development.

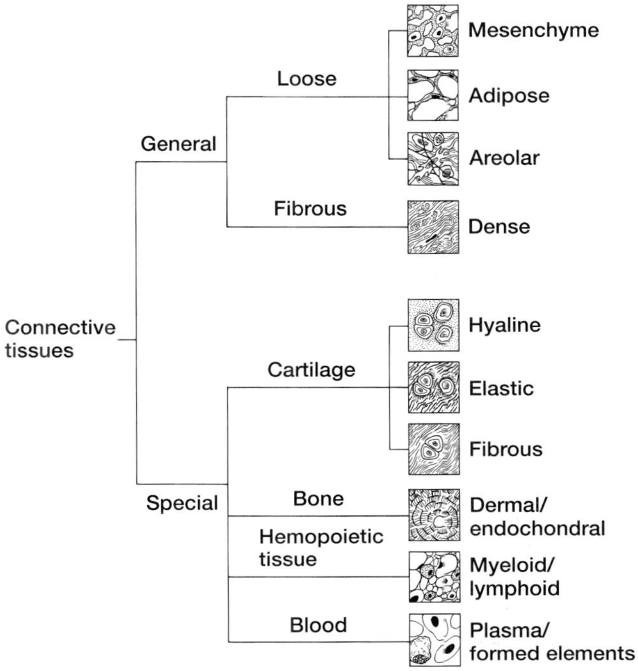

Figure 5.1
Categories of connective tissue. Bone, cartilage, fibrous tissue, adipose tissue, and blood are some of the body's connective tissues. Each type of connective tissue includes a distinctive cell type surrounded by an extracellular matrix.

Form & Function

Reach up with your hand and bend your ear. Let go, and your ear returns to its previous shape. Did this hurt? Will there be excessive bruising? *Imagine* doing this to a bone in your arm. Why would the bone react differently? Your ear contains elastic cartilage, with a matrix packed with elastic fibers that act like a rubber band, returning the ear to its original shape. In addition, your ear's cartilage lacks a direct blood and nerve supply, resulting in little to no pain. Bones, with a rigid calcium matrix and well supplied with blood and nerves, don't bend as easily and will be painful if bent too far.

→ Three types of cartilage are recognized primarily upon the basis of the type and abundance of protein fibers in the matrix. Examine representative cross sections of each of these tissues under the microscope and identify the structures indicated in figure 5.2.

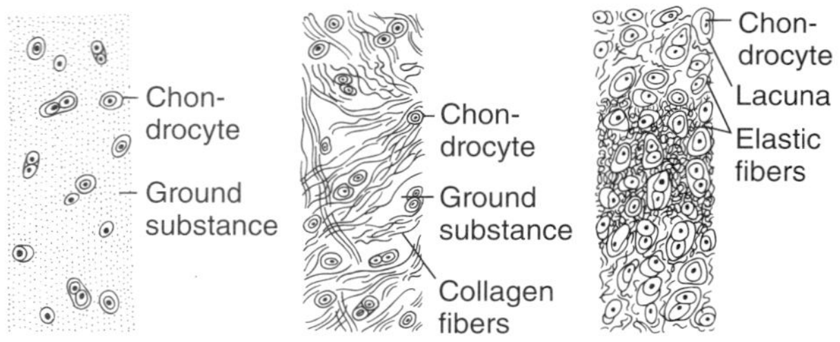

(a) Hyaline cartilage (b) Fibrocartilage (c) Elastic cartilage

Figure 5.2
Types of cartilage. The cartilage cell, or chondrocyte, is surrounded by a matrix composed of a ground substance and protein fibers. (a) Fibers are not apparent in the matrix of hyaline cartilage when it is viewed with a light microscope. (b) Collagen fibers are abundant in fibrocartilage, giving it mechanical resistance to tensile forces. (c) Elastin, the predominant protein fiber in elastic cartilage, makes it springy and flexible.

Hyaline Cartilage

→ Collagen fibers are present in the matrix surrounding chondrocytes, but never in sufficient abundance to be easily visualized with standard preparation in the microscope. Note the absence of blood vessels. Hyaline cartilage is the most common type of cartilage, found widely throughout the body. Hyaline cartilage is found at the tips of ribs, within developing bones, and at the ends of bones where it forms a smooth surface and thus assists the gliding motion of joints.

Fibrocartilage

→ Fibrocartilage is similar to hyaline cartilage except that the collagen fibers are present in much greater abundance so that their presence can usually be demonstrated routinely under microscopic examination. Chondrocytes are often lined up into short rows, formed because of the collagen fibers around them. The numerous strands of collagen give fibrocartilage its special physical property of being more resistant to various forces than other types of cartilage. Thus, fibrocartilage is found at sites where tensile and shear forces are thought to be significant, such as in the pubic symphysis and intervertebral disks.

Elastic Cartilage

→ As the name implies, elastic cartilage is flexible, a physical property attributable to the abundance of elastic fibers within the matrix. Chondrocytes are also found within the matrix. Elastic cartilage is found at sites where flexible support is required, such as the end of the nose, epiglottis, and pinna of the ear.

Bone

Bone is a calcified connective tissue, but one in which the calcium and other inorganic salts have been deposited in a regularly organized matrix around a central core. In compact bone this repeated unit of bone is the **osteon.** Unlike cartilage, bone is vascular and supplied directly with nerves. It is responsive to mechanical stresses. Thus, if its use increases, it undergoes **hypertrophy** (increase in bone density and deposition); if use decreases, it undergoes **atrophy** (decrease in bone density).

Classification of bone is done by several criteria. If position is used, we recognize **cortical bone** in the outer boundary or cortex of a bone and **medullary bone** that lies within the core. If gross visual appearance is used, then two types of bone are recognized: **cancellous** (= spongy) **bone,** which is porous, and **compact bone,** which appears dense to the naked eye (figure 5.3). A sheet of fibrous connective tissue surrounds the outside of a bone forming the **periosteum.**

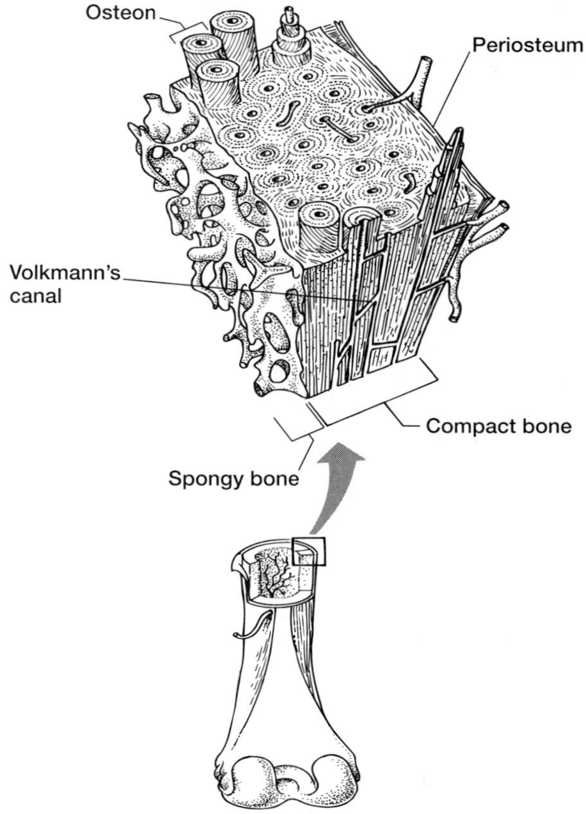

Figure 5.3
Bone architecture. Osteons make up compact bone. Each osteon is a series of concentric rings of osteocytes and their matrix. Nerves and blood vessels pass through a central canal within each osteon. Diagonal connections, known as Volkmann's canals, allow blood vessels to interconnect between osteons. As new osteons form, they usually override existing, older osteons as part of the ongoing dynamic process of bone remodeling.
After Krstic'.

→ Examine a cross section of developing endochondral bone. Identify cancellous bone composed of interconnected islands deep to compact bone composed of repeated osteon units (figure 5.3). Identify the thin periosteum surrounding the bone.

Form & Function

Vertebrates that reduce their usage of a limb experience a rapid decline in bone density (mass) in that appendage. The reduction of bone mass is usually measurable after as little as a week. For example, humans that injure an ankle joint favor the leg, use it less, and lose bone mass and muscle tone in the injured leg. Astronauts living in weightless space have similar consequences. The loss of gravity reduces the workload of the skeletal and muscular systems. Significant declines in bone mass routinely occur in their bodies after living for weeks or months in space. After returning to Earth, astronauts may not recover former bone mass if the loss in space was significant. This, of course, poses a serious medical problem that must be overcome if humans are to remain in gravity-free space vehicles for years at a time.

Table 5.1

Patterns of the Embryonic Origin Within the Vertebrate Skeleton.

Endochondral	Dermal
vertebrae	
bones in limbs*	
pelvic girdle	
pectoral girdle	pectoral girdle
(scapula, coracoid)	(all other bones)
skull	
chondrocranium, splanchnocranium	dermatocranium

*Except sesamoid bones (patella, pisiform).

If the embryonic origin is used to classify bone, then three types of bone, each with a distinctive embryonic history, are recognized. **Endochondral bone** development is perhaps the most elaborate. Basically, it involves the formation of a cartilage model of the future bone from mesenchyme and the subsequent replacement of this cartilage model by bone tissue. **Dermal bone** forms directly from mesenchyme without a cartilage precursor. **Sesamoid bones** form within tendons and are not preceded by a cartilage model. The patella bone of the knee and the pisiform bone of the wrist are examples.

Regardless of the method of embryonic development—endochondral, dermal, or sesamoid—the resulting bone may be cancellous or compact in appearance. Thus, the gross visual appearance of a bone will not tell you its method of embryonic origin. The skeleton is a composite of endochondral, dermal bones, and sesamoid bones. You should keep in mind which kind of development characterizes any given bone in the body.

Rather than memorize each bone and its developmental history in isolation, consider the general part of the skeletal system to which it belongs: vertebral column, skull, or appendicular skeleton (table 5.1). All bones of the vertebral column are endochondral, as are bones of the limbs, and all elements of the pelvic girdle. The pectoral girdle is a composite structure: scapula and coracoids are endochondral, but all other elements are dermal. The skull is also a composite structure: chondrocranium and splanchnocranium are endochondral; dermatocranium is dermal.

Table 5.2

Common Anatomical Terminology and Directions

Cavities
- **canal**—an extensive tubelike opening
- **fenestra**—an opening within a bony braincase
- **fissure**—a long narrow cleft
- **foramen**—a perforation or hole through a tissue wall
- **fossa**—a large hollowed cavity or depression
- **sinus**—a large cavity, usually within a bone

Projections
- **apophysis**— an outgrowth or bony protuberance
- **condyle**—a smooth articular end of a long bone
- **head**—general name for the proximal end of a long bone
- **transverse process**—a general term for a lateral process from a vertebra
- **tuberosity**—a large, roughened projection

Directions
- **lateral**—away from the midline of the body
- **medial**—toward the midline of the body
- **proximal**—toward the point of attachment of an appendage
- **distal**—away from the point of attachment of an appendage
- **dorsal**—toward the back or upper surface of the body
- **ventral**—toward the belly or lower surface of the body
- **cranial**—toward the head or front end of the body
- **caudal**—toward the tail or back end of the body

Divisions of the Skeletal System

You will find several ways to collectively group skeletal elements. Some persons group endochondral bones and cartilages associated with the mouth and pharynx into the **visceral skeleton.** All other endochondral bones and cartilages belong to the **somatic skeleton.** The somatic skeleton in turn is divided into the axial skeleton (skull and backbone) and appendicular skeleton (limbs and girdles). For convenience only, we divide laboratory material into **Postcranial** and **Cranial Skeleton.** Table 5.2 provides a quick reference for basic terminology used to describe skeletal and general anatomical structure. Quickly learn to use this terminology, as it will provide essential clues to the structures you seek to identify. You may also wish to review the planes of the body in chapter 1, figure 1.5.

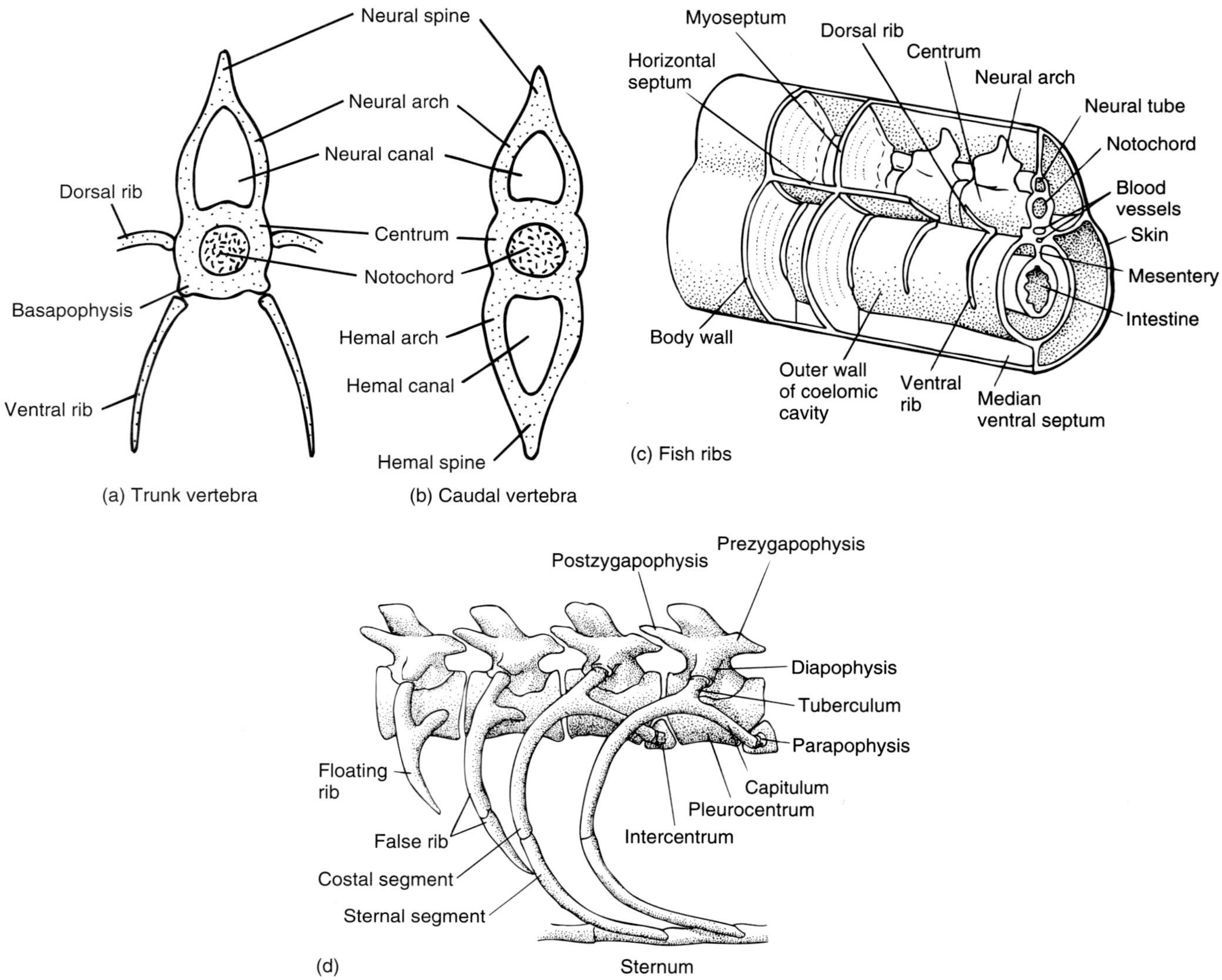

Figure 5.4
Ribs. (a) Cross section of trunk vertebra of a fish. (b) Cross section of caudal vertebra of a fish. (c) In fishes, dorsal ribs develop where myosepta intersect with the horizontal septum, and ventral ribs develop where myosepta meet the wall of the body cavity. (d) Amniote ribs. Ribs are named on the basis of their articulation with the sternum (true ribs), with each other (false ribs), or with nothing ventrally (floating ribs). Primitively, ribs are bicipital, having two heads, a capitulum and a tuberculum, that articulate respectively with the parapophysis on the intercentrum or the diapophysis on the neural arch. The body of the rib may differentiate into a dorsal part, the vertebral rib or costal segment, and a ventral part, the sternal rib or segment that articulates with the sternum.

Postcranial Skeleton

Vertebrae and Ribs

Two general phylogenetic trends occur in the **vertebral column** or backbone. First, the **notochord** becomes reduced as the vertebrae (segmental cartilaginous or bony elements) increase in prominence to surround and mostly replace the notochord. Second, the chains of vertebrae in the column differentiate (change in form and function), especially in later terrestrial vertebrates.

Typically each **vertebra** (pl. *vertebrae*) consists of a middorsal **neural arch** enclosing the **nerve cord,** a midventral **hemal arch** enclosing blood vessels, and a **centrum,** a spool-shaped body enclosing the notochord (figure 5.4a, b). Projecting beyond each arch is a **neural spine** and **hemal spine,** respectively.

Two types of segmental ribs, dorsal and ventral (figure 5.4a), occur primitively. Both form in **myosepta** separating muscles (figure 5.4c). **Ventral** or **subperitoneal ribs** lie beneath the axial muscles, just outside the peritoneum of the coelom. They are serially homologous with the hemal arch. **Dorsal ribs** form at the intersection of myosepta with a longitudinal sheet of fibrous connective tissue, the **horizontal skeletogenous septum.** Opinion varies as to the homology of fish and tetrapod ribs (figure 5.4d), although most now hold that tetrapod ribs derive from dorsal ribs of fish. Partly because of this uncertainty, the

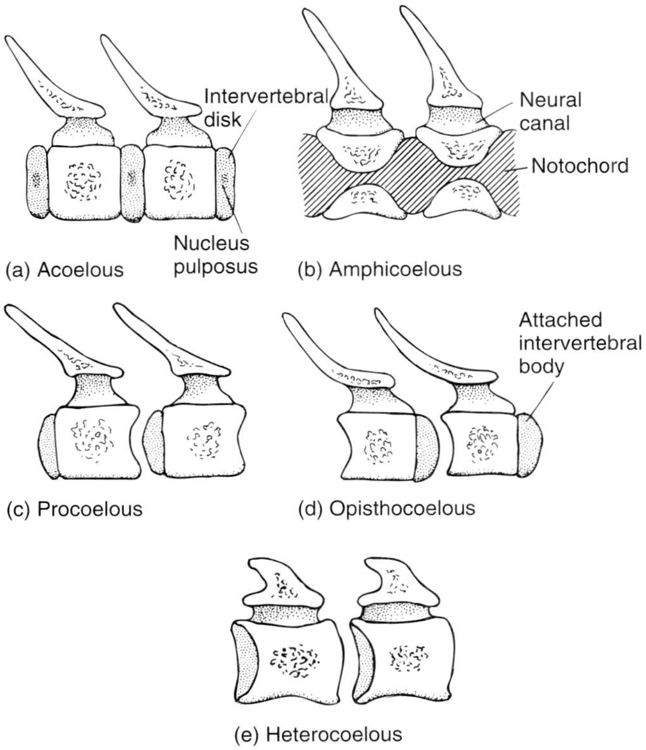

Figure 5.5
General centra shapes. Anterior is to the right. The shapes of articulating centra ends define specific anatomical types: (a) acoelous, both ends are flat; (b) amphicoelous, both ends are concave; (c) procoelous, anterior end is concave; (d) opisthocoelous, posterior end is concave; (e) heterocoelous, saddlelike articulating ends.
After George C. Kent and Larry Miller, Comparative Anatomy of the Vertebrates, 8th edition, 1997, McGraw-Hill Company, Inc., Dubuque, Iowa.

Table 5.3
The Structure of Centra

Shape of the ends of centra
- **amphicoelous**—both ends concave (amphi = "double"; coelo = "hollow")
- **acoelous (amphiplatyan)**—both ends flat (platy = "flat")
- **heterocoelous**—both ends saddle shaped (hetero = "different")
- **opisthocoelous**—anterior convex, posterior concave (opistho = "behind")
- **procoelous**—anterior concave, posterior convex (pro = "in front")

Common types of Apophyses (Processes) of centra
- **basapophyses**—(basapophysis, singular) paired ventrolateral processes on centrum, the likely remnants of the hemal arch bases now the articulation with a ventral rib (or remaining hemal arch)
- **diapophysis** and **parapophysis**—lateral projections from the centrum that articulate respectively with dorsal (tuberculum) and ventral (capitulum) heads of a dorsal rib
- **hypapophysis**—unpaired midventral processes from centra
- **zygapophysis**—articulations between successive vertebrae

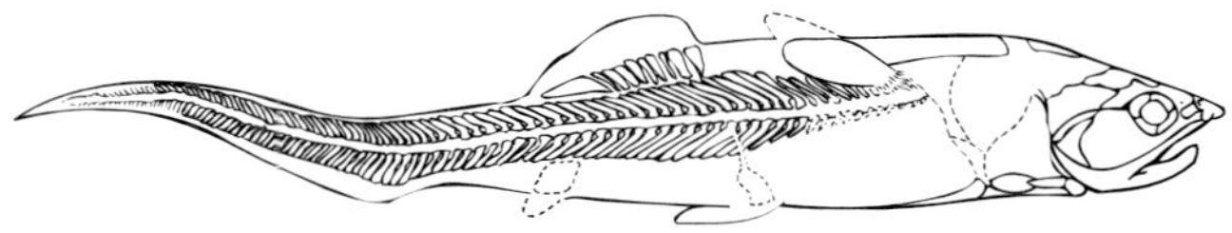

Figure 5.6
Axial skeleton of the placoderm *Coccosteus*, with prominent notochord supporting dorsal and ventral vertebral elements.
After J.A. Moy-Thomas, 1971, Palaeozoic Fishes.

anatomical descriptions usually do not attempt to specify whether a rib is dorsal or ventral, but term it simply "rib."

Since the function of the vertebral column in large part depends upon the shapes and articulations of vertebrae, much descriptive detail accompanies the centrum. Five principle centrum types are recognized on the basis of the shape of their ends (figure 5.5; table 5.3).

→ Generally, two regions and hence two vertebral types can be identified in the fish vertebral column: trunk and tail (or caudal) vertebrae.

Agnatha

→ **Ostracoderms** Ostracoderms are known only from fossils. Only trace impressions of vertebral elements have been found in fossils of some ostracoderm groups. These elements are probably small unossified pieces of vertebrae surrounding a prominent notochord. Thus, among ostracoderms, a strong notochord provided the central mechanical axis for the body.

→ **Cyclostomes** Lampreys possess vertebral elements, but these are small, cartilaginous structures resting dorsally upon a very prominent notochord that provides axial support for the body (see figure 3.3).

Placodermi

→ Some placoderms preserve evidence of a prominent notochord supporting ossified neural and hemal arches (figure 5.6).

Elasmobranchii—Shark

→ Examine segments of a shark vertebral column embedded in clear plastic blocks. Identify boundaries of each centrum (figure 5.7). Directly above each is the triangular neural arch that supports a low ridge, the neural spine. Between successive neural arches are inverted triangular elements, the interneural arches. Together neural and interneural arches enclose the neural canal through which passes the spinal cord. If you are looking at a vertebral section from the tail, then hemal arch and hemal spine are present. If the vertebral section comes from the trunk, then lateral projecting ridgelike basapophyses are present and neural and hemal spines are absent.

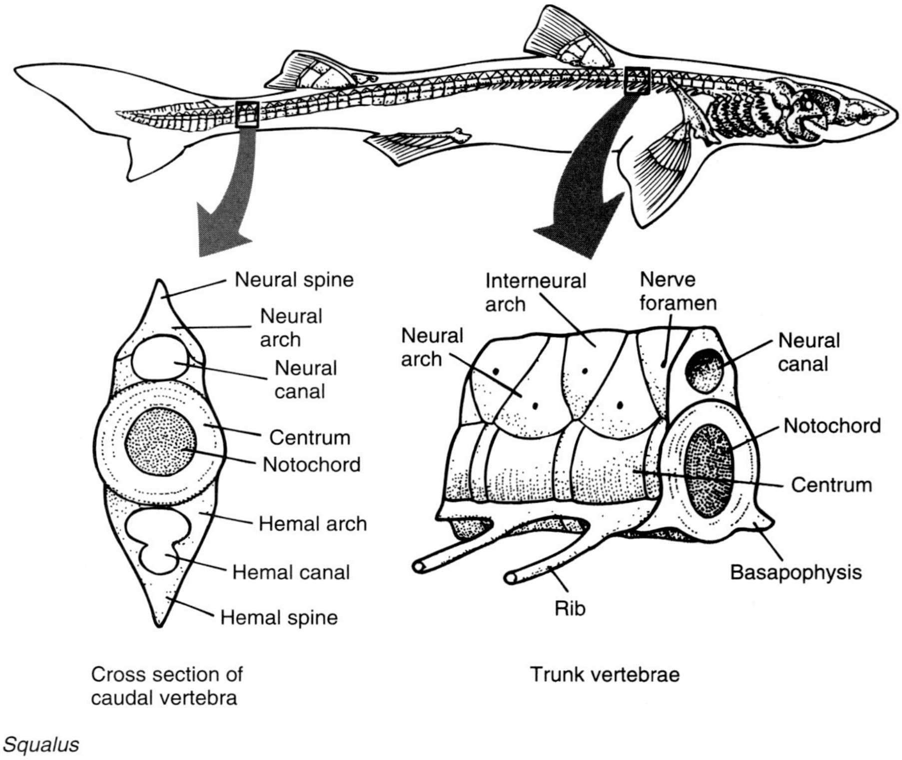

Figure 5.7
The axial skeleton in the shark *Squalus*. The vertebral elements tend to enlarge in elasmobranchs, surpassing the notochord as the major mechanical support for the body in modern sharks.

Within and through successive centra runs the notochordal canal, a canal constricted centrally, but expanded on the ends of each centrum so the notochord, filling this canal, likewise is constricted and expanded.

Examine a preserved shark skeleton in tall jars to identify the same specific features just located. In addition, examine cross sections of a shark tail to appreciate the vertebral column set in its natural environment of muscles and septa. Get your bearings by first locating the neural arch, notochord, and so on. From either side of the centrum, a fibrous connective tissue sheet runs directly outward to the skin. This is the horizontal skeletogenous septum and divides the body musculature into two regions. Muscles above this septum are epaxial muscles and below are hypaxial muscles. A dorsal skeletogenous septum from the neural spine upward to the skin of the back and a ventral skeletogenous septum from the hemal spine downward to the skin further divide muscles into left and right sides. The concentric whorls of muscle are myomeres each separated by myosepta.

Osteichthyes

→ **Chondrostei—Sturgeon** Examine vertebrae from a sturgeon. Note the prominent notochord with cartilaginous elements riding upon it in dorsal and ventral positions (figure 5.8a). Homologies are presently in doubt, although the neural spine, neural arch, and rib seem clear enough.

→ **Holostei—*Amia* (Bowfin)** Examine a mounted skeleton. The vertebrae are ossified. The notochord is not in evidence because it is much reduced in size and now is completely enclosed by the vertebrae. Follow ventral ribs posteriorly noting how they are serially homologous with hemal arches (figure 5.8b). Take a close look at vertebrae in the tail. Here two centra are present per segment: the intercentrum carries no arches, but the adjacent pleurocentrum articulates with neural and hemal arches. Try to find the forward point where the pattern of two centra ends and a single centrum per segment begins.

→ **Teleostei—Perch** Again consult the mounted specimen looking for what should begin to be familiar structures: centra, hemal arches and spines, neural arches and spines, and so on (figure 5.8c). In the trunk, dorsal ribs articulate with small ventrolateral projections, basapophyses. Note their serial homologies.

Now examine an isolated, single dried teleost vertebrae—centrum, spines, arches, and so on (figure 5.8d). If you have a tail vertebra, how do you determine dorsal and ventral (i.e., hemal and neural arches)? Compare with a mounted perch skeleton if necessary. Compare

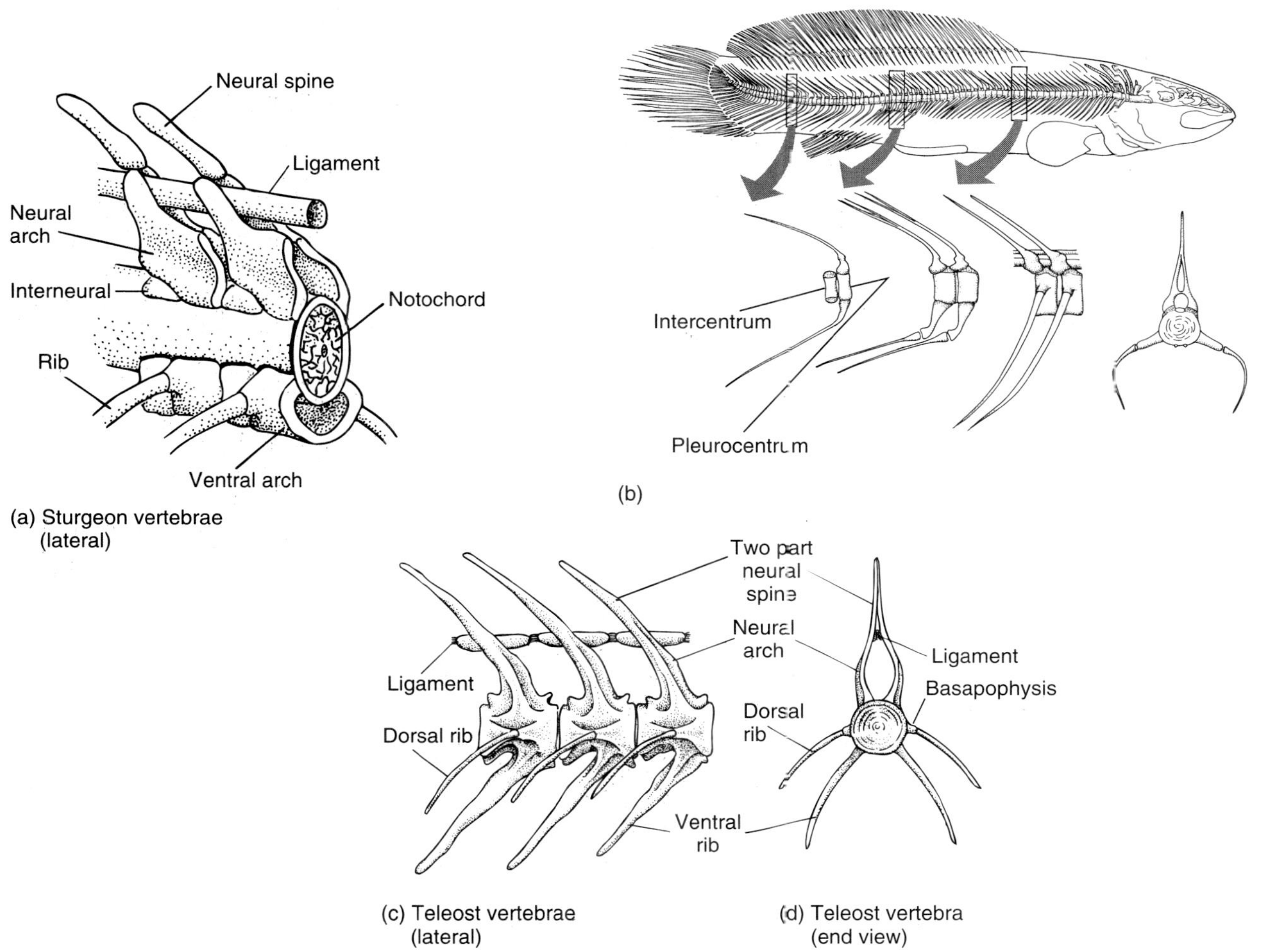

Figure 5.8
Actinopterygian vertebrae. (a) Sturgeon vertebrae, lateral view. (b) Axial skeleton of the bowfin *Amia calva*. (c) Teleost vertebrae, lateral view. (d) Teleost vertebra, end view.
(a, c) After Jollie; (b) after Jarvik.

relative sizes of neural and hemal canals. The centrum is biconcave, and thus is what type of vertebra? Note the pin-sized hole in the center of the centrum. Through this the constricted notochord passes, then upon exit, expands to partially fill the concavities between successive vertebrae in a fashion similar to sharks.

Amphibia

→ **Urodela—*Necturus*** Start with a mounted skeleton. Four sequential regions are present (figure 5.9a): the cervical (or neck) region consists of a single vertebra that joins the vertebral column to the skull; the long trunk region supports tiny wishbonelike ribs (figure 5.9c); the sacral region consists of one vertebra whose ribs join with girdles of the hind limbs; and the caudal region (figure 5.9b), whose vertebrae commonly possess hemal arches, lack ribs, and have short lateral projections called transverse processes.

Seen to best advantage in isolated, disarticulated vertebrae, ribs of the trunk are bicipital (possessing two heads). Where they articulate with the vertebrae, ribs fork into a dorsal head, the tuberculum, and a ventral head, the capitulum. These heads meet respective articulation points on the transverse processes: a diapophysis (dorsal) and a parapophysis (ventral).

Near the base of the neural spines each vertebra sends a pair of shelflike projections forward: prezygapophyses (articular face upward) to engage with paired posteriorly projecting postzygapophyses (articular face downward) of the adjacent anterior vertebra. These prevent twisting or wringing of the vertebral column.

→ **Anura—Frog** On mounted frog skeletons, identify regions of the vertebral column (figure 5.10). The cervical region consists of a single vertebra. The trunk region is longest, usually 7–10 vertebrae depending upon species; transverse processes, low neural spines, prezygapophyses and postzygapophyses are present. The single vertebra constituting the sacral region attaches to the long ilium (pl. ilia) of the pelvic girdle by transverse

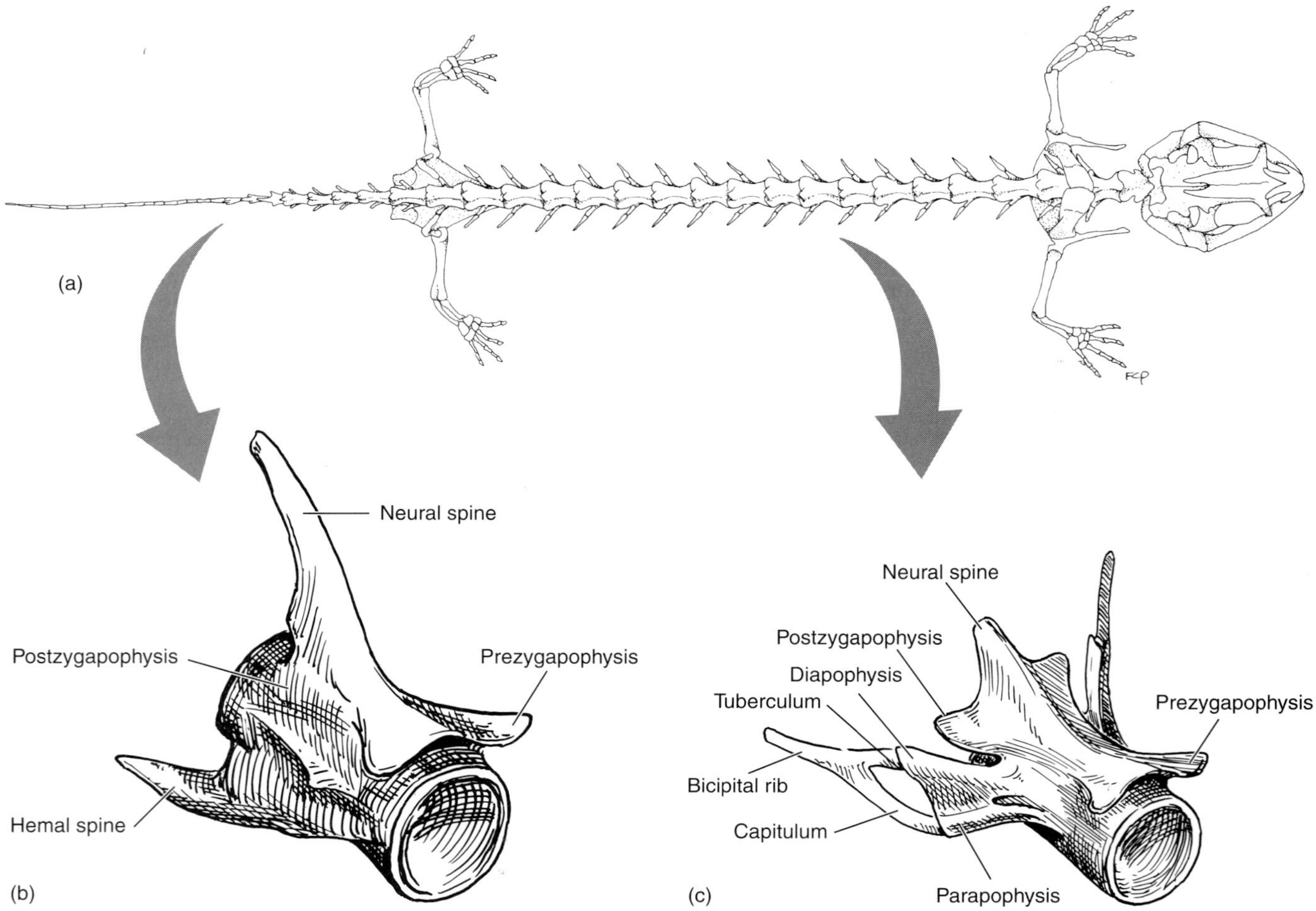

Figure 5.9
Skeleton of the mudpuppy, *Necturus*. (a) Skeleton, dorsal view. (b) Caudal vertebra. (c) Trunk vertebra with articulated ribs. Cranial is to the right in both (b) and (c).

Patterns & Connections

The pattern of two centra per body segment found in the tail of the bowfin, *Amia*, is not found in the modern tetrapods (amphibians, reptiles, birds, mammals). Each of these modern groups possesses but a single centrum per body segment. But why should we see this difference? Functionally, a vertebral column with fewer elements per body segment is more rigid and provides greater support. As tetrapods became increasingly adapted to life on land, the advantages of a firm vertebral column to carry the weight of the body became more important.

processes. The vertebral column continues beyond the sacrum as a single long splintlike element, the urostyle, the stout bone between the two ilia of the pelvic girdle. Caudal vertebrae are absent.

Reptilia

→ **Alligator (Mounted Skeleton)** The vertebral column divides into five regions: cervical, thoracic (long ribs), lumbar (without ribs), sacral, and caudal (figure 5.11a). Because of their unique structure, complex embryology, and specialized function, the first two cervical vertebrate receive their own names, atlas and axis, number 1 and 2 cervical vertebrae respectively (figure 5.11d). The atlas is a ring composed of four components: intercentrum, two bases of the neural arch, and part of the proatlas (an extra vertebra, now lost save for its contribution here to the reptile atlas). The axis has a neural arch and broad neural spine. Attached to its centrum and projecting forward like a spike is the odontoid process (phylogenetically an outgrowth of the atlas centrum) that will probably be obscured by adjacent vertebrae. A short pair of spinelike ribs belonging to the axis actually articulate with the odontoid process. The remaining cervical vertebrae are more conventional: neural arch and neural spine, prezygapophyses, postzygapophyses, and a single procoelous centra. Short, V-shaped ribs attach to transverse processes. Cervical and a first few thoracic vertebrae bear midventral bumps, hypapophyses.

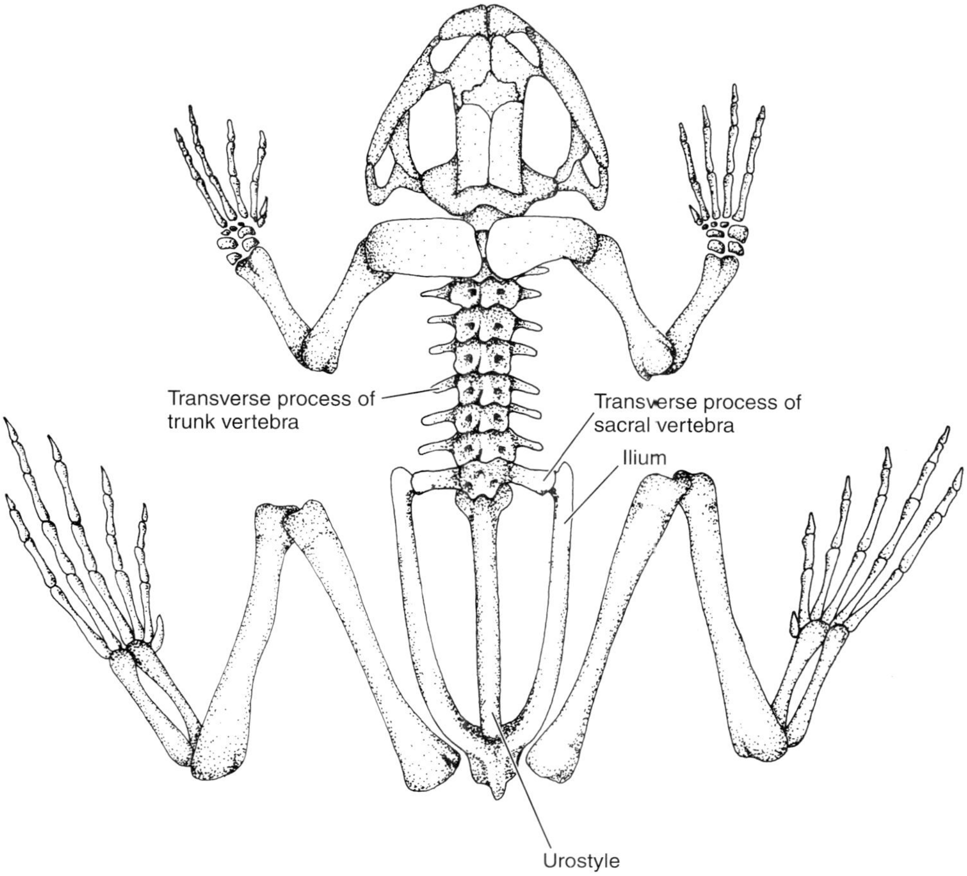

Figure 5.10
Skeleton of a bullfrog; dorsal view.

Progressing posteriorly, the first vertebra to bear long ribs reaching the midventral region is the first thoracic vertebra. Approximately nine others follow, all bearing ribs on the ends of transverse processes (figure 5.11c).

The following five lumbar vertebrae carry broad transverse processes, but no ribs (figure 5.11a).

Two vertebrae, each with strong transverse processes attached at their tips to the pelvic girdle, compose the sacral region (figure 5.11a).

The caudal vertebrae follow: neural arch, neural spine, transverse processes, hemal arches (missing on first caudal vertebra) that articulate almost between centra (figure 5.11b). Ribs are absent altogether.

Overall, note the prominence of neural spines and their basic similarity in shape and orientation throughout thoracic, lumbar, and most of the caudal region (figure 5.11a). Pre- and postzygapophyses occur through most regions. Cervical and thoracic ribs (phylogenetically dorsal ribs) are bicipital (tuberculum and capitulum), with respective articulations to a diapophysis and a parapophysis riding on transverse processes. The two heads define a transverse foramen. As successive arches might define a passageway, the successive transverse foramina define the vertebrarterial canal parallel to the column. Blood vessels to the head pass through this canal. Each rib has three parts: vertebral rib including the two heads, intermediate part, and terminally the sternal or costal cartilage that articulates midventrally with thin bone or dried cartilage (figure 5.11a,c). Depending upon the size of the individual, the first few sternal cartilages are likely to articulate with the sternum and the remainder of the costal cartilages with a midventral thin, dried membrane. Lying midventrally between the last sternal cartilage and the pelvic girdle are the gastralia (figure 5.11a). The gastralia are Y-shaped dermal bones of the belly, directly below the lumbar region and are often slightly obscured by dried cartilage.

→ **Turtle (Mounted Skeleton)** Identification of homologous parts of the vertebral column is complicated in the turtle because elements fuse with dermal bone to form the protective dorsal carapace. Cervical, trunk, sacral, and caudal regions compose the vertebral column (figure 5.12).

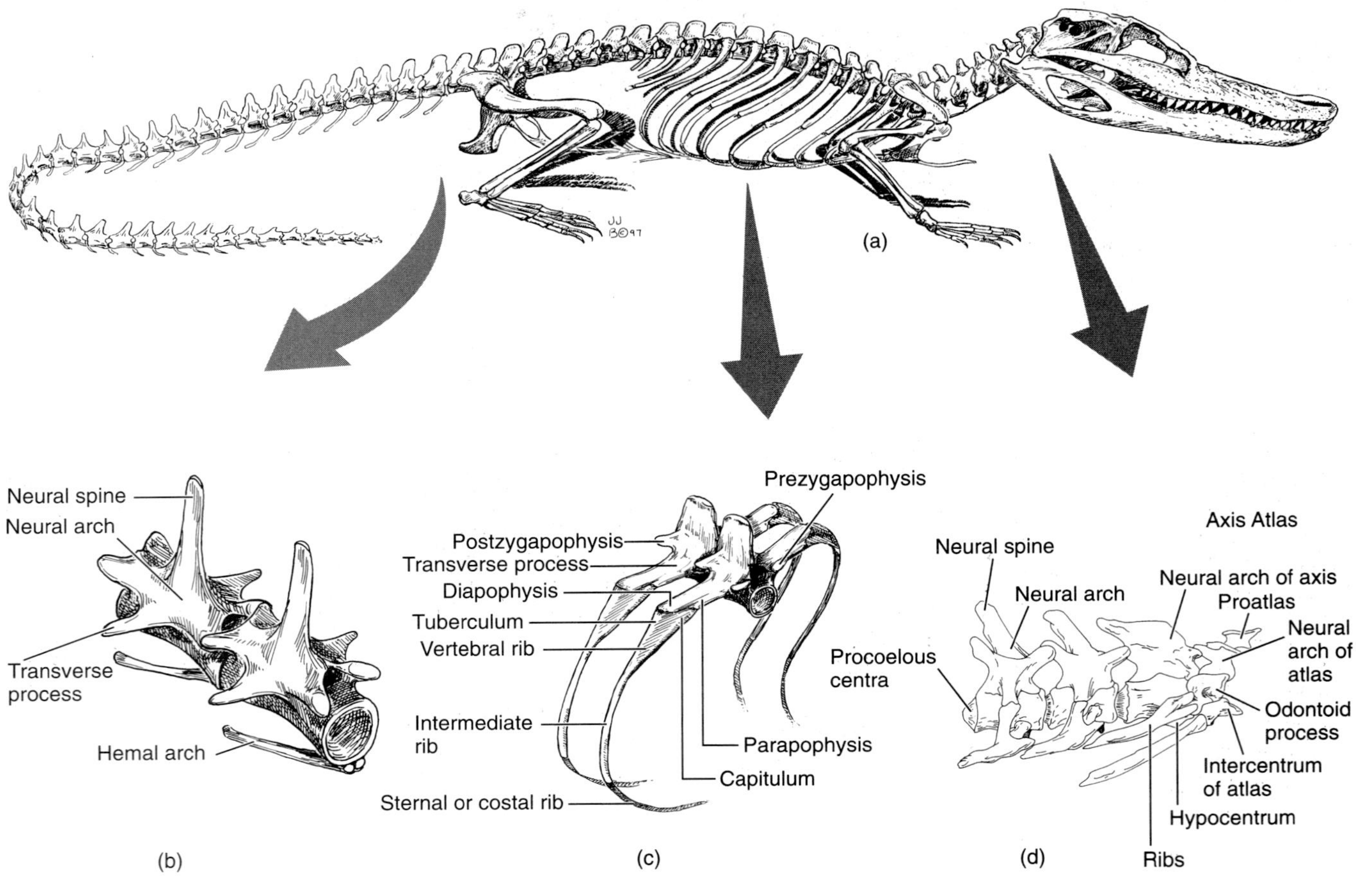

Figure 5.11
Skeleton of an alligator. (a) Lateral view of complete skeleton. (b) Two caudal vertebrae. (c) Oblique view of two thoracic vertebrae. (d) Lateral view of cervical vertebrae. Cranial is to the right in (b–d).
(d) Redrawn from E. S. Goodrich, *Studies on the Structure and Development of Vertebrates.*

An atlas and axis begin the cervical series of vertebrae. The remaining cervical vertebrae lack ribs.

Neural spines and neural arches of the trunk, sacral, and first caudal vertebrae are fused to the carapace. Ribs of trunk vertebrae, possessing but a single head (capitulum), distally broaden and fuse with bony plates of the carapace.

Two vertebrae compose the sacrum.

The first caudal vertebra fuses with the sacrum. Remaining caudal members possess neural arches, hemal arches, and transverse processes of decreasing prominence.

Aves—Pigeon or Chicken (Mounted Skeleton)

→ Fusion is the key feature of the bird skeleton. Slight differences in skeletal structures occur between species, but generally all regions of the vertebral column fuse except the first, the cervical region (figure 5.13a).

The cervical region contains numerous vertebrae (16 in chicken, 13–14 in the pigeon). They are heterocoelous, saddle-shaped centra, that give great flexibility to this region, a fact you should confirm on disarticulated neck vertebrae (figure 5.13b). The atlas is small and ring shaped. The axis bears an odontoid process. The following cervical vertebrae possess zygapophyses, low neural arches and spines, hypapophyses, and reduced ribs. The bicipital ribs are lateral swellings that bear sharp posteriorly directed spines. The base of each forked rib defines a transverse foramen and so contributes to the vertebrarterial canal. Capitulum and tuberculum fuse with parapophysis and diapophysis on the transverse process.

From their transverse processes, thoracic vertebrae support long ribs, which usually have two segments (figure 5.13a). In the middle of the ribs, the vertebral section usually carries a flat posterior projection, the uncinate process (also present in some reptiles) resting over the lateral face of the next successive rib. The sternal section is the distal part of the rib that articulates with the sternum. Besides possessing an incomplete rib, the first thoracic vertebra, in most species, fuses with the next few successive vertebrae. Note fusion of centra, spines, transverse processes, and zygapophyses.

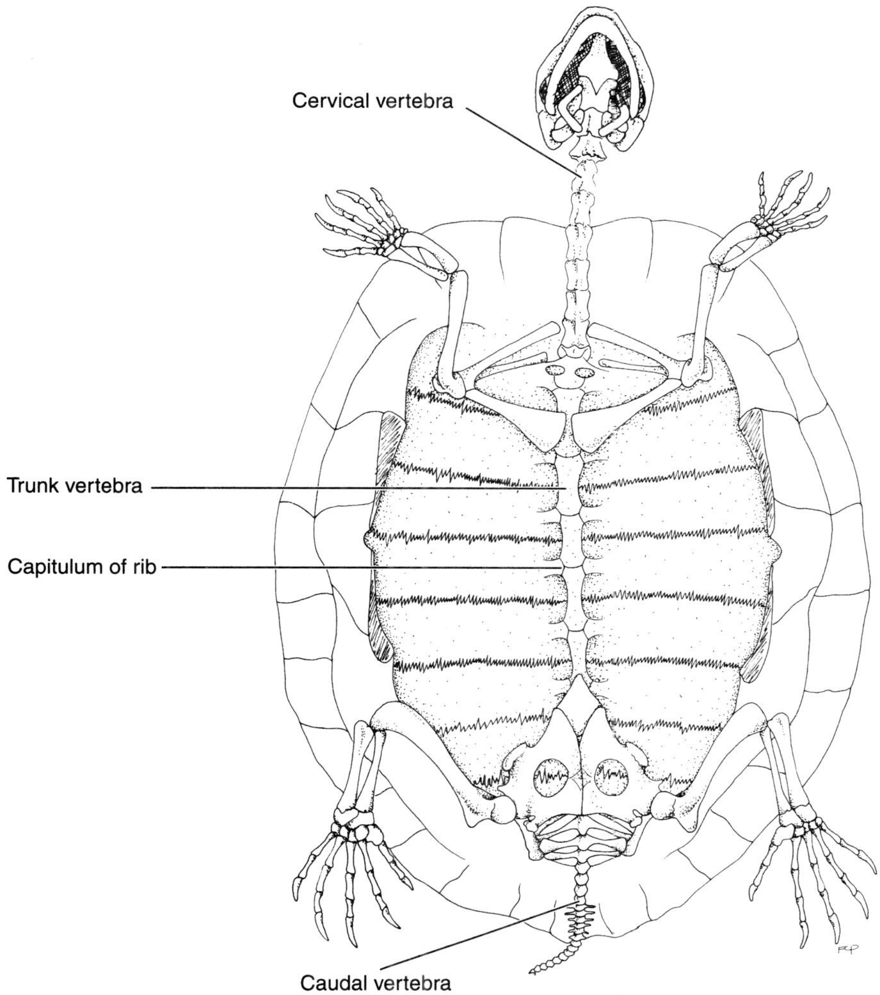

Figure 5.12
Skeleton of a painted turtle, ventral view. The plastron has been cut away to reveal the internal skeletal detail.

Form & Function

Birds are an interesting example of the close match between form and function within the vertebral column. Cervical vertebra are flexibly articulated to give the head great freedom of movement and reach when a bird preens its feathers or probes for food. However, most of the vertebrae in the middle and posterior part of the vertebral column are fused to each other (synsacrum) and to the three fused bones (innominate) of the pelvic girdle. This adds rigidity and establishes a firm and stable axis for control while a bird is in flight. Indirectly, this fusion of vertebral elements decreases the weight of the body because less muscle is required to control flexible, individual vertebrae.

The fusion of the last thoracic, all lumbar, all sacral, and first few caudal vertebrae forms a composite piece, the synsacrum (figure 5.13c). In turn, the synsacrum fuses with the large ilium, which itself is fused with other pelvic elements. The number of truly separate vertebrae involved can be counted by examining the ventral synsacrum (figure 5.13c) to count still distinct transverse processes. Only two vertebrae contributing to the synsacrum are truly sacral (except in ostriches, where three contribute).

A few free caudal vertebrae follow. The last several fuse into a small solid terminal piece, the pygostyle. Examine disarticulated vertebrae from throughout the vertebral column to better appreciate the structure of the bones indicated above.

Mammalia—Cat (Mounted and Disarticulated Skeletons)

→ Five distinct regions characterize the mammalian vertebral column: cervical, thoracic, lumbar, sacral, caudal (figure 5.14a). Between vertebrae in life lie circular pads of fibrous cartilage: the intervertebral cartilages or intervertebral disks. Centrally they contain a gelatinous material, the **nucleus pulposus,** the only adult derivative of the

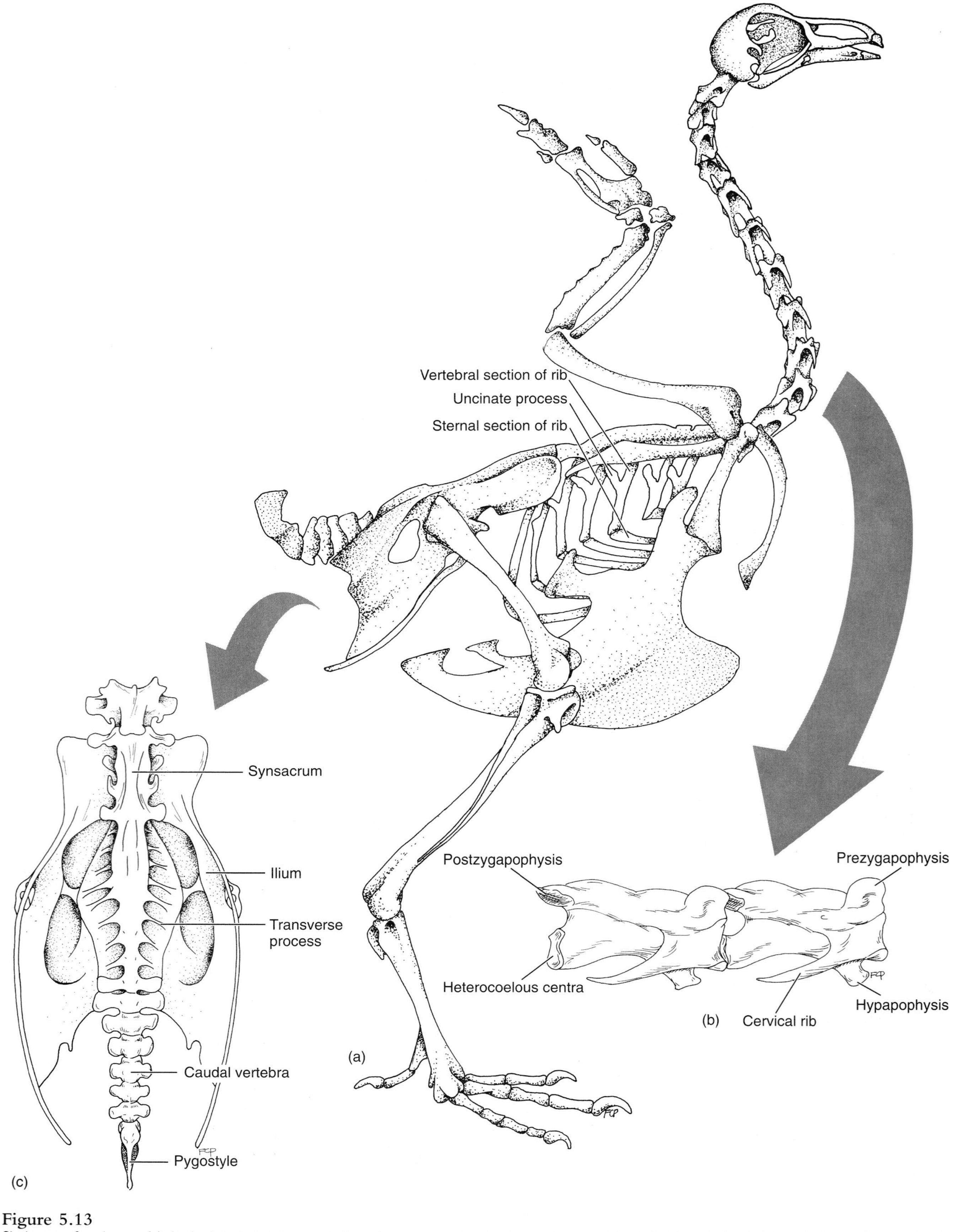

Figure 5.13
Skeleton of a pigeon. (a) Articulated skeleton with the left member of paired structures not included for clarity. (b) Cervical vertebrae, lateral view. (c) Synsacrum, ventral view; cranial is up.

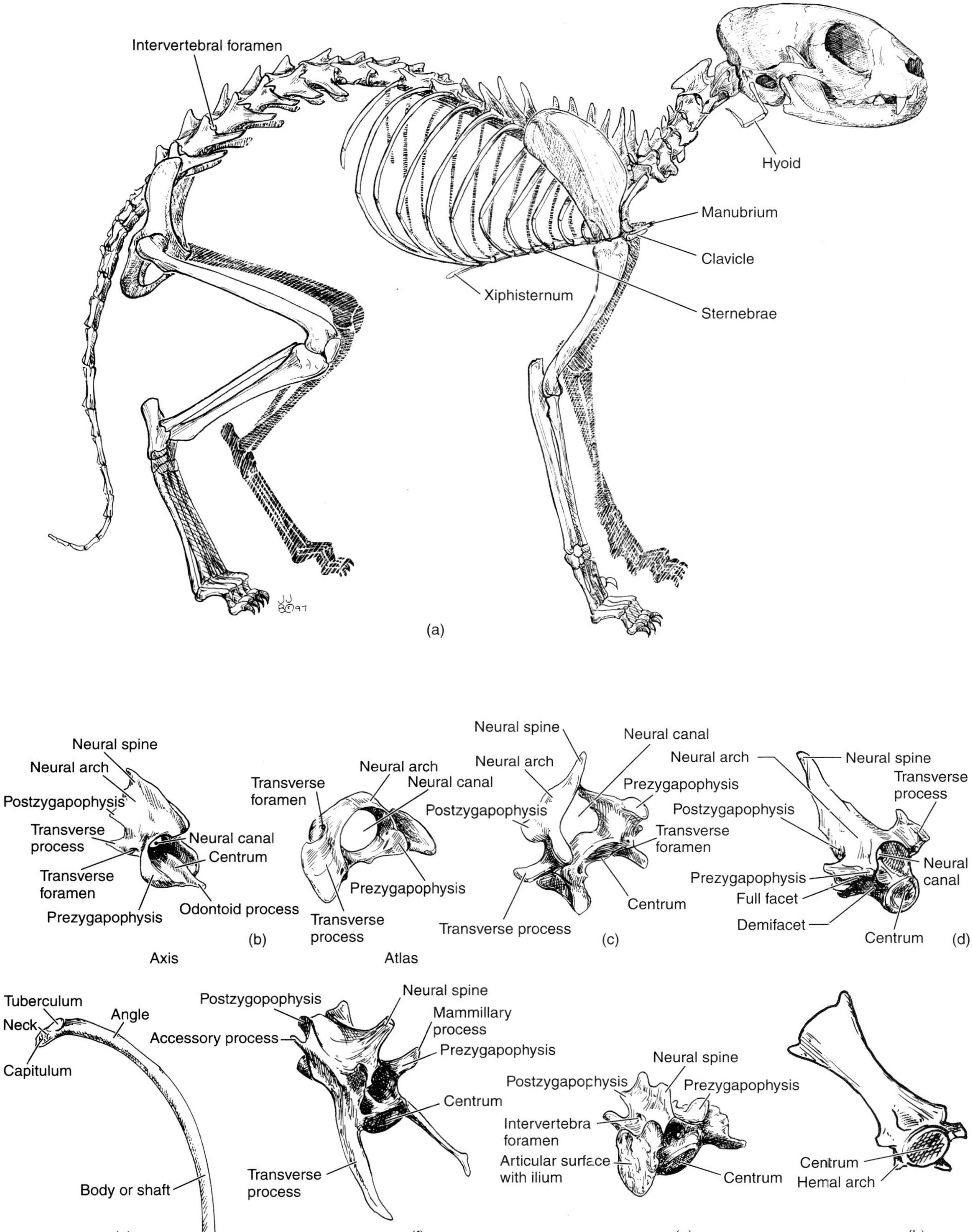

Figure 5.14
Axial column of a cat. (a) Complete skeleton, lateral view. (b) Axis (left) and atlas (right). All vertebrae (b–d, f–h) are in oblique view, with the cranial end to the right. (c) More caudal cervical vertebra. (d) Thoracic vertebra. (e) Rib. (f) Lumbar vertebra. (g) Sacral vertebra. (h) Caudal vertebra.

embryonic notochord in mammals. The cartilage of intervertebral disks shrinks and deforms when it dries. For this reason, these disks are usually absent on mounted skeletons prepared by routine methods. On a mounted skeleton, note that adjacent vertebrae form an opening between them near the base of each neural arch (figure 5.14a). This opening is the intervertebral foramen, through which spinal nerves exit. Each vertebra defines half this foramen, and so bears an indentation, the intervertebral notch.

The special interest humans have in mammalian anatomy has given rise to additional terms for previously learned morphological names. These special descriptive terms are given in brackets.

Now begin your examination of the individual, disarticulated vertebrae, observing regional changes in prominence of neural arch and spine, neural canal (vertebral foramen), zygapophyses (articular processes), transverse processes, and ribs. Often two regions of the neural arch are identified: pedicle (each foot of a neural arch where it stands on the centrum) and lamina (section between prezygapophysis and base of the neural spine).

→ **Cervical Vertebrae** The first two cervical vertebrae are the atlas and axis (figure 5.14b). Five more follow in almost all mammals. The atlas is ring shaped and carries lateral, paired, and winglike transverse processes. The base of each transverse process is pierced by a transverse foramen, a distinctive aperture in all cervical vertebrae. The neural arch lacks a spine. The base of the neural arch is perforated by holes for passage of nerves. Anteriorly, large scooped paired depressions, the articular facets, belonging to the prezygapophyses (= anterior articular processes) receive the occipital condyles of the skull.

Above its neural arch, the axis carries a broad neural spine (spinous process) that projects over the atlas. Posteriorly this spine forms the postzygapophyses. The odontoid process (dens) projects forward from the centrum into the interior base of the atlas. Lateral to the odontoid process are two swellings, the prezygapophyses (= anterior articular processes), which fit into the postzygapophyses (= posterior articular processes) of the atlas. Transverse processes pierced by transverse foramina lie along the sides of the centrum. Each remaining cervical vertebra possesses zygapophyses, a neural canal, a neural arch, a neural spine (spinous process), a well-defined centrum, a transverse processes, and transverse foramina (figure 5.14c).

→ **Thoracic Vertebrae** Note the tall, caudally directed neural spines (spinous processes), neural arches, zygapophyses (articular processes), short transverse processes, and centra (figure 5.14d). Locate the two facets that articulate with each bicipital rib, the full facet on the transverse processes (correctly diapophysis), and a demifacet on all centra where a capitulum articulates between centra. Between demifacet and prezygapophysis, the neural arch indents, the intervertebral notch.

→ **Ribs** Most ribs consist of a bony vertebral rib, and a bar of hyaline cartilage, the costal cartilage (or sternal rib) (figure 5.14a,e). Ribs are classified on the basis of their distal attachment:

true rib—the costal cartilage articulates separately with the sternum

false rib—the costal cartilage unites with adjacent costal cartilages before finally articulating with the sternum

free (or floating) rib—the distal end has no attachment to the sternum

On a mounted specimen, note the sites of rib contact with the vertebrae. The rib is bicipital (figure 5.14e). The capitulum (head) articulates with the centrum, while the more distal tuberculum (tubercle) articulates with the transverse process. The small depression on the centrum where the capitulum articulates is the facet. If articulation lies only partially on the centrum, then the half notch is termed a demifacet. The narrow part between the two heads is the neck, the rest of the rib the body or shaft; the point of sharpest bend just distal to the tuberculum is the rib angle.

→ **Lumbar Vertebrae** Note the change in orientation of neural spines, and forward directed transverse processes (pleurapophysis by some since it is thought to be formed by the fusion of rib and diapophysis) (figure 5.14f). Above the articular face of the prezygapophysis lies a small knob, the mammillary process (metapophysis). Below and lateral to the postzygapophysis is a spinelike accessory process (anapophysis).

→ **Sacrum** Individual sacral vertebrae fuse to form the sacrum so that individual transverse processes and ribs are indistinguishable (figure 5.14g). However, former boundaries can be determined by noting persisting individual neural spines and intervertebral foramina (dorsal and ventral sacral foramina). Three sacral vertebrae contribute in the cat and dog, four in the rabbit and pig, five in the horse and humans. Prezygapophyses and postzygapophyses occur on respective ends of the sacrum.

→ **Caudal Vertebrae** Note the loss in prominence of arches, transverse processes, and zygapophyses caudally until just centra remain (figure 5.14h). Tiny V-shaped hemal arches accompany some vertebrae, but are commonly lost during preparation of the skeleton. Numbers of caudal vertebrae vary with tail length. In humans, those present fuse into a single piece, the urostyle, or so as not to confuse with frogs, more commonly the coccyx.

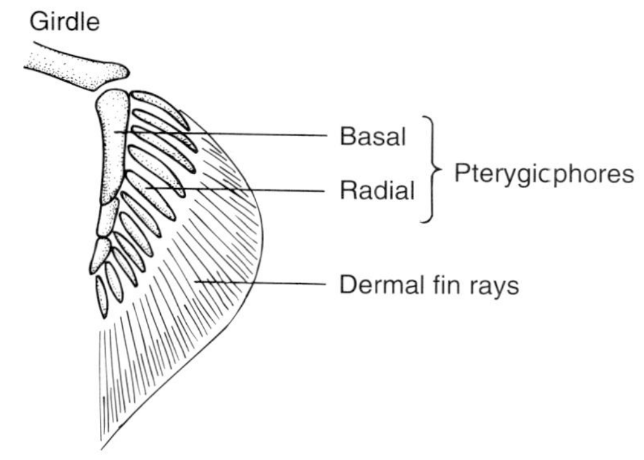

(a) Fin

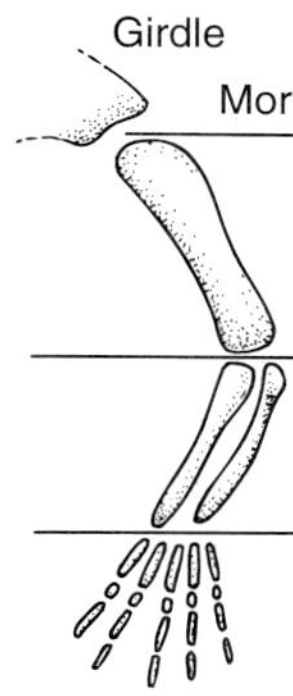

Morphological term	Forelimb	Hindlimb
Stylopodium	Upper arm	Thigh
Zeugopodium	Forearm	Shank (crus)
Autopodium	Manus (wrist-palm-fingers)	Pes (ankle-sole-toes)

(b) Limb

Figure 5.15
Basic components of the fin and limb. (a) The fin is composed of pterygiophores, basal and radials, and dermal fin rays. Fin rays are called lepidotrichia in bony fishes and ceratotrichia in elasmobranchs. (b) The limb, either fore- or hindlimb, includes three regions: stylopodium (upper arm/thigh), zeugopodium (forearm/shank), and autopodium (manus/pes).

Girdles and Limbs

Introduction

The girdles are parts of the skeleton that support the fins or limbs (figure 5.15), and serve as sites of muscle attachment. The **pectoral girdle** supports anterior appendages and the **pelvic girdle** posterior appendages. Tetrapod limbs evolved from fish fins. Many of the anatomical changes you will study in appendages and girdles reflect changing functional demands of locomotion and support. Primitive fins were supported by a series of skeletal rods, **pterygiophores,** some proximal members of which enlarged and fused in later forms to give rise to the major elements of the girdles.

Pelvic Girdle and Posterior Appendages

→ **Elasmobranchii—Shark** Consult an articulated skeleton for pectoral and pelvic girdle position on the vertebral column. The distal portion of the flat fin projects outward and is supported internally by slender dermal rays, the ceratotrichia, giving it a streaked appearance (figure 5.16b). At the base of the fin lie larger cartilaginous elements, pterygiophores. The fins in turn articulate with a straight, single cartilaginous pelvic girdle, the puboischiac bar. Notice, on each end of this bar, a lateral projection, the iliac process. In the shark, two groups of the pterygiophores can be recognized. The first group are the radials, an outer series of elements arranged like the diverging blades of a hand-held fan. The second group is composed of two basals, enlarged medial cartilages probably derived phylogenetically by the fusion of several radials. The largest of the two basals is the metapterygium, a long, stout medial cartilage supporting the series of radials. The smaller basal is the propterygium, positioned just cranial to the series of radials, and probably enlarged from an anterior radial. There are sexual differences in the pelvic girdle (figure 5.16b). Only the male possesses prominent claspers, inserted into the female to facilitate sperm transfer, that are also likely derived from fused radial cartilages.

> **Patterns & Connections**
>
> The pelvic girdles of tetrapods follow a basic pattern. In general, each girdle is formed by three pairs of bones; the pubic, iliac, and ischiac, seen laterally in a triangular configuration. At the top of the triangle is the iliac, usually identified as the dorsal portion of the pelvic girdle that makes contact with the sacral vertebrae. At one corner of the triangle are the pubic bones, usually directed ventrally and cranially. The ischiac bones are in the remaining corner, directed ventrally and caudally. Derivatives of this tripartite pattern exist, so stay alert to deviations from this pattern.

→ **Amphibia—*Necturus*** The pelvic girdle in urodeles is largely unossified. In *Necturus,* the single sacral vertebra bears, on its transverse process, a rib that articulates with the ilium, a bony rod joined ventrally to the puboischiac plate (figure 5.17a,b). This plate is shaped like an arrowhead directed anteriorly, and lacks any sutures to clearly delineate regions. Basically, the anterior, pointed half is termed the pubic cartilage, and the posterior half the ischiac cartilage. Within the ischiac cartilage a pair of semicircular centers of ossification form, termed ischia (figure 5.17a,b). The acetabulum is an open depression that receives the rounded proximal end of the femur. It is formed where the pubic cartilage, ischium, and ilium join together. Medial to the acetabulum in the pubic cartilage is the small obturator foramen through which the obturator nerve passes.

The limb, from proximal to distal, is divided into thigh, shank, ankle, and foot. In order, the femur articulates with two parallel bones. The tibia lies anterior and fibula posterior when the limb is held extended at right angles to the body. These two bones join distally with

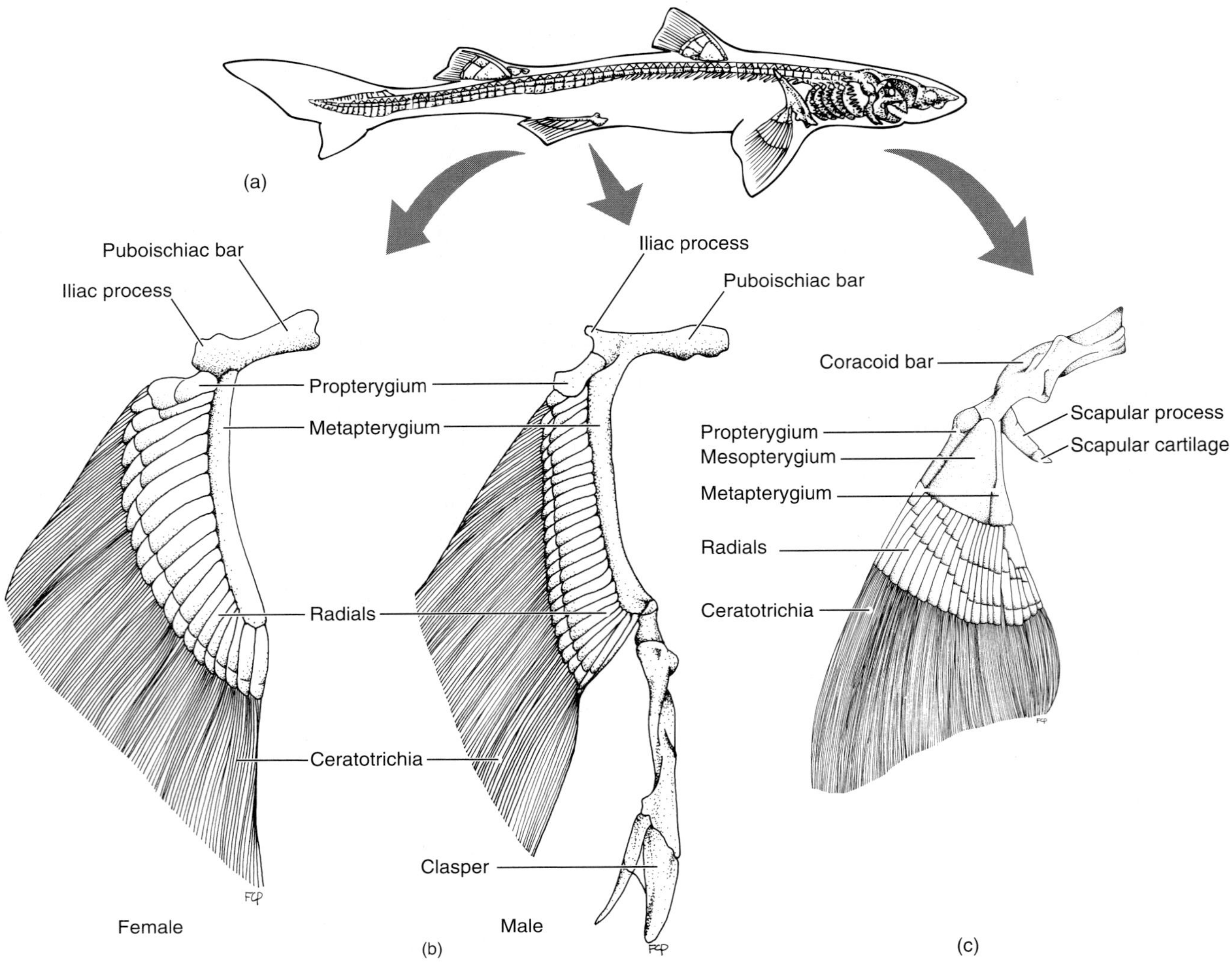

Figure 5.16
Right pelvic and right pectoral girdles of a shark. (a) Skeleton showing the general position of each girdle. (b) Female (left) and male (right) pelvic girdles, ventral view. Note the elongate clasper of the male used for sperm transfer. (c) Pectoral girdle (similar in males and females), ventral view.

six smaller bones, the tarsals, followed distally by rays. The first elongate bone in each ray is a metatarsal, the shorter two or three that end the ray are phalanges (sing., phalanx). Thus, tarsals form the ankle, metatarsals form the sole, and phalanges form the toes or digits of the foot. *Necturus* normally has but four toes on each foot, although most salamanders have four on the front feet and five on the back feet.

→ **Reptilia—Alligator** Examine a mounted skeleton of an alligator (figure 5.18a). The pelvic girdle is ossified in the adult, although added cartilaginous components may associate. The basic pelvic girdle is tripartite, consisting of three stout bones (figure 5.18b). The ilium joins dorsally with two sacral ribs. Ventrally it contributes with the ischium and pubis to form the acetabulum that receives the proximal head of the femur. The ischium is posterior and passes to the ventral midline where its expanded end joins the ischium of the opposite side in an ischiac symphysis. The pubis is anterior and also flattens and widens as it passes ventrally; here, where it touches its partner from the opposite side, a pubic symphysis is formed. Forward from this, it continues with gastralia, not formally part of the pelvic girdle.

Proximally the femur is expanded into a prominent knob, the head. Distally it articulates with a robust tibia (anterior) and slender fibula (posterior). Beyond this follow the tarsals, the two most prominent being the tibiale and fibulare, which articulate with the tibia and fibula, respectively. There are four rays in the alligator hind limb, each beginning with a long metatarsal and ending with one or more phalanges tipped with claws.

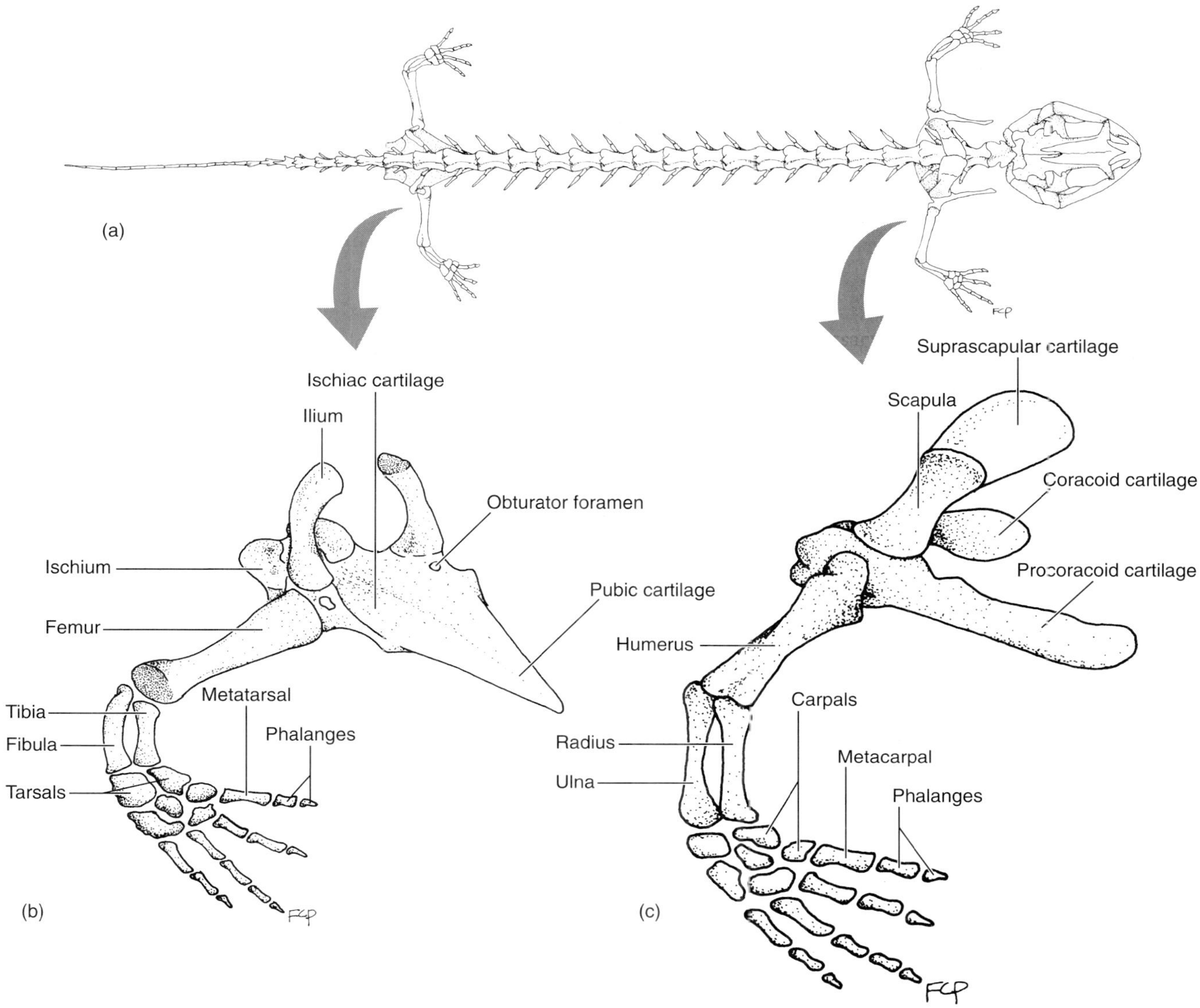

Figure 5.17
Right pelvic and right pectoral girdles of *Necturus*. (a) Skeleton showing the general position of each girdle. (b) Pelvic girdle, oblique view. (c) Pectoral girdle, oblique view.

→ **Reptilia—Turtle** Like an alligator, the pelvic girdle of the turtle is tripartite (figure 5.19b). The expanded dorsal end of the ilium joins with three or more sacral ribs depending upon species. The anterior pubis bears a lateral projection in its middle, the prepubic (or pectineal) process. At the ventral midline, the pubis of each side touches at the midline to form a pubic symphysis, which bears a forward directed epipubic cartilage from this symphysis. The ischium behind the pubis also bears in its middle a posteriorly directed process (unnamed) and enters into an ischiac symphysis with the ischium from the other side. Ischium and pubis form the edges of an opening in the ventral girdle, the large obturator foramen. In a disarticulated skeleton, note the bones that contribute to the acetabulum.

Present in the limb are the femur with prominent head and a pair of ventral flanges, tibia (larger) and fibula (smaller), tarsals, rays, and claws. Fibulare and tibiale can be identified by position to the fibula and tibia, respectively. The remaining tarsals can be simply termed tarsalia, which vary in size and number between species. Metatarsals and phalanges are also present.

→ **Aves—Chicken** Fusion of bones is a hallmark of the bird skeleton (figure 5.20a). As seen when examining the vertebral column, components of the last thoracic, lumbar, sacral, and first few caudal vertebrae unite into a single piece, the synsacrum (see figure 5.13c). Straddling each side and sutured to it is the paired broad hip bone, the innominate, part of the pelvic girdle (figure 5.13c).

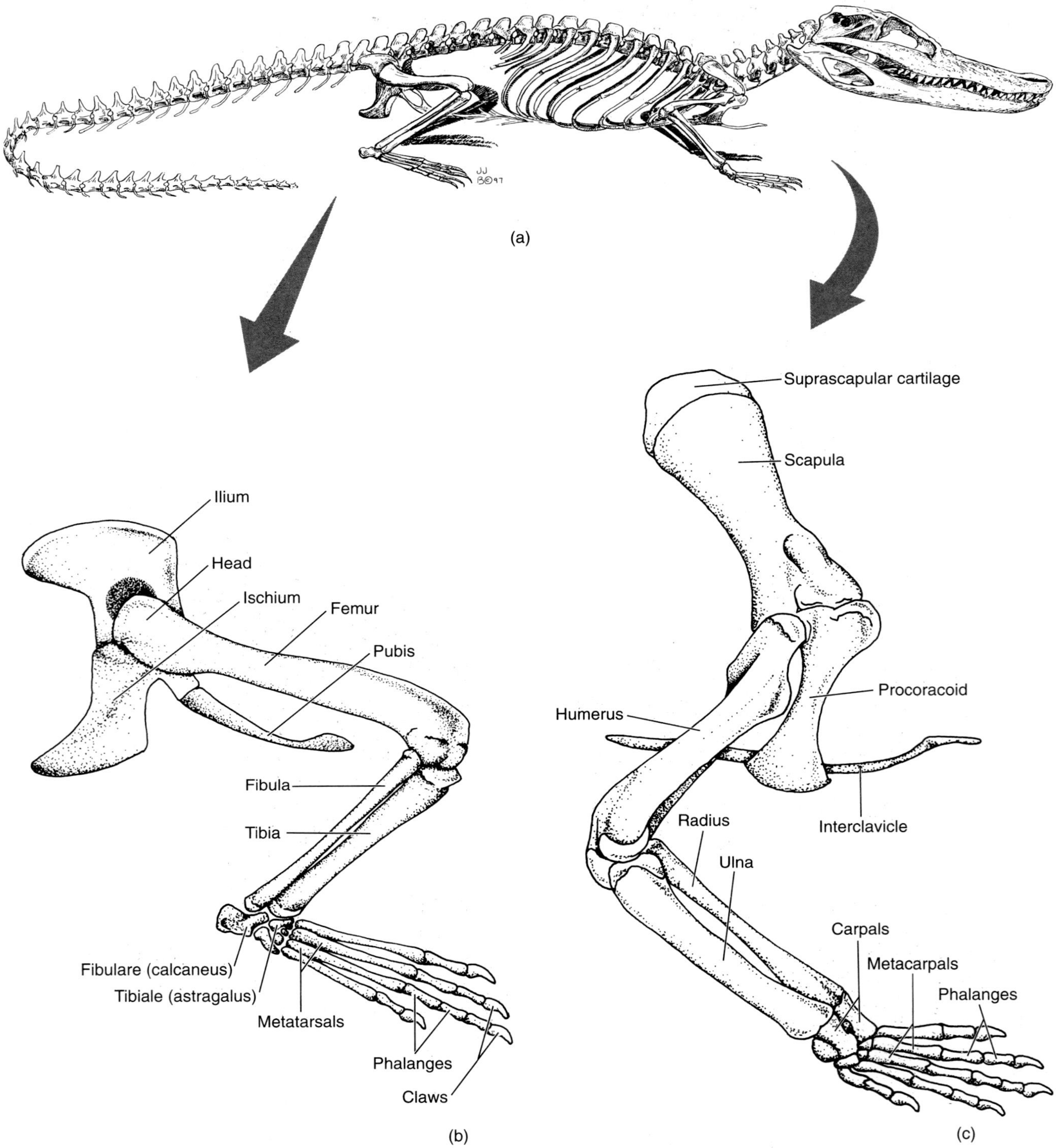

Figure 5.18
Right pelvic and right pectoral girdles of an alligator. (a) Skeleton showing the general position of each girdle. (b) Pelvic girdle, lateral view. (c) Pectoral girdle, lateral view.

Ilium, ischium, and pubis fuse to form it, thus obliterating fine boundaries between them in the adult bird. The perforated acetabulum marks the approximate point of their common union. The slender, riblike pubis runs backward from the acetabulum. The bladelike ilium extends forward and dorsally from the acetabulum and dorsally to articulate with the synsacrum. The anterior part of the ilium is concave, while its posterior half is convex. Between the posterior ilium and pubis lies the ischium. A low ridge marks the point of fusion of the

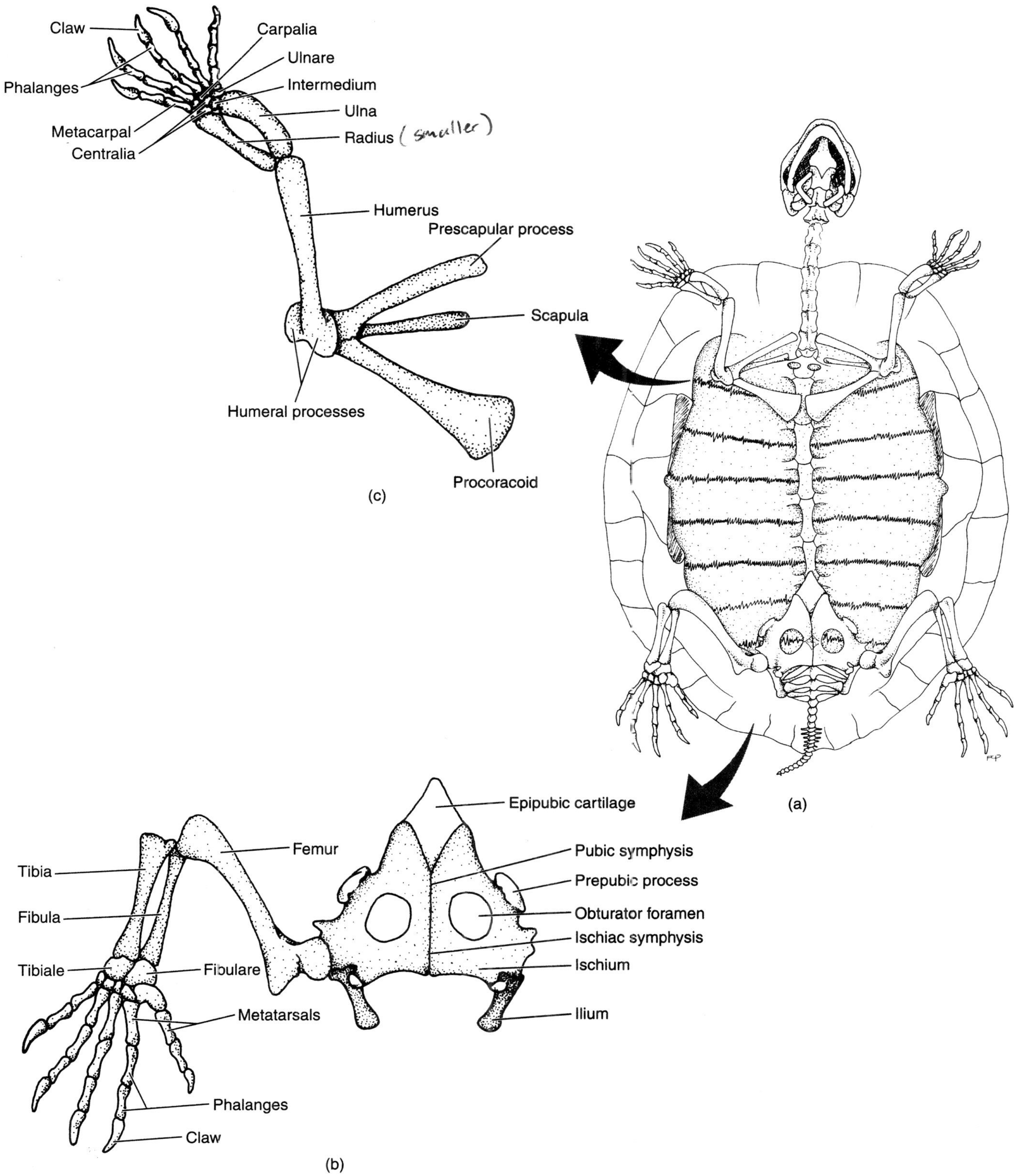

Figure 5.19
Right pelvic and right pectoral girdles of a turtle. (a) Skeleton showing the general position of each girdle. (b) Pelvic girdle, ventral view. (c) Pectoral girdle, ventral view.

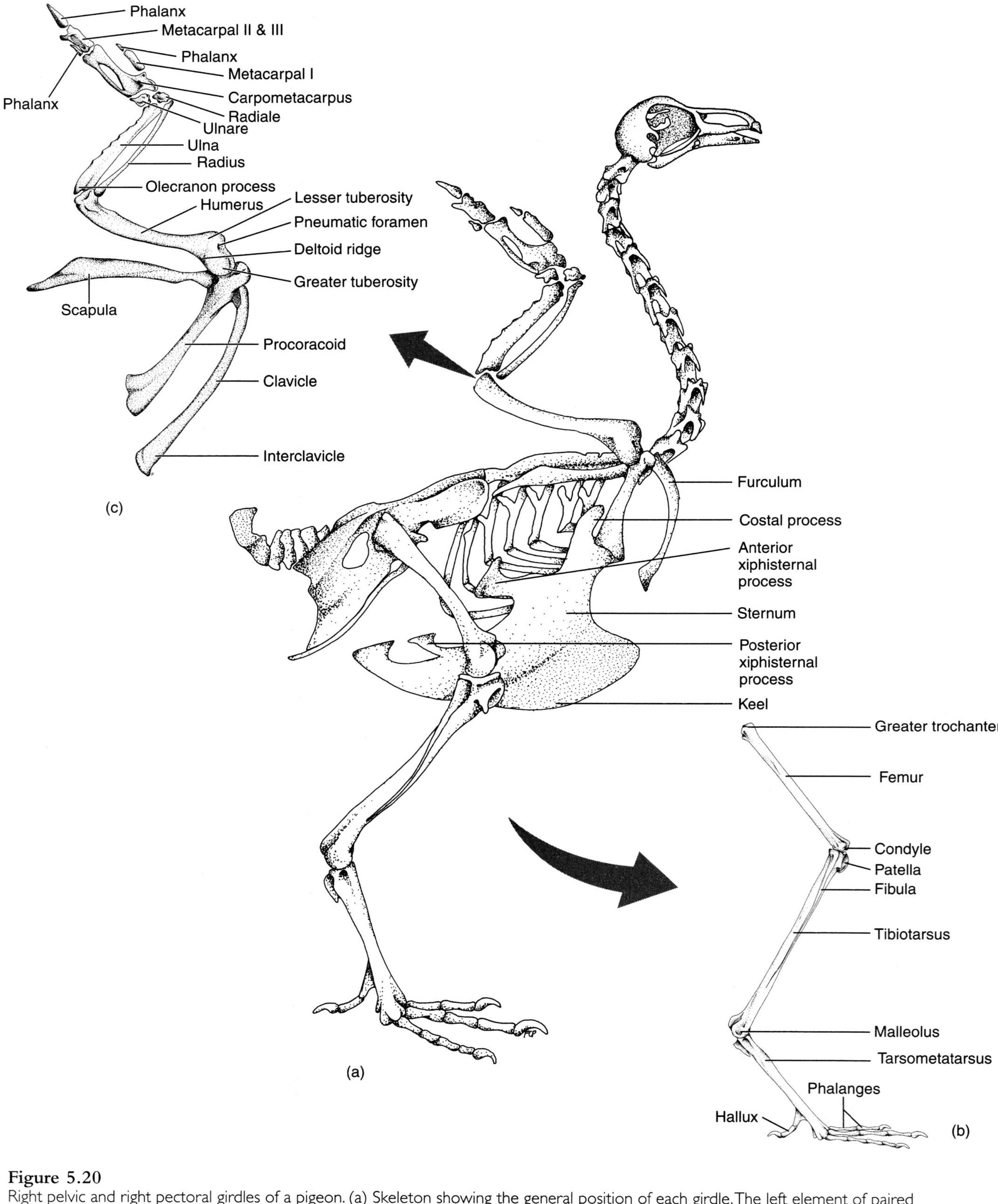

Figure 5.20
Right pelvic and right pectoral girdles of a pigeon. (a) Skeleton showing the general position of each girdle. The left element of paired structures is not shown for clarity. (b) Hindlimb, lateral view. (c) Partial pectoral girdle and forelimb, lateral view.

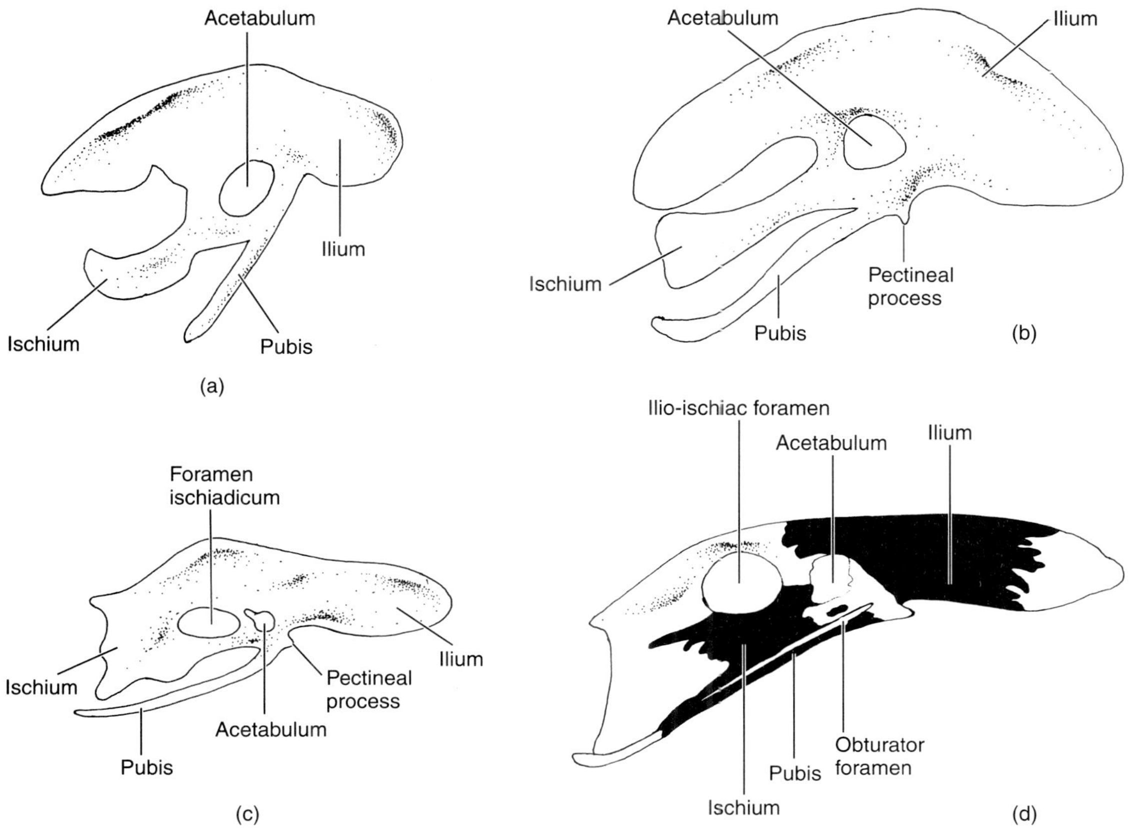

Figure 5.21
The development of the pelvic girdle in birds during their 21-day incubation period. Bone begins to replace cartilage after day 15. The acetabulum marks the junction of the three main components of the pelvic girdle: the ilium, pubis, and ischium. The origin of the name ilio-ischiac foramen is revealed by the two bones that develop around it. Note the migration of the pubis toward the ilium and the formation of the obturator foramen. (a) 7 days. (b) 8 days. (c) 13 days. (d) 18 days.
Redrawn with modification after Lebidinsky, 1913.

ischium with the posterior convex half of the ilium. Note that neither a pubic nor ischiac symphysis forms. Nestled between anterior points of union of pubis and ischium is the small, round, obturator foramen, and above this foramen and behind the acetabulum is the large, oval ilio-ischiac foramen. Consult embryonic stages in development of the bird pelvis (figure 5.21a–d) to clarify your understanding of the bony contributions to the innominate.

Fusion of skeletal elements continues to varying degrees in the hind limb (figure 5.20b). A lateral projection, the greater trochanter, is located on the lateral head of the femur. At its distal end, the femur carries raised, curved ridges, the condyles, and a depression between in which the patella or kneecap slides. In the shank, the large bone is the tibiotarsus, formed by fusion of the distal tibia with proximal tarsal bones. The small, splintlike bone accompanying the tibiotarsus is the reduced fibula. Proximally, the tibiotarsus has two condyles positioned opposite the condyles of the femur. Below these condyles on the tibiotarsus run two low ridges, or crests, to which muscles attach. Distally, the end of the bone swells into two ridges, or malleoli (sing. malleolus), that form a rocking articulation with the succeeding composite bone the tarsometatarsus. As the name suggests, several tarsals and metatarsals fuse. Hence, tarsal bones are not separate, but are incorporated into this or the preceding tibiotarsus. The ankle joint formed is thus within the tarsals, an intratarsal joint. The fifth ray is lost. Metatarsals first through fourth are incorporated into the tarsometatarsus and support respectively numbered toes with separate phalangeal series. The short first digit (sometimes called hallux) is on the medial side. Digits two through four are opposite it.

Embryonic Development of Bird Pelvic Girdle The pelvic girdle appears first as a mesenchymal condensation that forms a single triradiate cartilage, at least in the chick (figure 5.21). Within this single cartilage, three areas of ossification eventually form that are destined to be the three bones of the pelvis: namely, **ilium, ischium,** and **pubis.** In some species of birds, as in reptiles generally, the cartilaginous elements formed from mesenchymal precursors are initially separate. That is, each has its own center of **chondrogenesis.**

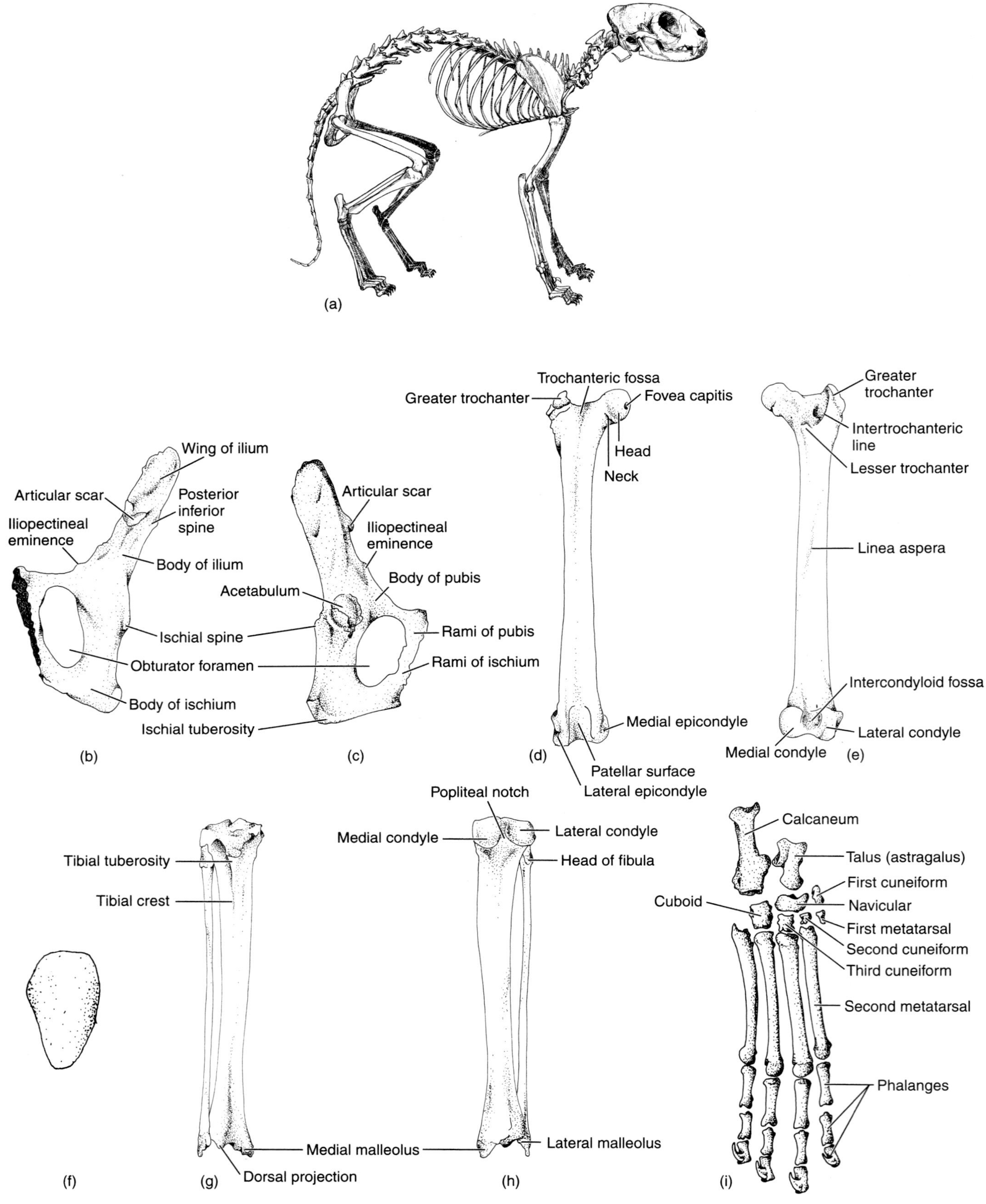

Figure 5.22
Right pectoral girdle and right forelimb of a cat. (a) Lateral view of cat skeleton. (b) Innominate bone, dorsal view. (c) Innominate bone, ventral view. (d) Femur, anterior view. (e) Femur, posterior view. (f) Patella. (g) Tibia and fibula, anterior view. (h) Tibia and fibula, posterior view. (i) Hind foot, dorsal view.

Later, as ossification begins, each cartilage will possess its own **center of ossification.** In the chick, the posterior ends of the ilium and ischium grow together, leaving only the **ilio-ischiac foramen** open between them. There is also a tendency for the pubic bones to fuse with the ventral edge of the ischium except anteriorly where the **obturator foramen** persists. The three bones of the pelvic girdle so united with one another are collectively the **innominate bone.**

→ **Mammalia—Cat** The three bones of the girdle fuse indistinguishably into the innominate bone (figure 5.22a–c). The large opening in the bone is the obturator foramen. The bladelike ilium, contributing to the anterior part of the innominate, articulates with the sacrum and has two major regions. The roughened region, the articular scar, is the site where the ilium joined to the sacrum. The body of the ilium lies behind the articular scar. The wing of the ilium lies in front of the scar and bears the crest of the ilium along its anterior rounded end. A small swelling, the posterior inferior spine, projects from the dorsal edge just above the scar. On this same edge, but posteriorly above the obturator foramen, protrudes a slightly more prominent process, the ischial spine. The iliopectineal eminence is a small rise along the ventral edge of the body of the ilium. The ischium contributes to the posterior girdle beyond the acetabulum. The end of the ischium is blunt and roughened, the ischiac tuberosity. The pubis lies below the acetabulum. Two general regions of the ischium and pubis are recognized: the bodies, which contribute to the acetabulum and also include most of the rest of the bones, and the rami, those remaining parts of each bone that define the ventral border of the obturator foramen. The rami meet along the ventral midline with rami of respective ischium and pubis of the opposite side to form the ischiac and pubic symphyses.

The femur has a hemispheric head pedestaled upon a short columnar neck that joins the rest of the long bone at about a 45° angle to its long axis (figure 5.22d,e). The fovea capitis is a shallow pit offset from the middle of the head. A greater trochanter projects lateral to the head. Between the base of the greater trochanter and the neck of the femur is a cavity, the trochanteric fossa. The lesser trochanter is a smaller knob below and behind the head. The low, posterior ridge between trochanters is the intertrochanteric line. The posterior linea aspera is a low ridge coursing in a slow spiral from the proximal end of the femur distally along the shaft. The two large, rounded articulating swellings at the distal end of the femur are lateral and medial condyles. Between them resides a pit, the intercondyloid fossa. The roughened sides of the condyles are lateral and medial epicondyles. A patella (figure 5.22f) is present at the knee and slides in a low depression on the anterior face of the distal femur, the patellar surface.

In the shank is a slender fibula and robust tibia (figure 5.22g,h). The articular surfaces of the proximal end of the tibia are condyles (lateral and medial). Posteriorly they form a projecting shelf, whose indented underside is the popliteal notch. Along the front of the tibia runs the raised tibial crest that begins in front of the condyles in a roughened tibial tuberosity. At the distal end of the tibia are two projections. The longer medial one is the medial malleolus. The shorter lateral process is the dorsal projection. The lateral malleolus forms the distal articulating end of the fibula. Note how the proximal end, the head of the fibula, is expanded and articulates with the tibia.

Morphological terms for ankle bones help correlate these elements with homologues in other vertebrates already studied. But these bones in the cat are the same as in humans, although of different shape, and so have more commonly been given the descriptive terms of their counterparts in human anatomy. These descriptive names will be used, with morphological equivalents in parentheses. The heel is a backward projection of the largest of the six ankle bones, the calcaneum[1] (fibulare) (figure 5.22i). The astragalus (talus) articulates below with the calcaneus, above with the malleoli of the tibia and fibula, and in front with the compressed, curved navicular (one or two centralia) reaching to the medial side of the foot. The cuboid (fourth and fifth tarsalia) lies in front of the calcaneus and articulates with the fourth and fifth metatarsals. Next to it, proceeding medially, is the third or lateral cuneiform (third tarsale) articulating with the third metatarsal. Next, the second or middle cuneiform (second tarsale) articulates with the second metatarsal, and the small first or medial cuneiform (first tarsale) articulates with a rudimentary first metatarsal. Phalanges are present. The first toe (hallux) is reduced, and each toe ends in a claw.

Pectoral Girdle, Sternum, and Appendages

Like the pelvic girdle, the pectoral girdle arose by fusion and enlargement of simple cartilaginous elements and underwent later modification reflecting changes in locomotion and support. However, unlike the pelvic girdle, the basic endochondral bones were joined by dermal bones to give rise to a composite pectoral girdle made up of both. Thus, in addition to noting changes in bone shape, size, and presence from group to group, also be aware of the type of bone involved.

→ **Elasmobranchii—Shark** The pectoral girdle is a semicircular, curved cartilage divided into regions (see figure 5.16c). The ventral portion between the bases of the fins is the coracoid bar. The continuations above the fin articulations are the scapular processes, tipped with separate cartilage points, the scapular cartilages.

Like the pelvic girdle, fins are supported by many dermal rays, ceratotrichia, and proximally by pterygiophores. Three large basal pterygiophores support the fin at its articulation with the girdle. The central one is the

[1]In the human literature, "calcaneum" is often spelled "calcaneus."

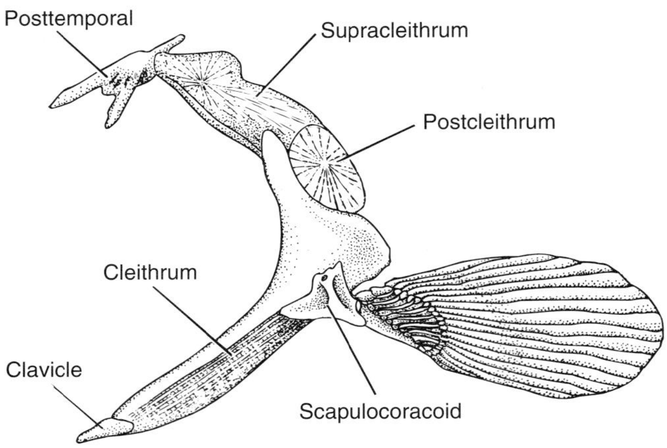

Figure 5.23
Right pectoral girdle of *Amia*, a primitive actinopterygian. Lateral view with cranial to the left.

large triangular mesopterygium. Behind it lies the slender metapterygium, and in front a small cartilage wedge, the propterygium. Distal to these lie rows of aligned radials. No sternum is present.

→ **Teleostei—Perch and *Amia*** During early bony fish evolution, dermal bones became applied to the endochondral girdle and compose the most prominent part. In fact, the cartilage bones are difficult to see in most dried fish skeletons. Although impractical to see on most specimens, a scapulocoracoid is usually present. The approximate position of the dried scapulocoracoid can be found at the point where the base of the fin attaches to the girdle (figure 5.23). Occasionally, specific ossified pterygiophores can be seen as well, although these are also impractical to see. Further, Amia, but not the perch, has a tiny sliverlike clavicle near the anterior, ventral end of the cleithrum (figure 5.23); however, this clavicle is also impractical to locate in commercially purchased specimens.

Examine a mounted skeleton to find the position of the pectoral fin and site of the pectoral girdle that supports it. Return now to an *Amia* skull. The pectoral girdle is covered by the large, disklike opercular bones at the back of the skull. The girdle attaches dorsally at the top of the skull and runs in an arched series of increasingly larger size to beneath the gill arches (figure 5.23). The gill arches support hairlike gill filaments that fill most of the posterior skull opening. The dorsal most posttemporal is the first bone in the series. Forked in most bony fish, here it occupies the upper, posterior corner of the skull and sends a strut from its undersurface down to meet a projection from the skull. Next follows the supracleithrum, which joins the extensive cleithrum in an overlapping articulation. Positioned on the caudal margin of this lap joint is a flakelike postcleithrum. No sternum is present.

→ **Amphibia—*Necturus*** Left and right sides overlap ventrally when in place, but do not fuse as occurs in the pelvic girdle (see figure 5.17c). Only the scapula is ossified. It is a short bone that bears a suprascapular cartilage above. Below the scapula is a concave depression, the glenoid fossa, that receives the head of the humerus. The ventral part of the girdle is the coracoid cartilage, a broad plate that enlarges like a disk toward the ventral midline where it may overlap with its partner. The slender anterior projection at a right angle to the coracoid cartilage is the procoracoid cartilage. The single humerus is the upper limb bone. Next, the forearm consists of the radius and ulna that lie, respectively, anterior and posterior when the limb is outstretched. Six small carpals follow and support four projecting rays. Formally, the proximal element of each ray is termed a metacarpal and the remaining two, phalanges.

Care should be taken in generalizing the *Necturus* pectoral girdle to other amphibians, especially fossilized forms, because it is in many ways specialized. In early amphibians, the girdle was more completely ossified and usually possessed an interclavicle, clavicle, and cleithrum.

Reptilia

In primitive reptiles, the pectoral girdle consists of the dermal (interclavicle, clavicle, and remnants of the cleithrum) and cartilaginous (scapula and procoracoid[2]) components. These have been lost or modified in the two specific reptiles examined here.

→ **Alligator** Clavicles are absent. A bladelike scapula tipped dorsally with a suprascapular cartilage meets a dumb-bell shaped procoracoid laterally to form a posterior glenoid fossa that receives the proximal end of the humerus (see figure 5.18c). The left and right procoracoids articulate medially with the sides of a thin oval sternum. A slender interclavicle is usually difficult to see because it is fused with the ventral side of the sternum, although the spear-shaped head of the interclavicle can be seen projecting forward from the anterior edge of the sternum.

The radius forms the medial margins and the larger ulna the lateral margins of the forearm. Carpals, metacarpals, and phalanges tipped with claws continue distally.

---

[2]In cotylosaurs there appear two ossified coracoids in the ventral part of the pectoral girdle. These are also found in some early synapsids and monotremes. The rear member of this pair enlarged at the expense of its partner to become the coracoid (or sometimes posterior coracoid) of modern placental mammals. The front member alone was inherited by modern reptiles and birds. Although some call this front member only "coracoid," this invites confusion with a different bone mammals possess in their pectoral girdles. Thus, formally in reptiles and birds it is the precoracoid, anterior coracoid, or even procoracoid.

Patterns & Connections

Examine again the *Amia* pectoral girdle. The scapulocoracoid and proximal fin elements are endochondral but the other bones of the girdle are dermal. The pectoral girdles of later vertebrates retain this basic composition, but the dermal elements tend to be reduced as the endochondral elements become more prominent. In primitive synapsids, a new endochondral bone appears, the posterior coracoid, to join with the phylogenetically older scapulocoracoid bone. The three persist into primitive mammals. In marsupials and placental mammals, only the scapula and posterior coracoid (called just coracoid) persist. In modern reptiles and birds, the scapula and anterior coracoid (or procoracoid) persist.

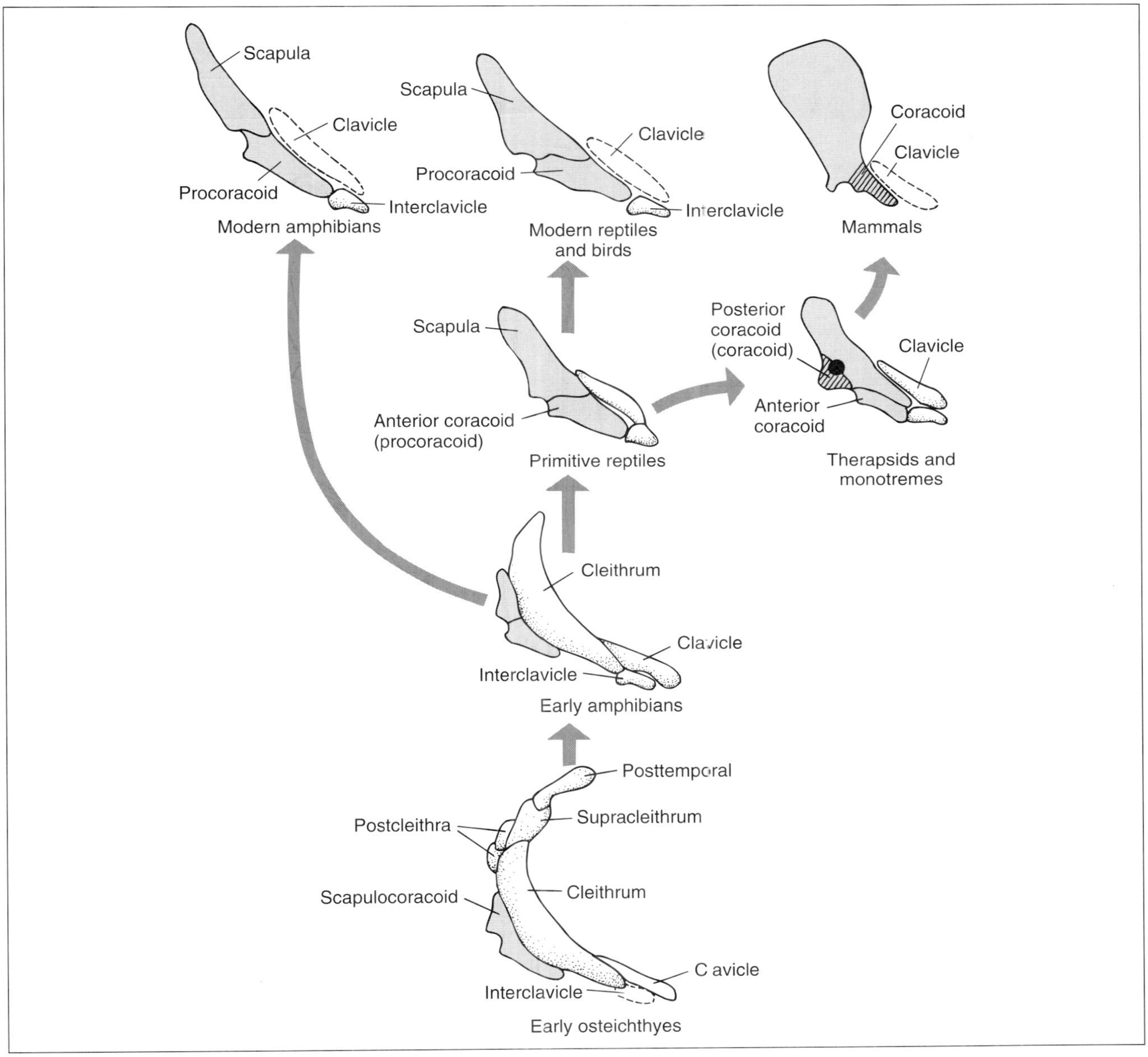

Box Figure 5.1
Summary of pectoral girdle evolution. Notice that dermal elements of the girdle tend to be lost and endochondral elements tend to assume a greater role. In primitive therapsids, a third endochondral bone appears, the posterior coracoid, to join with the phylogenetically older scapula and anterior coracoid bones. The three persist into primitive mammals. In marsupials and placental mammals, only the scapula and posterior coracoid (called just coracoid) persist. In modern reptiles and birds, the scapula and anterior coracoid (or procoracoid) persist.

→ **Turtle** A sternum is absent, although a plastron is present in its place. The outer surface of the plastron is covered by cornified material derived from the epidermis. On the inner surface of the plastron, dermal bones are fused along zigzag sutures. Anteriorly are two, small medial bones called epiplastra, likely derived from the clavicles. Behind them lies the single central entoplastron from the interclavicle. The larger plates that follow may be gastralia.

Two bones form the pectoral girdle, which lies, unlike in other vertebrates, within the rib cage (figure 5.19c). The flat lower bone is the procoracoid. In the glenoid fossa, it joins the scapula, an L-shaped bone reaching at one end dorsally to the carapace. The lower leg of the L-shaped scapula is the prescapular process (acromion by some) that lies in the same plane as the procoracoid. Forelimb bones are as you would expect. A single proximal humerus articulates proximally with the girdle and distally with the ulna and radius. When the limb is extended laterally, the ulna is more caudal than the radius. On either side of the head of the humerus are two triangular shaped projections termed simply humeral processes. The bones are bowed to allow limb retraction and to accommodate restricted locomotion. The distal end of the ulna makes contact with two small bones: the ulnare (outer) and intermedium (inner). The two centralia reside between these two bones and the five carpalia, which each support a metacarpal and two or three phalanges.

→ **Aves—Chicken or Turkey** Three bones compose the pectoral girdle: a long, slender bone over the top of the rib cage, the scapula; a robust bone running downward to the sternum, the procoracoid; and anterior to this the furculum or wishbone (see figure 5.20a,c). The furculum is a composite bone formed by the fusion of two clavicles (forks) joined ventrally to the single interclavicle (compressed plate).

The large sternum, between ventral tips of the ribs, bears a large ventral projection, the keel or carina for attachment of flight muscles (see figure 5.20a). At its anterior end, the sternum bears two large depressions or facets in which rest the procoracoids. Medially, between these two facets arises a single projecting plate, the rostrum.[3] Like a pair of curved horns, paired costal processes (see note 3) project anteriorly and dorsally from the sternum; they depart at a point between the procoracoid and a series of small depressions for articulation with ribs, the rib facets. Behind these rib facets, a branched process sends an anterior xiphisternal process upward, and a slender posterior xiphisternal process (see note 3) arching backward above the sternum.

The glenoid fossa, formed by scapula and procoracoid, receives the head of the humerus (see figure 5.20c). The greater tuberosity, on the ventral side of the head of the humerus, bears a large hole at its base, the pneumatic foramen, portal of entrance into air sacs that fill much of the long bone's core. The lesser tuberosity, on the dorsal side of the head, is continuous with a peaked, lateral projection, the deltoid ridge. The curved ulna, larger than the straight slender radius, possesses an elbow or olecranon process, a short projection at its articulation with the humerus. Carpals fuse and some are lost in the wrist, but there is no consensus as to which contribute to the two small bones, ulnare and radiale, that articulate with the respective forearm bones of the adult chicken. Other wrist bones fuse with metacarpals to form the carpometacarpus, the large composite bone that follows next distally. Long metacarpals II and III are united. At their bases a third metacarpal (I) persists that supports the small first digit. Metacarpal II is the most prominent, and its two well developed phalanges continue beyond the distal end of the carpometacarpus. Fused metacarpals I and III each support a small phalanx.

→ **Mammalia—Cat** The pectoral girdle of mammals typically includes the smallest number of elements, usually just a clavicle and scapula. The clavicle is small and resides in shoulder muscle. It will be located when muscles are dissected, but should be located now on a mounted skeleton where it is often wired near the joint between the forelimb and girdle. The scapula is a flat, large, and not quite triangular bone bearing a ventral cup-shaped depression, the glenoid fossa (figure 5.24a–c). A prominent spine divides the lateral surface of the scapula into two regions: the anterior supraspinous fossa and posterior infraspinous fossa. The entire, undivided medial surface of the scapula facing the rib cage is called the subscapular fossa. The short anterior edge of the scapula is the anterior border, the straight, posterior edge is the axillary border, and the curved upper edge is the vertebral border. The middle roughened border of the spine is the tuberosity. Ventral to the tuberosity, the spine forms a sharp, posterior flat projection, the metacromion. The ventral end of the spine terminates in a downward pointed acromion, receiving one end of the clavicle. The small hooklike coracoid process, remnant of the coracoid bone, forms embryologically as a separate center of ossification. It can be found fused with the anterior rim of the glenoid fossa.

The chain of eight sternebrae constitute the sternum and run in tandem between the ventral tips of the ribs (figure 5.14a). The first sternebra in the series is the manubrium, and the last is the xiphisternum (often with a terminal sternebra xiphoid or ensiform cartilage). The six sternebrae between these constitute the body of the sternum.

---

[3] In the pigeon, these processes are less prominent. The posterior xiphisternal process fuses with the sternum, but its hooklike tip can be seen as a small forward projection at the posterior end of the sternum.

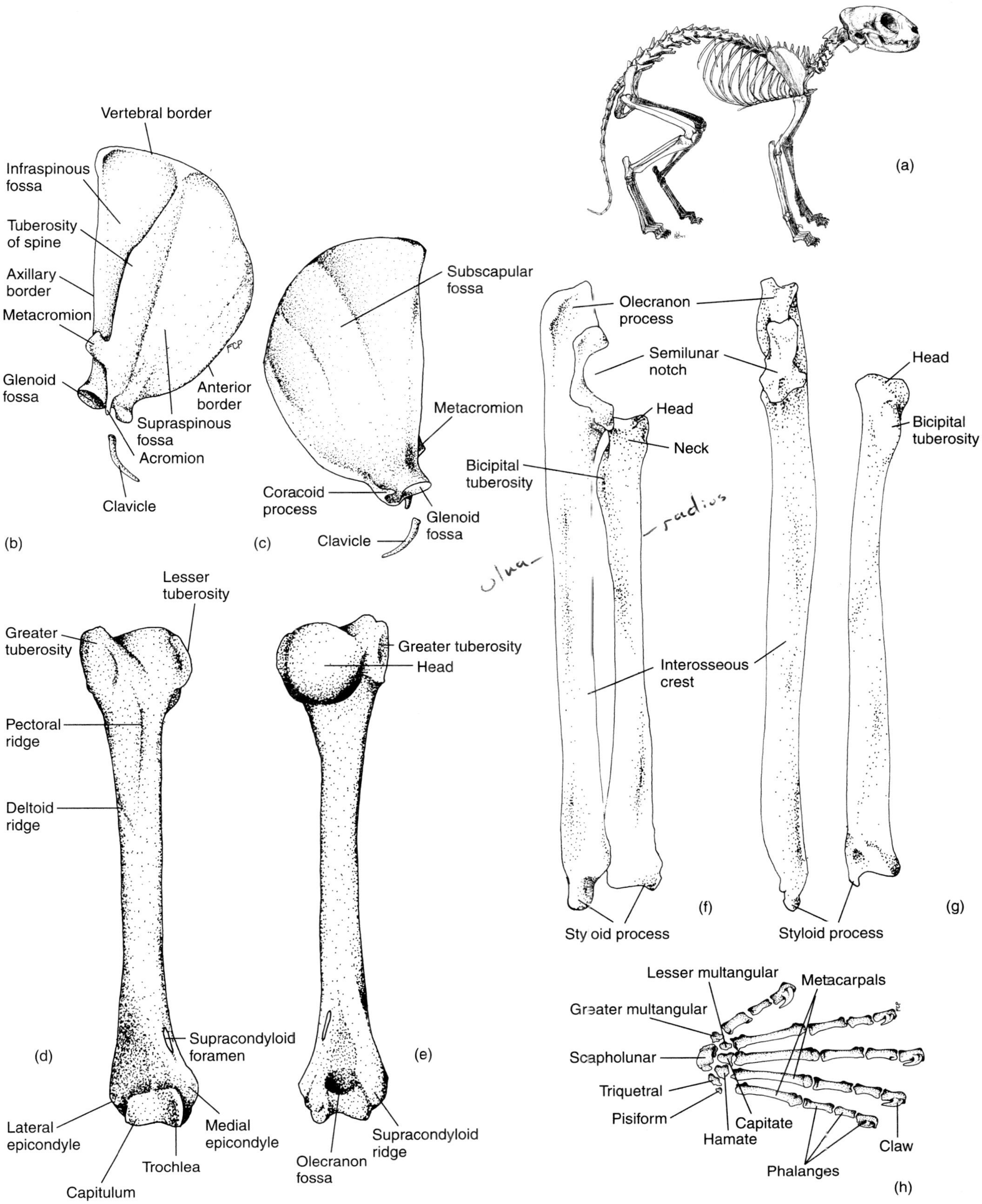

Figure 5.24
Right scapula, clavicle, and forelimb skeleton of a cat. (a) Lateral view of cat skeleton. (b) Scapula and clavicle, lateral view. (c) Scapula and clavicle, medial view. (d) Humerus, anterior view. (e) Humerus, posterior view. (f) Articulated radius and ulna, lateral view. (g) Radius and ulna, anterior view. (h) Fore foot, dorsal view.

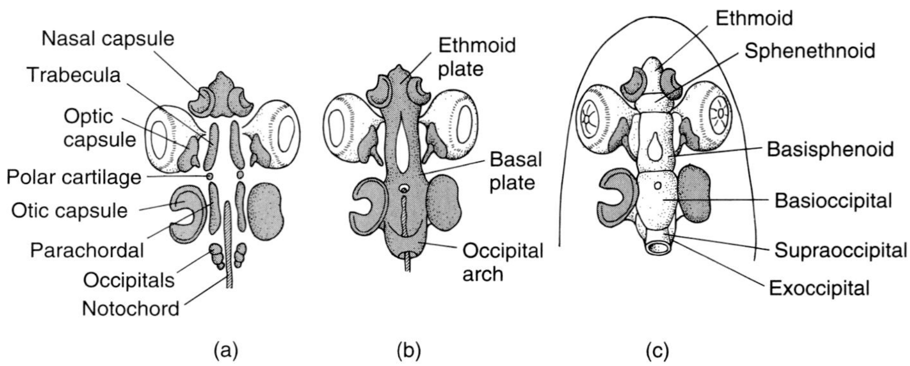

Figure 5.25
Embryonic development of the chondrocranium. Cartilage (shaded) appears first but in most vertebrates is replaced by bone (not shaded) later in development. The chondrocranium includes these cartilaginous elements that form the base and back of the skull together with the supportive capsules around sensory organs. Early condensation of mesenchymal cells differentiates into cartilage (a) that grows and fuses together to produce the basic ethmoid, basal, and occipital regions (b) that later ossify (c), forming basic bones and sensory capsules.
After de Beer.

The forelimb bones are those as expected, although a few distal elements are modified for the cat's locomotor requirements, and the forearm bones cross to direct the toes forward. The head of the humerus projects medially and is bounded on its sides by the greater (lateral) and lesser (medial) tuberosities (figure 5.24d,e). Named for the general muscle groups that insert below the greater tuberosity, the low deltoid (posterior) and pectoral (lateral) ridges appear like seams on the bone. The distal humerus is smooth and rounded where it articulates with forearm bones: the swollen medial part of the articular surface is the trochlea, and the lateral articular swelling is the capitulum. On the side and above each of these are the epicondyles, medial and lateral, respectively. The slit above the medial epicondyle is the supracondyloid foramen. Running out of the lateral epicondyle along the side of the humerus is the supracondyloid ridge. When the forelimb is extended, the elbow rests in the olecranon fossa lying between humeral epicondyles.

In the forearm, the ulna is deeply indented, forming the semilunar notch, located where the ulna articulates with the trochlear surface of the humerus (figure 5.24f,g). Behind the semilunar notch is the elbow or olecranon process. The distal end of the ulna terminates in a pointed styloid process. The interosseous crest runs along the shaft of the ulna. The cup-shaped head of the radius articulates with the capitulum of the humerus. Supporting the head is a short neck. The knoblike swelling on the side and just adjacent to the neck is the bicipital tuberosity. The radius and ulna each ends distally in its own small projection, a styloid process.

In the wrist, the scapholunar (formed from the fusion of the radiale and intermedium) rides on the end of the radius, and the triquetral (ulnare) rides on the ulna (figure 5.24h). Lateral to the triquetral is the small pisiform bone, a sesamoid bone. Lying in a transverse row distal to these elements are, from medial to lateral, the greater multangular (first carpale), lesser multangular (second carpale), capitate (third carpale), and hamate (fourth and fifth carpale). Five metacarpals, their phalanges and terminal claws are present distally.

Skull

Introduction

The vertebrate skull receives three phylogenetic contributions: the **chondrocranium,** which, if ossified, is often known as the **neurocranium,** the **dermatocranium,** and the **splanchnocranium.**

Chondrocranium

Details of its embryonic formation differ between species, but generally, head mesenchyme condenses into cartilage elements (figure 5.25), such as the **parachordals** (next to notochord), **trabeculae** (anterior to parachordals), the **occipitals** (posterior to the parachordals), and **cartilaginous capsules,** which surround chief sense organs (e.g., nasal, optic, otic). These elements fuse to form the embryonic chondrocranium. In elasmobranchs, this cartilaginous skull persists, develops a roof, and becomes the adult braincase. However, in most vertebrates, these cartilaginous plates are transitory. Centers of ossification soon appear (not necessarily in the same sites as the original cartilage elements), replace the cartilage, and give rise to specific bones of the mature skull. These bones include the **occipitals** (basi-, ex-, supra-), **sphenoids, mesethmoid, otic capsule** (prootic, opisthotic, epiotic), **optic capsule** (birds, reptiles), and **nasal capsule.**

Splanchnocranium

The splanchnocranium generally supports the gills and offers attachment for the respiratory muscles. Elements of the splanchnocranium contribute to the jaws and **hyoid apparatus** of gnathostomes (figure 5.26).

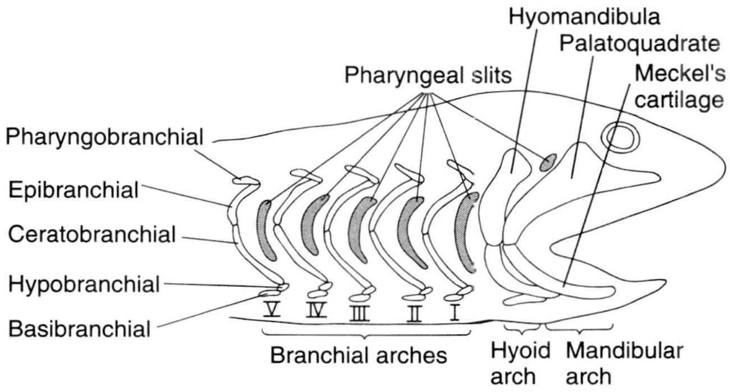

Figure 5.26
Primitive splanchnocranium. Seven arches are shown. Up to five elements compose an arch on each side, beginning with the pharyngobranchial dorsally and in sequence to the basibranchials most ventrally. The first two complete arches are named: mandibular arch for the first and hyoid arch for the second that supports it. The characteristic five-arch elements are reduced to just two in the mandibular arch, the palatoquadrate and Meckel's cartilage. The large hyomandibula, derived from an epibranchial element, is the most prominent component of the next arch, the hyoid arch. Behind the hyoid arch are variable numbers of branchial arches I, II, and so on.

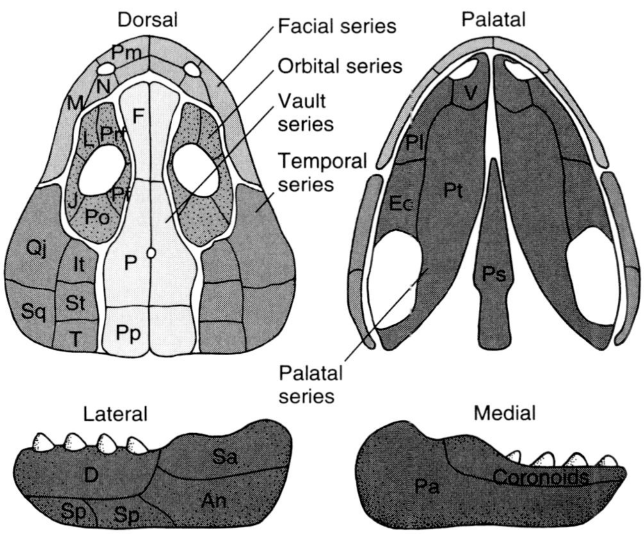

Figure 5.27
Major bones of the dermatocranium. Sets of dermal bones form the facial series surrounding the nostril. The orbital series encircles the eye, and the temporal series composes the lateral wall behind the eye. The vault series, the roofing bones, run across the top of the skull above the brain. Covering the top of the mouth is the palatal series of bones. Meckel's cartilage (not shown) is encased in the mandibular series of the lower jaw. Abbreviations: angular (An), dentary (D), ectopterygoid (Ec), frontal (F), intertemporal (It), jugal (J), lacrimal (L), maxilla (M), nasal (N), parietal (P), prearticular (Pa), palatine (Pl), premaxilla (Pm), postorbital (Po), postparietal (Pp), prefrontal (Prf), parasphenoid (Ps), pterygoid (Pt), quadratojugal (Qj), surangular (Sa), splenial (Sp), squamosal (Sq), supratemporal (St), tabular (T), and vomer (V).

Dermatocranium

The dermatocranium consists of dermal bones that contribute to the skull. Phylogenetically these bones first appear as enlarged bony armor in ostracoderms, sink inward, and become applied to the chondrocranium and splanchnocranium. They also contribute to the pectoral girdle. The dermal skull may contain a considerable series of bones joined firmly at sutures in order to box in the brain and other skull elements. As a convenience, we can group these series and recognize the most common bones in each (figure 5.27).

The Composite Skull

Within the embryonic chondrocranium, centers of ossification appear destined to form specific endochondral bones of the base of the skull. The specific bones so formed vary somewhat between groups, and they often in turn fuse to form composite bones (e.g., temporal bone, occipital bone). But, the more common bones to form within the chondrocranium are shown in figure 5.28a. The nasal capsule often remains unossified, except in mammals where it forms the coiled **turbinate bones** (= turbinales, = nasal conchae).

Bones derived from the splanchnocranium are also endochondral (figure 5.28b,c). The **palatoquadrate** element of the first, **mandibular arch** gives rise to the **quadrate** and **epipterygoid. Meckel's cartilage** gives rise to the **articular.** The **hyomandibula** and its homologues (**collumella** and later **stapes**) arise from the second, hyoid arch. The ventral elements of the splanchnocranium contribute to the **hyoid bone.**

Generally, dermal bones come to encase the endochondral elements of the skull (figure 5.28d). The six regions of dermal bones identified earlier (see figure 5.27) form natural groupings that make learning their names easier. These groups are the **facial, vault, orbital, temporal, palatal,** and **mandibular** series listed in table 5.4 with their components.

Cranial Skeleton

Elasmobranchii—Shark

→ The shark skull is entirely cartilaginous, being derived from the chondrocranium with gill arch contributions from the splanchnocranium. Consequently, it does not air-dry well for inspection. Thus plastic or wax models are usually the handheld "skulls" actually examined. This is sufficient for most structures, but be sure to inspect fluid-preserved, bottled shark skulls for some structures broken or missing from the models.

Begin with a model of the shark chondrocranium. The skull forms by a dramatic enlargement of the embryonic chondrocranium, although this is hard to appreciate until the more modest chondrocranium of other vertebrates is studied. (After having examined the *Necturus* skull, this enlargement in sharks will be more apparent.)

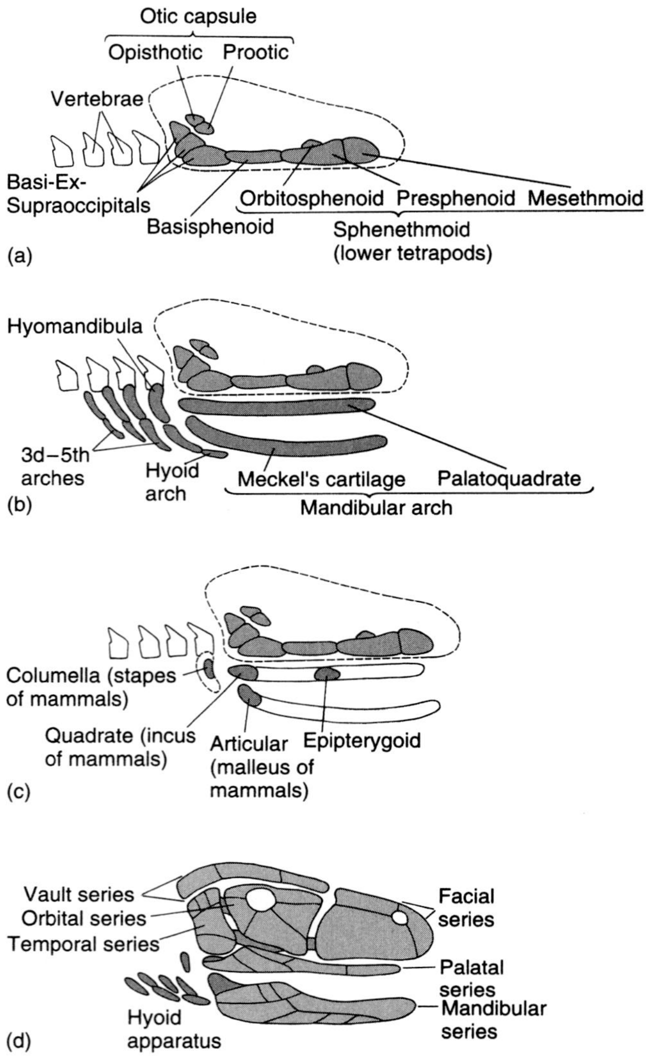

Figure 5.28
Contributions to the skull. The chondrocranium (a) establishes a supportive platform that is joined by contributions from the splanchnocranium (b), in particular the epipterygoid (c). Other parts of the splanchnocranium give rise to the articular, quadrate, and hyomandibular, as well as to the hyoid apparatus (d). The dermatocranium encases most of the chondrocranium together with contributions from the splanchnocranium.

Dorsal Surface

→ The prominent prowlike scoop projecting forward is the rostrum, its rounded floor the precerebral cavity (figure 5.29a). The base of the rostrum lies around the sides and base of an opening, the precerebral fenestra, through which one can peer into the cranial cavity, occupied in life by the brain. Lateral to the base of the rostrum are the paired nasal capsules. They imprint poorly into models. Thus, what in life are marble-sized swellings each pierced anteriorly by two slitlike nares, usually turn out to be missing in wax models. Often only a remnant of the base of the walled capsules remains. Behind the nasal capsules, the lateral edge of the dorsal skull first forms an outward process, the antorbital process. This edge then dips sharply inward, forming the supraorbital crest, and finally swings laterally again as a prominent projection, the postorbital process. The supraorbital crest is thus the edge of the dorsal skull between bases of the antorbital and postorbital processes, and in life lies over the median half of the eyeball.

Many holes pierce the dorsal surface of the skull. Two openings lie along the midline. Anteriorly, the single median hole behind the rostrum and opening to the cranial cavity directly below is the epiphyseal foramen, filled in life by the epiphysis or pineal gland. Posteriorly, a large medial depression with bilaterally paired openings at its base can be seen. The border of the opening into the depression is called the endolymphatic fossa. On the floor of the depression, the small (anterior) pore is the endolymphatic foramen, and the large (posterior) pore is the perilymphatic foramen, which opens to a deep lateral tunnel leading into the auditory or otic capsule. Along the base of the supraorbital crest run, bilaterally, a series of small pores. In front of them is a large pore opening into the orbit, and in front of it is a final opening at the base of the nasal capsule. Collectively these are generally termed foramina, through which branches of cranial nerves (trigeminal, V, and facial, VII) exit.

Table 5.4

Six Series of Bones Forming the Dermatocranium

Facial series	nasal, premaxilla, maxilla
Vault series	frontal, parietal, postparietal
Orbital series	prefrontal, postfrontal, postorbital, jugal, lacrimal
Temporal series	intertemporal, supratemporal, tabular, squamosal, quadratojugal
Palatal series	vomer, palatine, ectopterygoid, pterygoid, parasphenoid
Mandibular series	dentary, splenials, coronoids, angular, prearticular, surangular

Ventral Surface

→ The rostral carina is a keel running the midventral length of the rostrum (figure 5.29b). On either side of its base are two prominent openings: the rostral fenestrae and lateral to these the nasal capsules (or their remnants). A lateral projection of the chondrocranium forms the anterior wall of the orbit, the anterior orbital shelf. Dorsally, it carries the antorbital process. Ventrally, the anterior orbital shelf ends in an unnamed

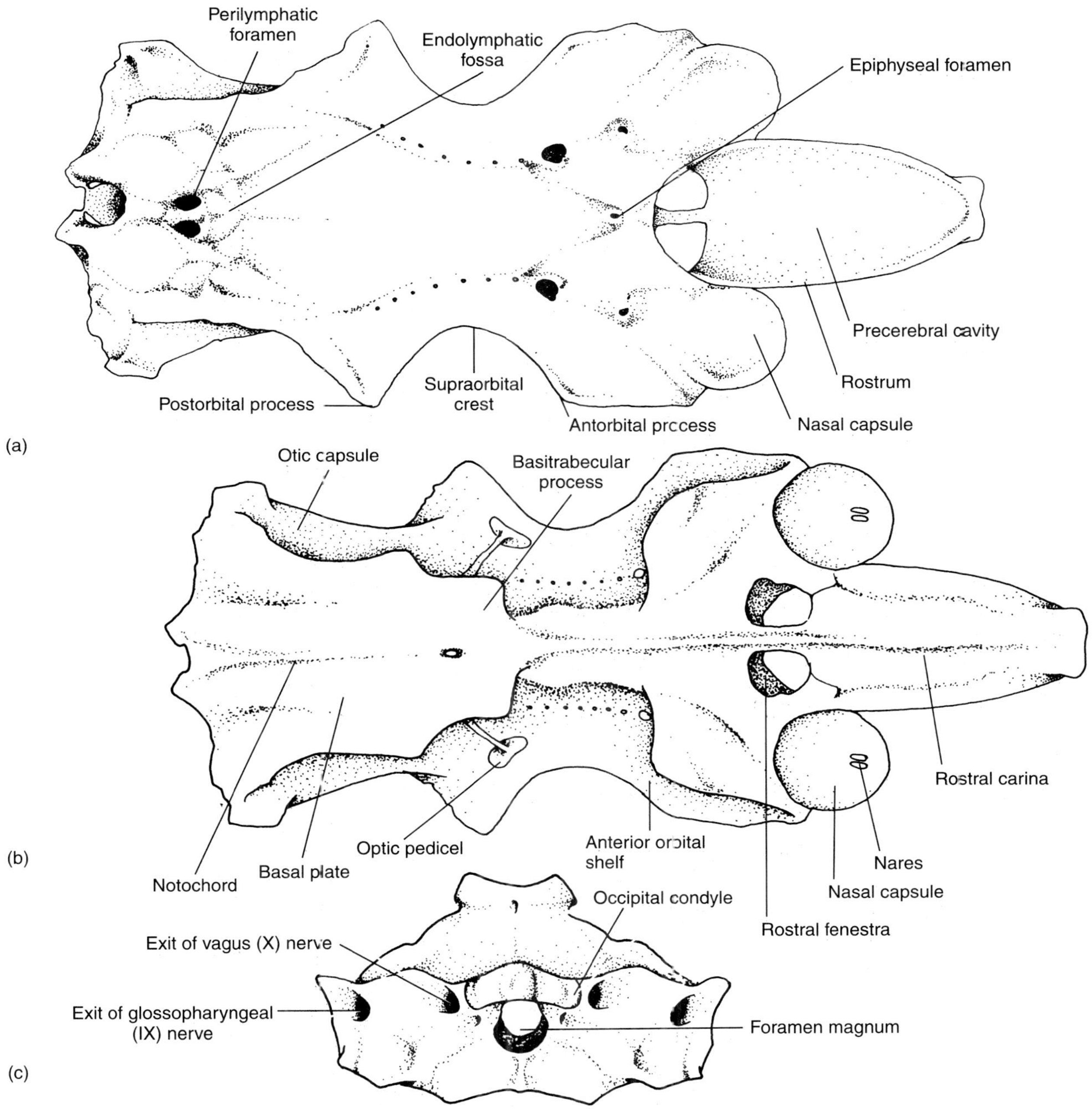

Figure 5.29
Chondrocranium and splanchnocranium of the shark. (a) Chondrocranium, dorsal view (b) Chondrocranium, ventral view. (c) Chondrocranium, posterior view. See page 70 for figure 5.29 d and e.

pointed process. The chondrocranium narrows between the orbits, then posteriorly swells into the paired basitrabecular processes. Immediately behind these processes is the single medial foramen for entrance of the internal carotid arteries. The posterior chondrocranium slopes away from the basitrabecular processes to widen into a flat area, the basal plate, bounded laterally by the otic capsules. The median swelling in the basal plate is the notochord.

Return to a preserved specimen and identify nasal capsules. Also locate, near the rear of the orbit, a thin stalk with a cup-shaped end. This is the optic pedicel, resembling a small golf tee in some respects.

Posterior End

→ The single, medial large opening is the foramen magnum (figure 5.29c). Next, lateral to it is the foramen for exit of the vagus nerve (X) and further laterally the foramen

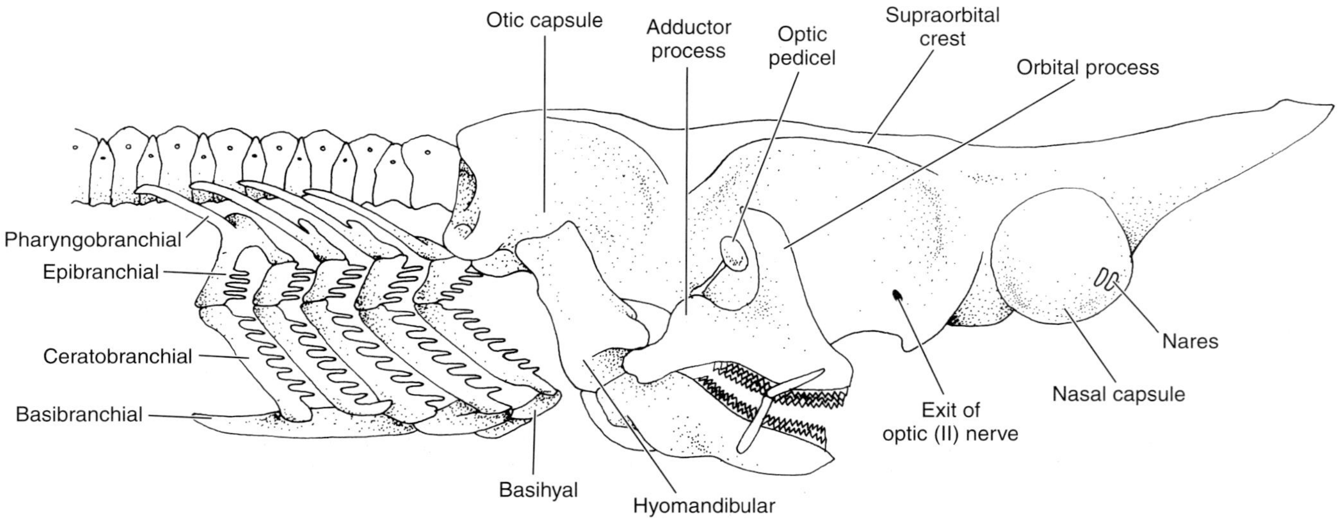

(d)

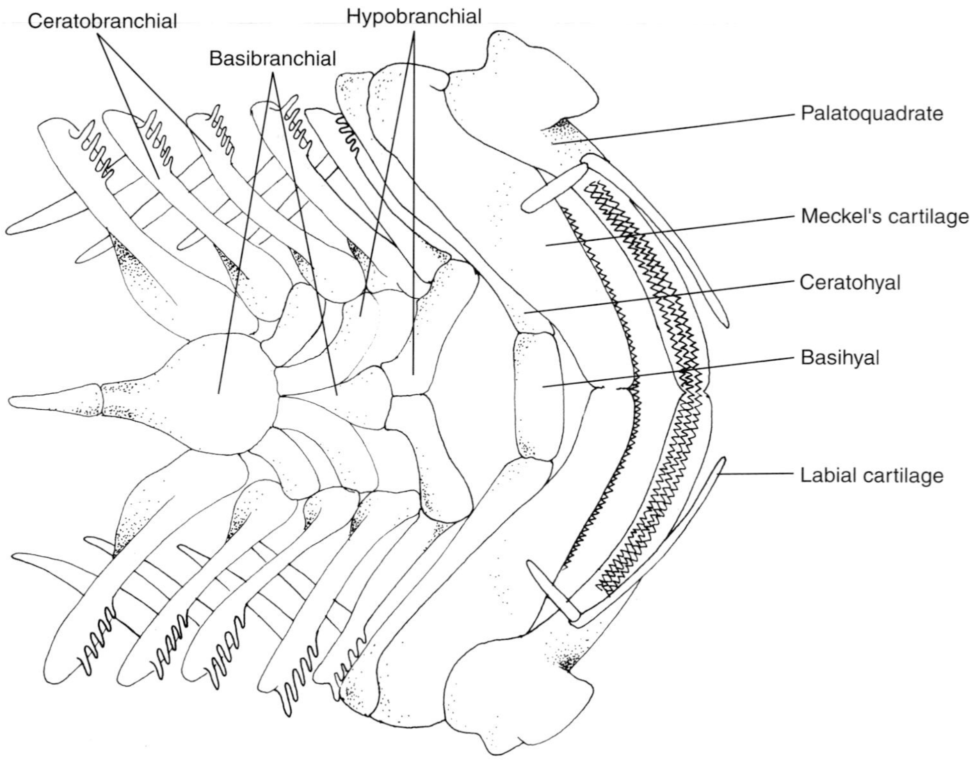

(e)

Figure 5.29
Continued. (d) Chondrocranium with attached splanchnocranium, lateral view. (e) Splanchnocranium, ventral view.

through which the glossopharyngeal nerve (IX) passes. Below the foramen magnum is the cupped surface receiving articulation of the first vertebra and lateral to it paired sharp processes, the occipital condyles.

Lateral Surface

→ The major cavity is the orbit, which in life holds the eyeball. Note the projecting postorbital process, antorbital process, and shelflike supraorbital crest between them (figure 5.29d). The back wall of the orbit is pierced by many pores for nerves and blood vessels. The two largest are the openings for the optic nerve (anterior) and trigeminal nerve (posterior, directly below postorbital process). An optic pedicel, identified previously in the ventral view, projects forward from this same wall in front of the trigeminal foramen, but is usually broken off in skull models. Try to remember to look for the optic pedicel later in the course when you dissect the muscles of the eye.

Form & Function

All vertebrates, except agnathans, have jaws. Vertebrates with jaws are called gnathostomes (jaw mouth). Some biologists mark the advent of vertebrate jaws, evolving through modification of the anterior gill arches, as one of the most important transitions in vertebrate evolution. Powerful muscles, which function in closing the jaws, are derivatives of the branchial arch musculature. It is not surprising, then, that with the evolution of jaws, gnathostomes experienced a dietary shift away from suspension feeding of their ostracoderm ancestors to more active feeding modes and the ability to ingest larger food items. With a change in diet also came a more active lifestyle, like that of carnivorous sharks and bowfins.

Splanchnocranium

→ Study a preserved specimen with gill arches in place (figure 5.29d). Seven gill arches are present in series, although the first two are highly modified. The last five (3–7) in sequence, known as gill or branchial arches, are similar. Each of these arches consists of up to five elements. Dorsal to ventral, these elements are: pharyngobranchial, a posteriorly directed splintlike element that lies next to the vertebral column; epibranchial, a stout piece that together with the next, longer ceratobranchial, bear fingerlike gill rays to support gills; and a hypobranchial, represented by three short pieces on the anterior three branchial arches. The basibranchials, seen to best advantage in ventral view (figure 5.29e), are unpaired. Two exist between the tips of the gill arches, 3–7.

The second arch, or hyoid arch, is modified. A single, midventral piece, the basihyal, articulates on each end with a slender bar, the ceratohyal. Above each of these is the robust hyomandibular. The hyomandibular articulates with the otic region of the skull. The hyomandibular acts as a major site of attachment of the visceral skeleton to the chondrocranium, and is thus said to serve as suspensor of the lower jaw.

The first gill arch, or mandibular arch, is the most modified member of the gill arch series, and forms the tooth bearing upper and lower shark jaws. The paired ventral elements forming the ventral half of the mandibular arch are the large Meckel's cartilages. Rows of teeth run along their inner margins. Often preserved and still attached to Meckel's cartilage is a slender antennaelike process, the labial cartilage. The dorsal half of the first arch consists of a pair of substantial pieces, the palatoquadrates, which form the upper jaw of a shark. The anterior end of each palatoquadrate supports the orbital process, projecting upward along the back wall of the orbit. A shorter, adductor process (= quadrate process) for muscle attachment projects dorsally from the rear of the palatoquadrate.

Osteichthyes—Amia (Bowfin)

Dermatocranium

→ Gill arches and gill rays are a dried mass occupying the center of the skull, seen best in posterior view. The chondrocranium is present but obscured by the extensive covering of dermal bones.

Internally, on the roof of the mouth three pairs of bones bear teeth: ectopterygoid, palatine, and vomer. However, it is impractical to identify these three bones in dried specimens. The external row of upper teeth is carried on two pairs of bones: the premaxillae below the anterior nostrils, and the maxillae, which project laterally along the mouth (figure 5.30a,b). Riding atop each maxilla is the supramaxillary. Between the anterior nostrils lies the unpaired rostral (mesethmoid), often obscured in skeletal preparations by dried flesh. Just lateral to the rostral are the paired nasals. Running laterally and posteriorly from the nasals toward the eye is a path of bones. Lateral to the nasal is the antorbital, then the larger lacrimal, and finally the jugal that defines the lower boundary of the orbit. Behind the orbit are two large bones: the postorbitals (infraorbito-suborbitals). Behind them lies a slender bone running vertically, the hyomandibula. Along its posterior margin are three bones: dorsally the platelike opercular, joined ventrally to the subopercular, and below these the interopercular. Like blades of an oriental fan, the branchiostegal bones unfurl in a series across the ventral throat region. In front of them is the single shieldlike gular, which is bordered laterally by the lower jaw.

The lower jaw is made up of three lateral dermal bones. The dentary, located at the back of the jaw, bears the external row of lower teeth. Just posterior to the dentary at the back of the jaw is the triangular, angular bone. Dorsal to the angular is the small sliver of a bone, the surangular (= supraangular). Finally, the quadrate is deeply recessed below the hyomandibula where it extends laterally to articulate with the medial, posterior corner of the angular.

On the dorsal surface of the skull, five pairs of bones follow the nasals defined by irregular suture lines (figure 5.30b). Behind the nasals lie the fused internasals, followed caudally by the paired parietals. Behind the parietals are the postparietals, and lateral to them and above the postorbitals are the supratemporals. Finally, at the back of the skull behind the postparietals are the extrascapulars.

Amphibia—Necturus

Dermatocranium

→ A dissection microscope may assist you in the study of these small skulls. In the dried *Necturus* skull notice the position of the premaxilla, frontal, parietal, and quadrate across the dorsal surface (figure 5.31a). Ventrally, the

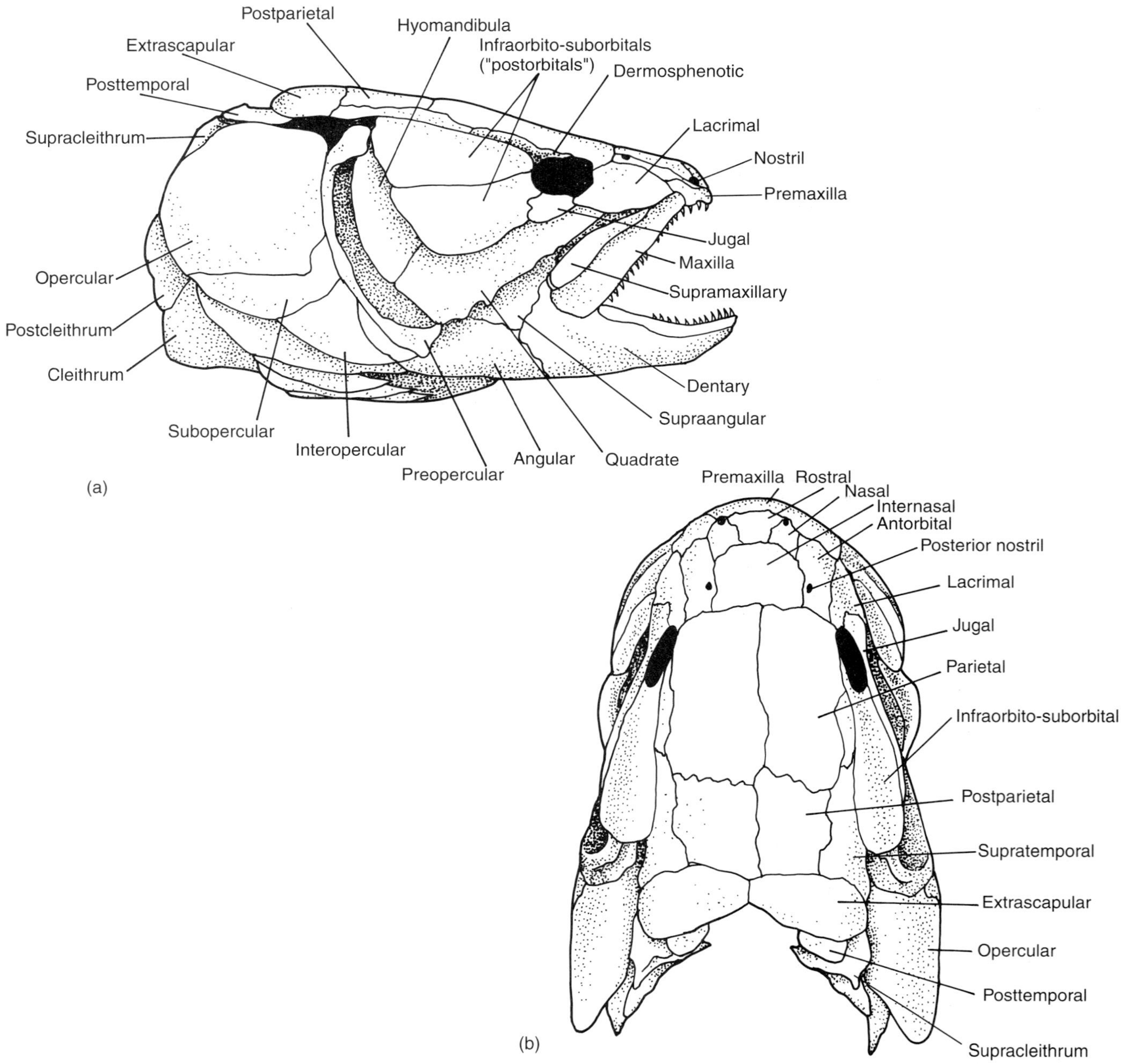

Figure 5.30
Skull of the bowfin, *Amia*, a chondrostean. Lateral (a) and dorsal (b) views.

premaxilla, vomer, pterygoid (= palatopterygoid), and parasphenoid can be seen (figure 5.31b). The bone covering the outer side of the lower jaw is the dentary (figure 5.31c).

Chondrocranium

→ An isolated chondrocranium embedded in plastic offers an opportunity to identify this part of the skull without dermal bones obscuring the study. Once examined, however, try to visualize how it fits within the casing of the dermatocranium (figure 5.31a).

Compare the chondrocranium to figure 5.31a. Note that trabeculae send forth lateral projections, the antorbital processes, and fuse anteriorly via an ethmoid plate from which two short trabecular horns project. The foramen magnum is surrounded ventrally by the tiny basal plate, laterally by the ossified exoccipitals and bulging otic capsules, and above by the synotic tectum. Within the otic capsule two sites of ossification occur: prootic (anteriorly) and opisthotic (posteriorly). The lateral quadrate cartilage ossifies, forming the quadrate bone. Again be sure to note how the chondrocranium fits with the dermatocranium.

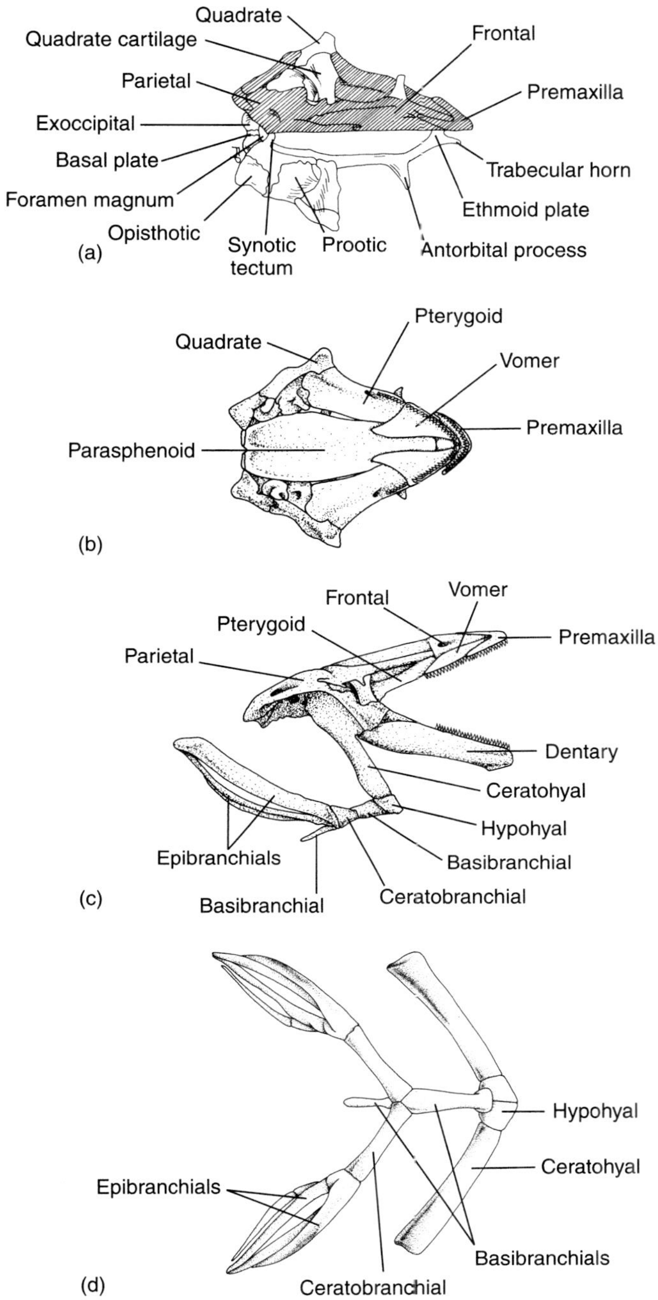

Figure 5.31
Skull of *Necturus.* (a) Dorsal view. Dermatocranium (top, shaded) and chondrocranium (lower, unshaded). (b) Ventral view of skull with jaw removed. (c) Lateral view of skull with splanchnocranium attached. (d) Splanchnocranium, ventral view.

Splanchnocranium

→ Compare figure 5.31c to a preserved or embedded skull with gill arches intact. Identify the corresponding elements in the various gill arches as indicated below and in figure 5.31d:

a) hyoid arch representatives: hypohyal, ceratohyal
b) 3rd, 4th, and 5th gill arches: basibranchials, ceratobranchials, and epibranchials

> **Form & Function**
>
> The palpebral bone rides upon the anteriomedial surface of the orbit. Look at an alligator skull from a lateral view and note how the nostrils and eyes are raised on the head. Consider the living animal, swimming just below the surface, with only its external nares and eyes above the water. In this position, the palpebral functions like a "cowcatcher" on a train, colliding with debris floating on the surface before it strikes a damaging cut to the eye.

Reptilia

Alligator

→ Note that the face is drawn out into a flattened snout. Second, the temporal region of the skull is pierced by two large paired openings: the infratemporal and supratemporal fenestrae.

→ **Dorsal Surface** The paired supratemporal fenestra are best viewed dorsally on the elevated top of the skull behind the orbits (figure 5.32a). The edge of the shelf to the side of each fenestra is the supratemporal arcade. Lateral and the next level below is the infratemporal arcade. The opening in this shelf is the infratemporal fenestra, just behind the orbit. The paired external nares open at the front of the snout. The premaxillae at the tip of the snout, and the maxillae that follow laterally are the only two pair of upper jawbones that bear teeth. Down the dorsal midline, beginning behind the external nares, run the paired nasals, followed in succession by the unpaired frontal, and finally the unpaired parietal between the supratemporal fenestrae. The lateral wall of the supratemporal fenestra is formed by the postorbital (anterior) and squamosal (posterior). These two particular bones lying between the infra- and supratemporal fenestrae are what make this a diapsid skull. Processes of the postorbital and jugal below join to form a postorbital bar defining the rear of the orbit. The long, stout jugal anteriorly meets the maxilla and medially the lacrimal, penetrated by a lacrimal canal best seen where it opens to the orbit (figure 5.32c). Between lacrimal and frontal lies the prefrontal, completing the circle of bones that define the orbit. One additional bone, the palpebral (= superciliary), lies within the anterior orbit attached to the prefrontal. However, the palpebral may have been lost in preparation of the skull. The quadrates project laterally at the back of the skull. Their ventral rounded surfaces, condyles, articulate with the lower jaw. Compressed between quadrates and jugal is the quadratojugal.

→ **Ventral Surface** The roof of the mouth forms by the inward growth of processes that collectively join along the midline. These processes form the secondary palate separating air and food passageways (figure 5.32b). That part

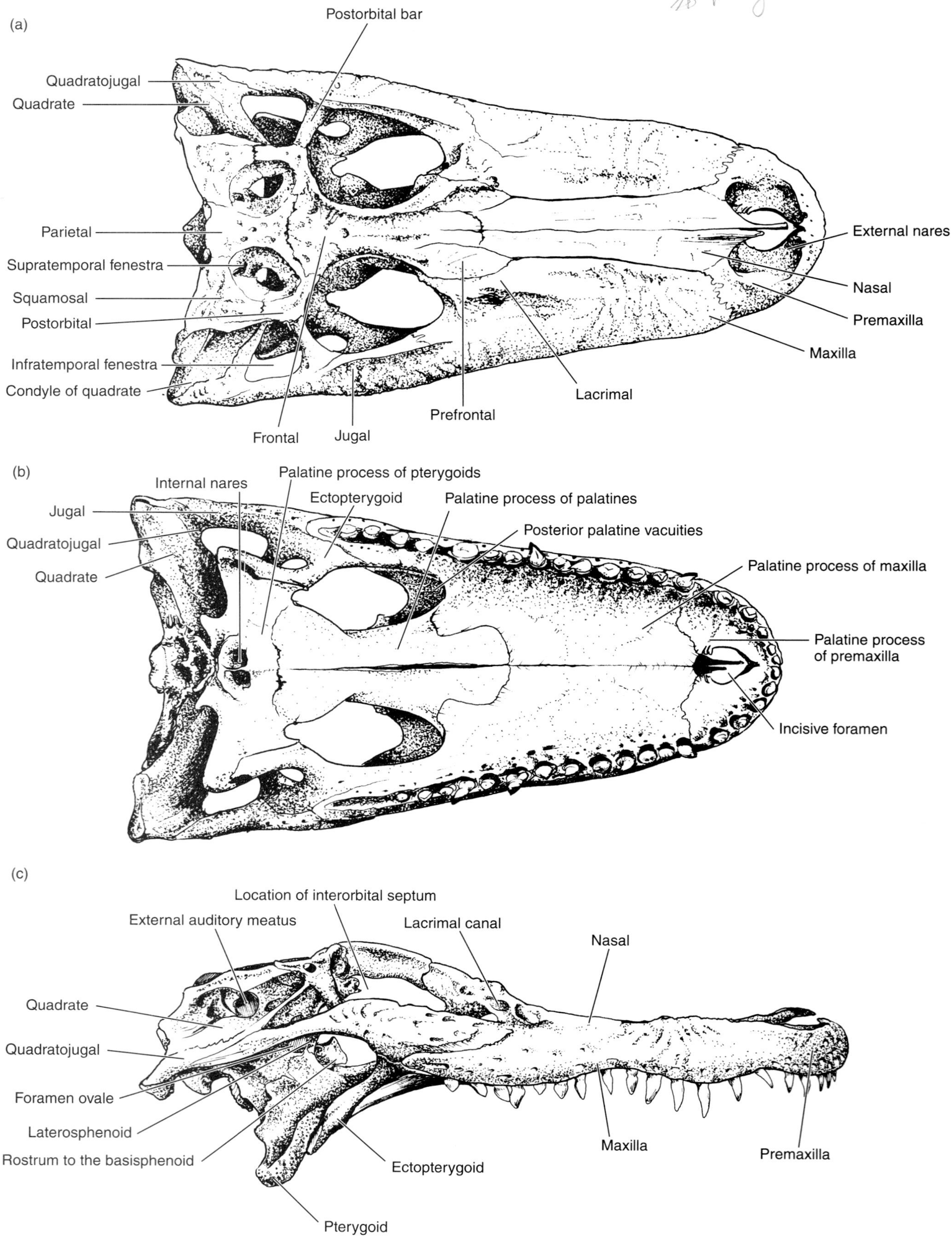

Figure 5.32
Skull of alligator. (a) Dorsal view. (b) Ventral view. (c) Lateral view. See page 75 for figures 5.32 d, e, and f.

(d)

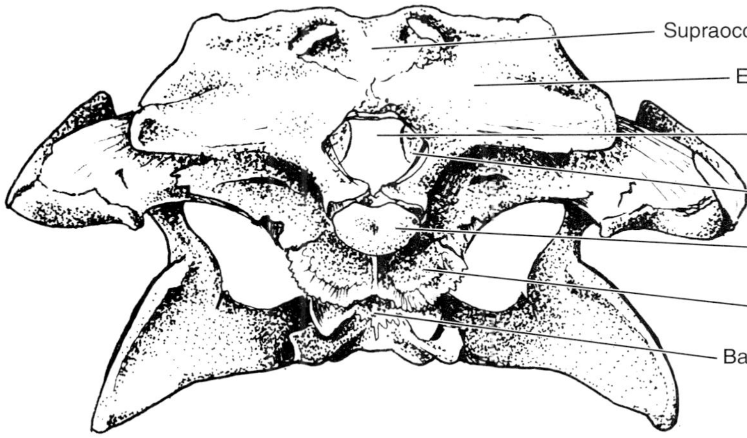

(e)

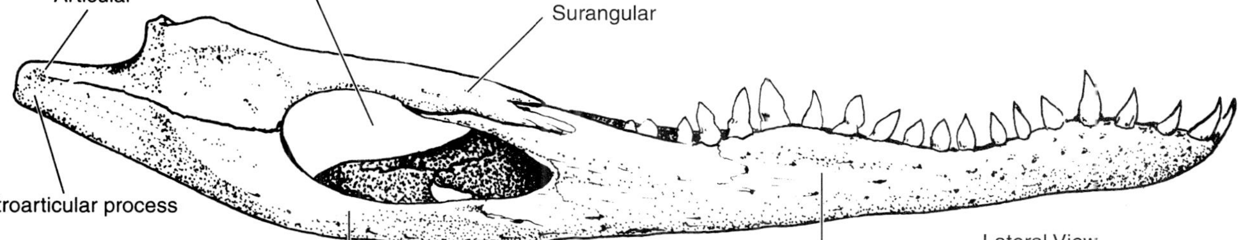

(f)

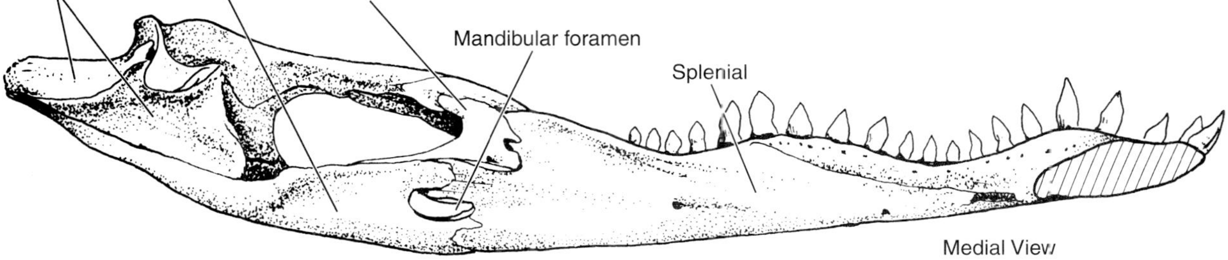

Figure 5.32
Continued. (d) Posterior view. (e) Lateral view of jaw. (f) Medial view of jaw.

of the premaxilla, maxilla, palatine, and pterygoid[4] contributing to the palate is called its palatine process. Thus, the palatine processes of the premaxillae contribute to the roof of the mouth and also form the margins of the small unpaired incisive foramen (= anterior palatine vacuity) covered in life by an epithelium. The posterior medial openings are the internal nares that open within the palatine processes of the pterygoids. The two large oval openings beneath the orbits are posterior palatine vacuities. Between them lie the palatine processes of the palatines. Follow the remaining regions of bones contributing to the secondary palate to identify their associations. Between posterior maxilla and lateral margins of the pterygoid lies the ectopterygoid. Note position and associations of jugal, quadratojugal, and quadrate.

[4]Collectively, these four, paired bones plus the ectopterygoids constitute the maxillary arch in the alligator.

Patterns & Connections

Look again at the alligator skull that you have been studying. Notice the long, extended snout. A longer face can evolve in two ways: (1) by adding new bones at the end of the snout, or (2) by stretching the components that are already present. Long snouts in alligators have evolved by reshaping the ancestral bones and not by the addition of new elements. Evolutionary remodeling of existing structure, and not new construction, is a common theme in morphology. Look for other examples as you compare and study the different skulls.

→ **Lateral Surface** The space between the orbits, in life, is closed by the cartilaginous interorbital septum (figure 5.32c). Although the septum is usually gone in prepared specimens, the paddlelike process projecting in posteriorly can be seen. This process is the rostrum to the basisphenoid, most of which is otherwise concealed beneath the pterygoid. Above and bordering the rostrum is the irregularly shaped and thin laterosphenoid (= pleurosphenoid) that reaches up to the undersides of the frontal and postorbital. The laterosphenoid forms the anterior rim and the quadrate the posterior rim of an opening, the foramen ovale, through which the large trigeminal nerve (V) exits from the cranial cavity. The two laterosphenoids do not meet, thus leaving an irregular medial cleft between them through which pass the optic nerves and olfactory tracts. Tucked under the supratemporal arcade is a large, deep pit, the external auditory meatus, down the core of which runs a slender, needlelike bone, the columella (frequently missing in specimens).

→ **Posterior Surface** Peering through the foramen magnum, the two lateral bulges, tympanic bullae, can be identified (figure 5.32d). The occipital ring includes the paired winglike exoccipitals lateral to the foramen magnum, the unpaired supraoccipital (dorsal), and basioccipital (ventral), which bears a single occipital condyle (with processes of exoccipitals lending some support). Between basioccipital and pterygoids is squeezed the basisphenoid. We shall not name them here, but do at least notice the several paired foramina on the posterior face or surface of the skull. These include a slitlike pair of foramina laterally between exoccipital and quadrate, and four pairs (two large, two small) on the exoccipital closer to the midline. Through the lower set of larger foramina adjacent to the occipital condyle, the internal carotid arteries pass to enter the skull and distribute to the brain. The other foramina are openings through which cranial nerves exit from the cranial cavity. Note the single, unpaired foramen between basioccipital and crowded basisphenoid.

→ **Lower Jaw** Each half of the lower jaw is a ramus (pl. rami) composed of six bones (figure 5.32e,f). Only the dentary bears teeth. The two dentaries unite anteriorly to form a cartilaginous symphysis. In lateral view (figure 5.32e), the large oval opening is the external mandibular fenestra bounded above by the surangular (= supraangular) and below by the angular. The articular, a cartilage bone derived from the posterior of Meckel's cartilage, occupies the rear of the ramus. The articular bone ossifies in the posterior end of Meckel's cartilage and is thus the only cartilage bone of the lower jaw. The articular contributes to the retroarticular process (retro = behind) extending behind the articular at the back of the jaw. The large splenial lies along the inner anterior half of the lower jaw. Posteriorly, it joins the angular along the ventral edge of the jaw. Directly between them opens the small internal mandibular foramen (= Meckelian fenestra). Above this foramen sits the crescent-shaped coronoid (= prearticular). The coronoid forms the anterior border and the articular the posterior border of the mandibular adductor fossa. In life, Meckel's cartilage lies along the bottom of this fossa. This cartilage continues forward to the mandibular symphysis within the appropriately named Meckelian (= mandibular) canal, formed laterally by the dentary and medially by the splenial bones.

→ **Hyoid Apparatus** The hyoid apparatus (see mounted skeleton) derives from parts of the hyoid arch and remaining gill arches. The broad cartilaginous plate constitutes the body of the hyoid, and the pair of processes constitutes the horns (= cornua).

Turtle

→ The following description is of an alligator snapping turtle. Although the pattern of bones is similar in related species, the path of suture lines will vary from one species to another.

Two major landmarks on the skull are the external nares (fused into a single aperture) and the two orbits (figure 5.33a–c). The small paired premaxillae lie at the very tip of the snout below the external nares. The maxilla follows laterally, contributing to the side of the external naris and anterior border of the orbit. Often a large cornified beak covers the entire upper jaw, obscuring the sutures between the premaxilla and maxilla. Behind the maxilla follows the long, broad jugal (= zygomatic), whose anterior end contributes to the lower, posterior rim of the orbit. The rear of the jugal joins the quadratojugal. A ventral process of the quadratojugal meets the quadrate below a deep depression, the otic notch. In life, this notch holds the external auditory meatus, which is covered externally by the tympanum of the ear. Note that in this specimen, as in most modern turtles, the nasal and lacrimal bones are absent. In their expected positions instead reside the bones interpreted to be the prefrontals, identified next.

Behind the external naris and between the orbits lie the prefrontals (figure 5.33a), followed posteriorly, in

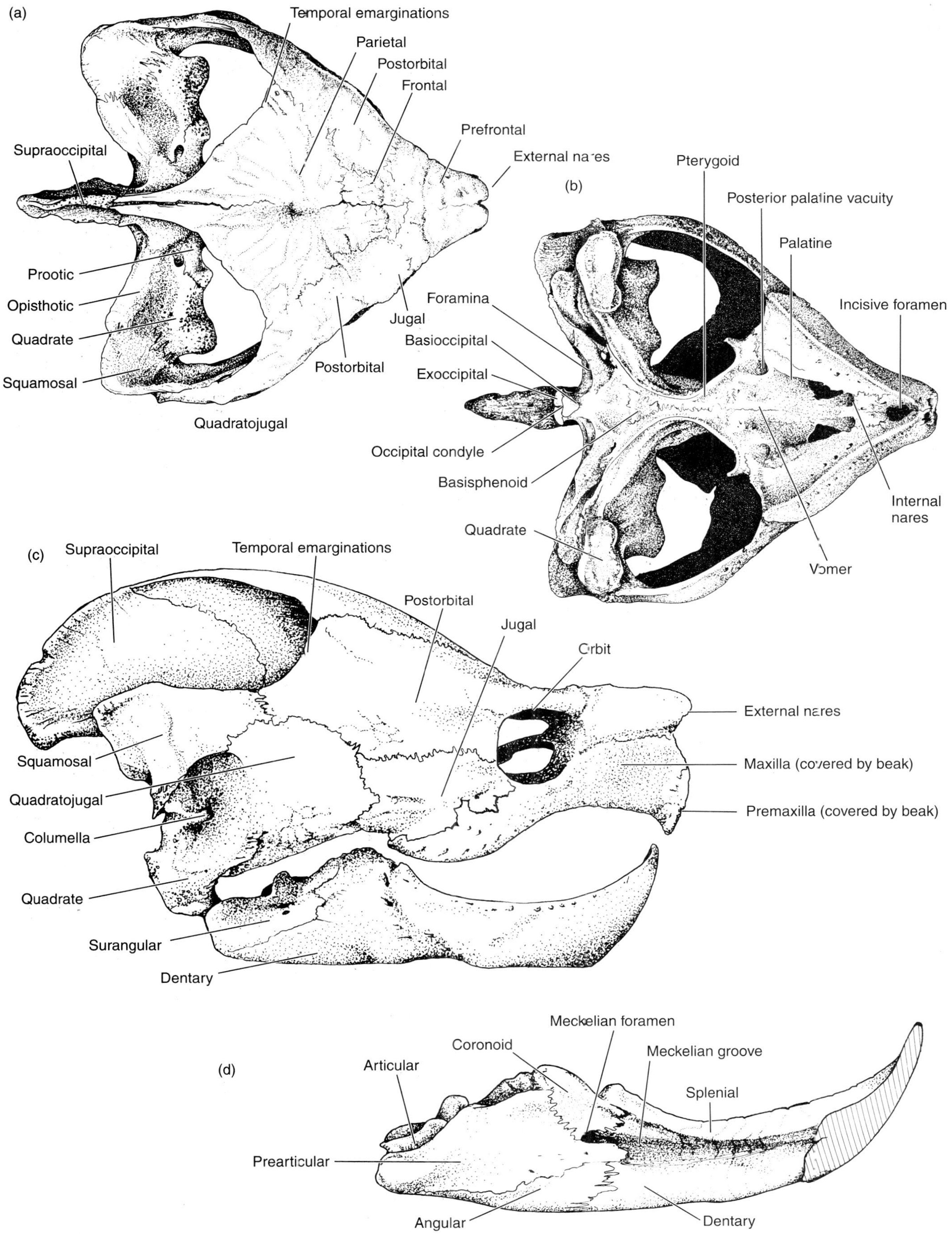

Figure 5.33
Skull of turtle. (a) Dorsal view. (b) Ventral view. (c) Lateral view of skull and jaw. (d) Medial view of jaw.

Patterns & Connections

Look around the room at the heads of your classmates. Notice how much of their heads are dedicated to housing their brains. Next return to the *Necturus,* alligator, and turtle skulls. If you haven't done so already, figure out precisely where the brain is located. It soon becomes clear that the skulls of these animals are primarily dedicated to many roles other than housing the brain. Look at these three skulls more closely to identify the various regions of the skulls not associated with the brain and examine their associated functions.

turn, by frontals (sometimes fused) and then parietals. At the posterior midline, the parietals narrow into a sharp backward directed process completed at its tip by the dorsal blade of the supraoccipital. Lateral to the frontal and parietal lies the postorbital, which defines the rear of the orbit (figure 5.33c). The squamosal follows, forming the outer posterior projecting corner of the skull above the otic notch. Some specimens retain the needlelike columella deep within the auditory meatus.

In the alligator snapping turtle in figure 5.33, the posterior edges of the parietal and squamosal bones form the rims of two enormous notches pushed forward from the back of the skull. Because these notches are open, they are not properly temporal fenestrae but rather temporal emarginations. In life, adductor muscles passing to the lower jaw bulge from these great forward notches. These adductor muscles rest on a broad laterally projecting shelf, the otic capsule. Otic bones, described below, roof over this capsule and fuse broadly with the lateral squamosal and quadrate.

→ **Posterior Skull** Above the foramen magnum resides the supraoccipital, already identified by its contribution to the middorsal posterior projection of the skull (figure 5.33a–c). Note that the occipital condyle is formed by three separate contributing bones (figure 5.33b). The lower, central part arises on the basioccipital; the lateral two contributions are made by the paired exoccipitals. The body of each exoccipital is pierced by small foramina. Dorsally, the exoccipital swings upward to form the lower, outer margins of the foramen magnum. Lateral to the exoccipital and running across the medial base of the otic capsule lie two bones: the prootic (lateral to the base of the parietal), and the opisthotic (posterior and adjacent to the supraoccipital) (figure 5.33a).

→ **Ventral Surface** In front of the basioccipital lies the unpaired triangular basisphenoid (figure 5.33b). Lateral and forward to it lie the broad pair of pterygoids bearing crescent-shaped emarginations on their lateral borders. Even more anterior lie the palatines. The posterior ends of the palatines contribute to the medial rim of small, lateral paired openings adjacent to the pterygoids, the small, slanted posterior palatine vacuities. You may need to tilt the snout downward to notice these vacuities. Anteriorly, between palatines, lies the unpaired vomer. The center of the vomer rises to form interior openings, internal nares, leading to a short canal that opens to the external nares (not to the orbit). Just in front of the internal nares, the premaxillae cooperate in forming a small medial opening, the incisive foramen (= anterior palatine vacuity), perhaps covered by the cornified beak.

→ **Lower Jaw** The lower jaw is composed of two bones, the two mandibles, left and right (figure 5.33c,d). They meet at the midline in a symphysis, a type of joint where bones appear fused to each other. The following description pertains to either mandible.

Articulating with the rounded, ventral end of the quadrate is the cup-shaped articular. Lateral to and above the articular is the slender surangular (= supraangular), forming the outer rear corner of the jaw. The large dentary on the anterior edge of the surangular forms almost all of the outer side of the jaw. The very slender angular runs along the rear, ventral edge of the jaw. Often, in some turtle species, articular, surangular, and angular fuse indistinguishably in the adult.

On the medial surface, the triangular prearticular forms the posterior half of the inner jaw surface. The edge of its posterior corner contributes to the medial articular surface that meets the quadrate. The anterior corner of the prearticular runs under a small foramen, the Meckelian foramen, and its anterior dorsal corner supports a small cap of bone, the coronoid. Anteriorly, the inner surface bears a longitudinal furrow, the Meckelian groove, filled in life by Meckel's cartilage. Often a dried remnant persists. Above this groove is the splenial, but its sutures are difficult to define.

In the turtle skull, follow the line of action of the jaw adductor muscle from the back of the skull downward to its attachment to the lower jaw primarily in the region of the coronoid.

Aves

→ The following description is of a chicken skull. Although the pattern of bones is similar in related birds, the path of suture lines will vary between species.

Bones of the bird skull fuse before hatching and many sutures helping to delineate elements are lost. The bird skull is also quite specialized, and this has led to considerable alteration in bone structure.

The occipital ring that forms the foramen magnum includes the fused supraoccipital (dorsally), exoccipitals (laterally), and the basioccipital (ventrally), which also bears the occipital condyle (figure 5.34a–d). The premaxilla lies where you would expect, medially in front of the nasal apertures, and forms most of the beak. The nasal process of the premaxilla defines the dorsal rims and the maxillary processes the ventral rims of the nasal

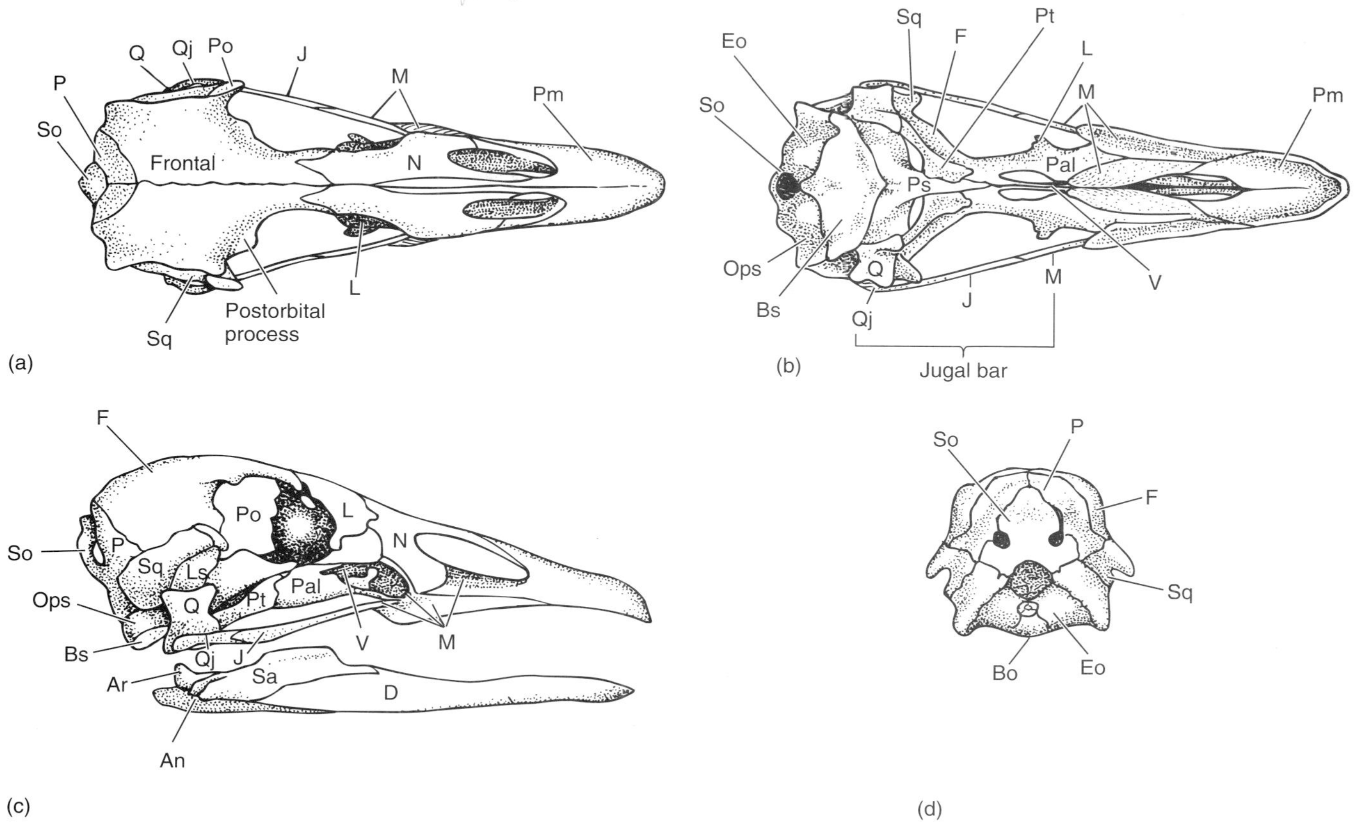

Figure 5.34
Bird skull. In the adult bird, sutures between skull bones fuse to obliterate identifiable borders. Dorsal (a), ventral (b), lateral (c), and posterior (d) views of the skull of a young gosling (*Anser*) before bones fuse. Abbreviations: angular (An), articular (Ar), basioccipital (Bo), basisphenoid (Bs), dentary (D), exoccipital (Eo), frontal (F), jugal (J), lacrimal (L), laterosphenoid (Ls), maxilla (M), nasal (N), opisthootic (Ops), parietal (P), palatine (Pal), premaxilla (Pm), postorbital (Po), parasphenoid (Ps), pterygoid (Pt), quadrate (Q), quadratojugal (Qj), surangular (Sa), supraoccipital (So), squamosal (Sq), and vomer (V).

apertures. The nasal bones lie behind and define the posterior rim of this large opening. The lacrimal projects ventrally from the dorsal side of the nasal, defining the anterior edge of the orbit. The frontals lie between the orbits and form the roof of the anterior skull. Behind them they fuse with the parietals to form the posterior vault of the cranium. Laterally, the parietals meet the squamosal bones. The ventral edge of the squamosal articulates with the dorsal end of the forked quadrate. Laterally the ventral end of the quadrate supports the end of the slender jugal bar (= arch) passing forward beneath the orbit (figure 5.34b). Three bones actually contribute to the jugal bar: quadratojugal (posterior), jugal (middle), and maxilla (anterior where the arch joins the nasal bone).

The lower jaw contains six or seven fused bones (figure 5.34c). The articular articulates with the ventral end of the quadrate. The surangular followed anteriorly by the dentary form the lateral face of the lower jaw. It is impractical to see, but splenial, angular, supraangular, prearticular, and occasionally coronoid contribute to the dermal casing as well.

Roof of Mouth

→ Between the quadrates lie the paired pterygoids (figure 5.34b). They angle medially and anteriorly where they join the paired palatines.

Sclerotic ring (eye) and hyoid bones (throat) are usually out of place. Note on the diagram the bones contributing to the hyoid.

Mammalia—Cat

→ Generally the skull divides into two regions: the facial region bearing eyes and nose, and the cranial region behind, enclosing the brain and the middle and inner ears (figure 5.35a,b). The cheekbone, or zygomatic arch, bows laterally between these two regions. The external nares in life are separated by a cartilaginous septum. On each side of the head a postorbital process projects down from the roof of the skull and up from the zygomatic arch (figure 5.35a). These processes incompletely separate the orbit (anterior) from the temporal fossa behind. At the posterior base of the zygomatic arch is an opening, the external auditory meatus (figure 5.35b). This

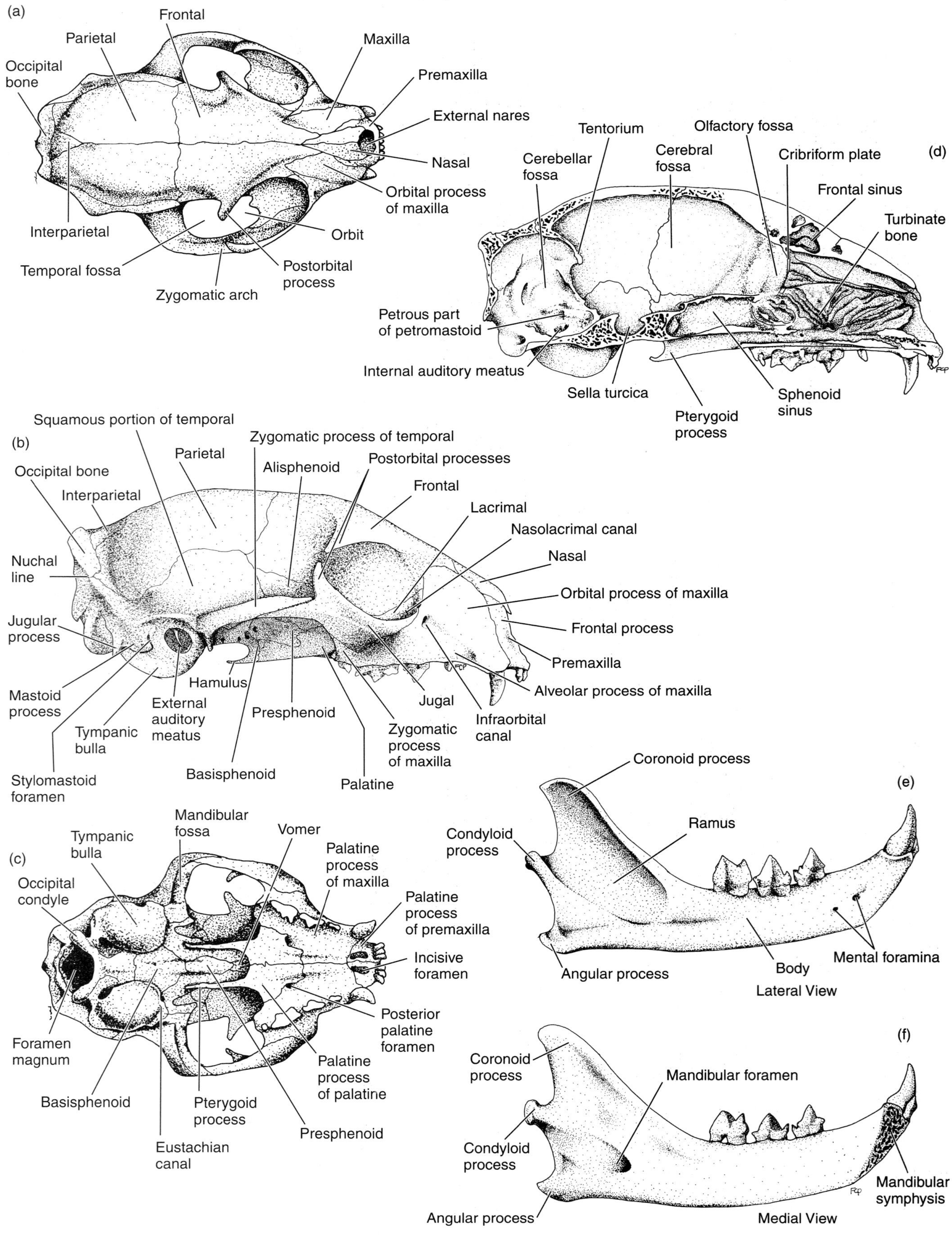

Figure 5.35
Skull of a cat. (a) Dorsal view. (b) Lateral view without the lower jaw. (c) Ventral view without the lower jaw. (d) Sagittal view. (e) Lateral view of jaw. (f) Medial view of jaw.

opening continues into a hollow cavity, the middle ear, contained in the prominent smooth swelling, the tympanic bulla. Immediately behind the external auditory meatus is a pin-sized opening, the stylomastoid foramen partially covered by a small ventrally directed knob, the mastoid process. About the same size, at the back of the tympanic bulla, projects the ventrally directed jugular process. Connecting the bullas is a ridge arching up across the back of the skull, the nuchal line, to which neck muscles attach.

Dorsal Surface

→ The premaxillae bear teeth and lie at the front of the face. They send their frontal processes upward along the sides of the nares (figure 5.35a). Directly above the external nares lie the nasals. Just posteriorly lie the frontals, each bearing a lateral postorbital process. Behind the frontals are the parietals, which form most of the roof over the cranial cavity. Look closely to find a small medial triangular bone, the interparietal, at the posterior ends of the parietals before the back of the skull starts up into the nuchal line. The occipital bone forms the back of the skull, including the foramen magnum and paired occipital condyles.

Lateral Surface

→ In the upper jaw, only the premaxillae and maxillae bear teeth (figure 5.35b). The part of the maxilla bearing teeth is called the alveolar process. The upward bladelike extension of the maxilla in front of the orbit is the orbital (= frontal) process, and the medial shelf contributing to the roof of the mouth is the palatine process. Finally, the lateral projection of the maxilla that begins the zygomatic arch is the zygomatic process. Penetrating the anterior base of the zygomatic process is the infraorbital canal for passage of blood vessels and nerves from the orbit. The zygomatic process supports the jugal (= malar) that bears the upward directed postorbital process. The slender posterior end of the jugal forms a scarf joint (slanted, overlapping) with the zygomatic process of the temporal bone, which completes the zygomatic arch. The smooth depression on the ventral base of the zygomatic process, the mandibular fossa, receives the lower jaw. The zygomatic process projects from a plate of bone lying above the tympanic bulla and below the parietal, the squamous portion of the temporal. The temporal bone is a composite bone in which the squamosal, otic elements, and tympanic bulla are fused.

Most of the back wall of the orbit forms by a downward growth of the frontal. The ventral edge of the frontal meets a series of bones at the base of the orbit. The most anterior of this series is the lacrimal, positioned behind the orbital process of the maxilla, and with a pore, the nasolacrimal canal, leading to the nasal cavity. Next, posteriorly lies the palatine bearing a palatine process that contributes to the roof of the mouth.

Patterns & Connections

Place a cat and alligator skull side by side in front of you. In both animals, compare the series of bones that form the top of the skull. Begin with the nasal bone immediately caudal to the external nares and continue to the foramen magnum. Notice that the names and sequences are generally the same. These similarities reflect part of the morphological pattern common to many vertebrate skulls. Now compare the ventral series of bones from foramen magnum to incisive foramen. By comparing similar regions of vertebrate skulls, you can start to see consistent patterns and make connections that make learning these names easier.

The bone is pierced by two openings: the sphenopalatine foramen (large, upper) and posterior palatine canal (pin sized, lower).

The sharp, posteriorly directed process that projects ventrally below the basisphenoid is the paired hamulus. Between the frontal bone and the tail of the palatine can be found the unpaired presphenoid, and behind it the single and larger basisphenoid. A dorsal, winglike projection of the basisphenoid, the alisphenoid, rises laterally between frontal and temporal bones. A row of four foramina run across the presphenoid and basisphenoid. From anterior to posterior, they are the optic foramen (in the presphenoid), the orbital fissure (between presphenoid and basisphenoid), foramen rotundum (basisphenoid), and foramen ovale (basisphenoid). Branches of various cranial nerves exit through these foramina. Just medial and posterior to the foramen ovale is the Eustachian canal, which runs over the anterior end of the tympanic bulla, slanting outward to connect with the middle ear.

Ventral View and Disarticulated Skull

→ Use a whole skull and a disarticulated skull, preferably simultaneously to work through the tightly interlocking bones along the base of the skull (figure 5.35c). The palatine processes of the premaxillae, maxillae, and palatines cooperate in the formation of the hard palate continued caudally in the life by a fleshy soft palate. The arched internal nares open at the back of the hard palate. The paired, oval incisive foramina (= anterior palatine foramina) lie in the premaxillae. The posterior palatine foramina are a pair of pin holes in the suture between palatine processes of the maxillary and palatine bones. Down the midline run the unpaired vomer dorsal to and partially hidden by the palatine processes of the palatines, followed posteriorly by the slender presphenoid, the broader basisphenoid, and finally occipital bones between the tympanic bullas. The pterygoid in some mammals (e.g., dogs) is a separate bone. In the cat, it fuses with the basisphenoid as the pterygoid process, with some delineating sutures difficult to see. Generally, it is a triangular process whose long base runs along the posterior half of the presphenoid and anterior body of the basisphenoid. It flares laterally, bearing the slender ventrally directed hamulus at its apex. Its anterior edge meets the posterior end of the palatine along an almost obliterated outward suture even with about the middle of the presphenoid. The presphenoid sends a winglike process into the orbit that carries the optic foramen (first in series of four foramina). This process is thought to be homologous to the orbitosphenoid of lower vertebrates, although here in the cat it is properly a part of the composite sphenoid. Compare the other disarticulated bones with those in place.

Sagittal View

→ Bones already identified should be further followed in this revealing preparation. The frontal sinus is the spacious cavity within the frontal bone (figure 5.35d). Delicate shell-thin turbinate bones (= nasal conchae) fill the nasal cavity. There are three compartments to the cranial cavity. The largest is the extensive cerebral fossa formed mostly by frontal, parietal, and temporal bones. The more posterior cerebellar fossa is partially separated from the cerebral fossa by a downward septum called the tentorium. Anteriorly, and continuous with the cerebral fossa, is the small projecting olfactory fossa, just ventral to the frontal sinuses. The cribriform plate is the part of the ethmoid through which tiny numerous olfactory nerves pass toward the brain. The medial depression pressed into the top of the basisphenoid is the sella turcica, which houses the pituitary. The presphenoid contains a dorsal cavity, the sphenoid sinus.

In the ventral rear of the skull, the petromastoid (= periotic) forms from fused otic bones and itself contributes to the temporal bone. The petromastoid bears an outer lateral mastoid process projecting over the external surface on the side of the tympanic bulla. Internally, the petrous part of the petromastoid rests on top of the bulla and bears an opening, the internal auditory meatus, through which the auditory nerve (VIII) passes to reach the small inner ear cavity contained within.

Lower Jaw

→ The mandible consists of only a single pair of bones, the dentaries, fused anteriorly into a mandibular symphysis (figure 5.35e,f). The longitudinal part of the dentary bearing teeth is the body, and the caudal part is the ramus. The coronoid process extends upward from the rear of the dentary. The condyloid process expands transversely into the rounded articulating surface of the mandible, which fits into the mandibular fossa at the base of the zygomatic arch. Below the condyloid process, the rear corner of the dentary projects as the angular process. Caudally, on the inner surface of the dentary, the mandibular foramen permits entrance of a nerve that passes forward but within the bone and finally exits on the outer, anterior end through two mental foramina.

> **Patterns & Connections**
>
> You may have learned previously that the smallest bones in your body are the incus, malleus, and stapes located in your middle ear. Its true. But where evolutionarily did these bones come from? These are not found as a trio in the fish, amphibians, reptiles, or birds that you have studied. You will recall that the jaws of *Necturus*, an alligator, a turtle, and a bird articulate with the quadrate bone at the rear of the skull (a pattern that makes learning this bone a bit easier). Further, you found the articular bone in the lower jaws of these animals but not in mammalian jaws. And finally, the columella in these same animals was already part of the middle ear. The quadrate (incus) and articular (malleus), have joined the columella (stapes) in the middle ear of mammals, where they have assumed a specialized role in sound conduction. Mammals hear with "old" bones in relatively new positions!

Hyoid apparatus

→ On a mounted skeleton, locate the hyoid apparatus (see figure 5.14a). This is formed by remnants of the hyoid and other gill arches. The single transverse bar is the body, and paired projecting extensions are the horns: anterior (long) and posterior (short).

Teeth

Teeth are unique among vertebrate animals in both form and function. They are usually capped with **enamel,** a mineralized coat found only in vertebrates. Teeth help catch and hold prey and can offer strong opposing surfaces that can crush hard shells of prey.

In mammals and a few other vertebrates, **mechanical digestion** begins in the mouth. After each bite, the tongue and cheeks collect food and place it between the upper and lower tooth rows. The teeth break down the bolus mechanically, reducing it to smaller chunks to make swallowing easier. By breaking the large bolus into many smaller pieces, **mastication** (chewing) also increases the surface exposed to chemical digestion. Even in vertebrates that do not chew their food, sharp teeth puncture the surface of the prey, creating sites through which digestive enzymes penetrate when food reaches the gut. For vertebrates that feed on insects and other arthropods, punctures through the chitinous exoskeleton are especially important in giving proteolytic enzymes access to the digestible tissues within.

Tooth Anatomy

The part of the tooth projecting above the gum line, or **gingiva,** is the **crown;** the region below is the **base.** If the base fits into a hole, or **socket** (alveolus), within the jaw bone, the base is referred to as a **root.** Within the crown, the pulp

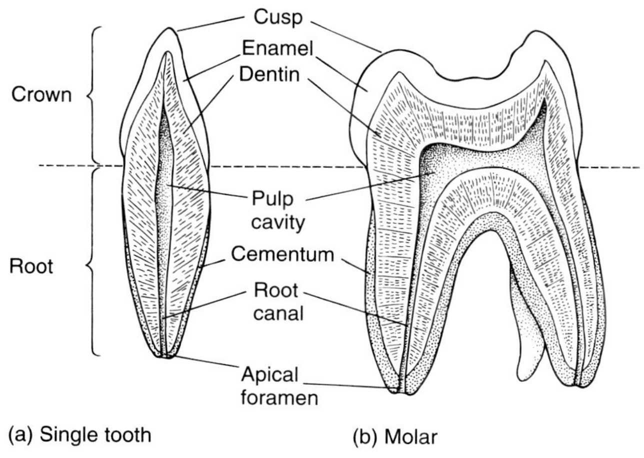

Figure 5.36
Tooth structure. (a) Tooth with single root. (b) Molar tooth with three roots.

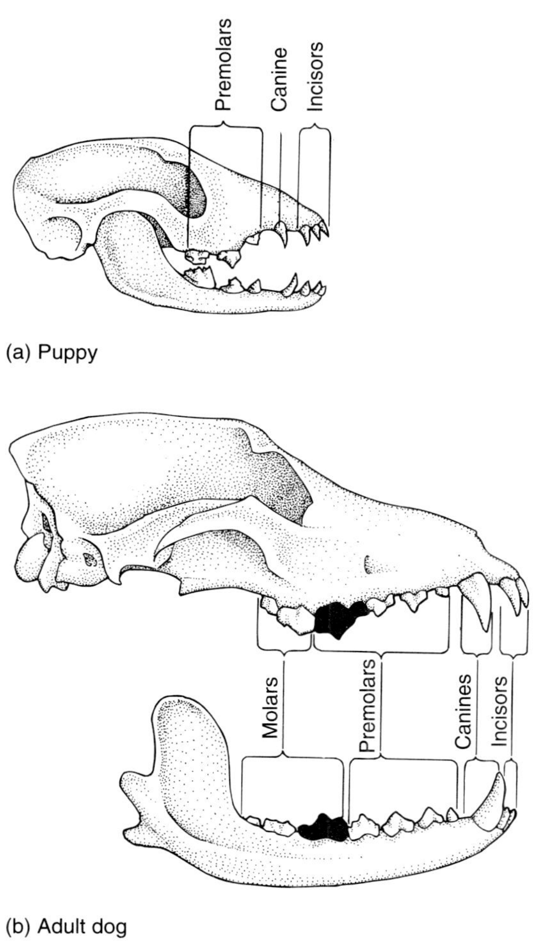

Figure 5.37
Deciduous (a) and permanent teeth (b) in a dog. The carnassials (shaded teeth) are specialized carnivore teeth derived from last premolar (upper) and first molar (lower).

cavity narrows when it enters the root, forming the **root canal,** and opens at the tip of the root as the **apical foramen.** Mucous connective tissue, or **pulp,** fills the pulp cavity and root canal to support blood vessels and nerves that enter the tooth via the apical foramen. The **occlusal surface** of the crown makes contact with opposing teeth. The **cusps** are tiny, raised peaks or ridges on the occlusal surface (figure 5.36).

In lower vertebrates, teeth are usually **homodont,** similar in general appearance throughout the mouth. Modern turtles and birds lack teeth altogether, but mammals have **heterodont** teeth that differ in general appearance throughout the mouth (figure 5.37). Most lower vertebrates have **polyphyodont** dentition; that is, their teeth are continuously replaced. A polyphyodont pattern of replacement ensures rejuvenation of teeth if wear or breakage diminish their function. However, most mammals are **diphyodont,** with just two sets of teeth. The first set, the **deciduous dentition,** or "milk teeth," appears during early life (figure 5.37). It consists of incisors, canines, and premolars, but no molars. As a mammal matures, these are shed and replaced by the **permanent dentition,** consisting of a second set of incisors, canines, and premolars and now molars (figure 5.37).

Sharks—Homodont and Polyphyodont Teeth

→ Examine a set of dried shark jaws. The teeth in some predators, such as sharks, have sharp, knifelike cutting edges along their sides to help pierce skin (figure 5.38a). For slicing chunks from flesh, these edges are further serrated, like those on a bread knife, to cut the tough skin. Notice the many rows of teeth that roll dorsally and forward to replace those routinely lost along the anterior margin.

Amphibia/Salamander—Homodont and Polyphyodont

In larval salamanders, most teeth are pointed cones, but the teeth of metamorphosed adults often show specializations (figure 5.38b). The crowns in some species are bicuspid, having two cusps, and the crown sits upon a basal pedicel to which it is attached by collagenous fibers.

→ Examine a <u>Necturus</u> skull under a dissection microscope and examine the type of teeth. Are the teeth in *Necturus* pointed cones or bicuspid?

Reptiles/Alligator—Heterodont and Polyphyodont Teeth

→ Examine an alligator's teeth still embedded within the upper and lower jaws. These sharp, pointed teeth vary in size. They are well adapted to grasping but are relatively

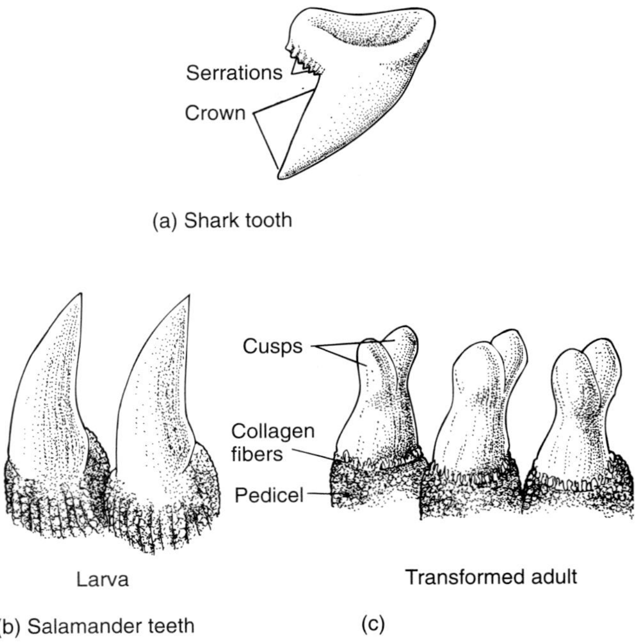

Figure 5.38
Specializations of teeth. (a) Shark tooth (blacknose shark, *Carcharhinus acronotus*). The pointed crown has a nearly smooth edge for piercing prey; the base is serrated for cutting flesh. This tooth can be more than ten times larger than the salamander teeth in (b) and (c). (b) Teeth before (larval) and after (adult) metamorphosis in the northwestern salamander (*Ambystoma gracile*). Larval teeth are pointed. Those of the transformed adult have divided cusps that articulate with a basal pedicel. The cusps are thought to inflect with the struggling prey, thus discouraging its escape from the mouth.

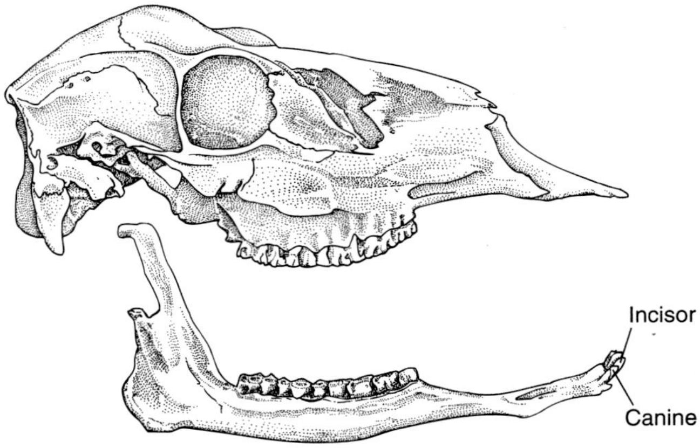

Figure 5.39
Skull of the mule deer (*Odocoileus hemionus*) with lower jaw detached. The dental formula for the upper row is 0-0-3-3, and the formula for the lower is 3-1-3-3. The absence of upper incisors and canines is normal, and is indicated in the dental formula by zeros. The lower incisors and canines are present, and the canines are adjacent to the incisors at the front of the mandible.

ineffective at chewing or cutting the integument of most prey. A tooth socket (alveolus) may be examined in places along the jaw where teeth have fallen out.

Mammal—Heterodont and Diphyodont Teeth

In mammals, the teeth not only capture or clip food but they are also specialized to chew it, producing a complex and distinctive dentition. In fact, the dentition in different groups is so distinctive that it is often the basis for identifying living animals and fossil species. Not surprisingly, an elaborate terminology has grown up to describe the particular features of mammalian teeth.

The heterodont dentition of mammals includes four types of teeth within the mouth: **incisors** at the front, **canines** next to them, **premolars** along the sides of the mouth, and **molars** at the back. The number of each type differ among groups of mammals. The **dental formula** is a short-hand expression of the number of each kind of tooth on one side of the head for a taxonomic group. For example, the dental formula of the coyote (*Canis latrans*) is:

I 3/3, C 1/1, PM 4/4, M 2/3.

This means that there are three upper and three lower incisors (I), one upper and one lower canine (C), four upper and four lower premolars (PM), and two upper and three lower molars (M), 21 per side or 42 total teeth. Sometimes the dental formula is written as 3-1-4-2/3-1-4-3, the first four numbers indicating the upper teeth and the second four the lower teeth (in this example, for the coyote). The dental formula for the mule deer (*Odocoileus hemionus*) is 0-0-3-3/3-1-3-3 (figure 5.39). Notice that the missing upper incisors and canines are indicated by zeros.

The cusp patterns of mammalian cheek teeth are so distinctive that they are also used to identify species. Cusps are termed **cones.** Major cones (figure 5.40) are identified by adding the prefixes *proto-* (first), *para-* (beside), *meta-* (beyond), *hypo-* (under), or ento- (within); minor cusps are indicated by the suffix *-ul(e).*

→ Examine the teeth in the upper and lower jaws of a cat. The dental formula for a cat is 3-1-3-1/3-1-2-1. Using this formula, calculate the number of each type of tooth and identify them in the cat. Notice the gap, or diastema, between the canines and molars. Notice that the last premolar and first molar, the carnassials, form the primary shearing surface along the lateral side of the jaw. These shearing forces can be best appreciated by articulating a jaw to the skull and noting the precise alignment, similar to scissor blades in form and function.

Examine other mammalian skulls before you. When available, carefully place the lower jaw in position and rotate it slowly to close the mouth. Notice how closely the teeth mesh in a very precise occlusion. Determine the dental formula for each of these skulls and verify them with your instructor.

Form & Function

The teeth of vertebrates are one of the best examples of the tight relationship between form and function. A close examination of the teeth of an unknown skull will provide insight into the diet of the animal. This is how paleontologists can make informed hypotheses about the diets of dinosaurs. This figure indicates the relationship between tooth shape and diet in three mammals. (a) A carnivore skull showing position of carnassial teeth (shaded). Carnassials function like scissors to slice through soft but pliable foods. (b) Artiodactyl skull showing the position of the grinding teeth (shaded). Corrugated occlusal surfaces of these teeth help grind fibrous foods. (c) Primate skull showing the position of the compression teeth (shaded) that pulverize hard foods.

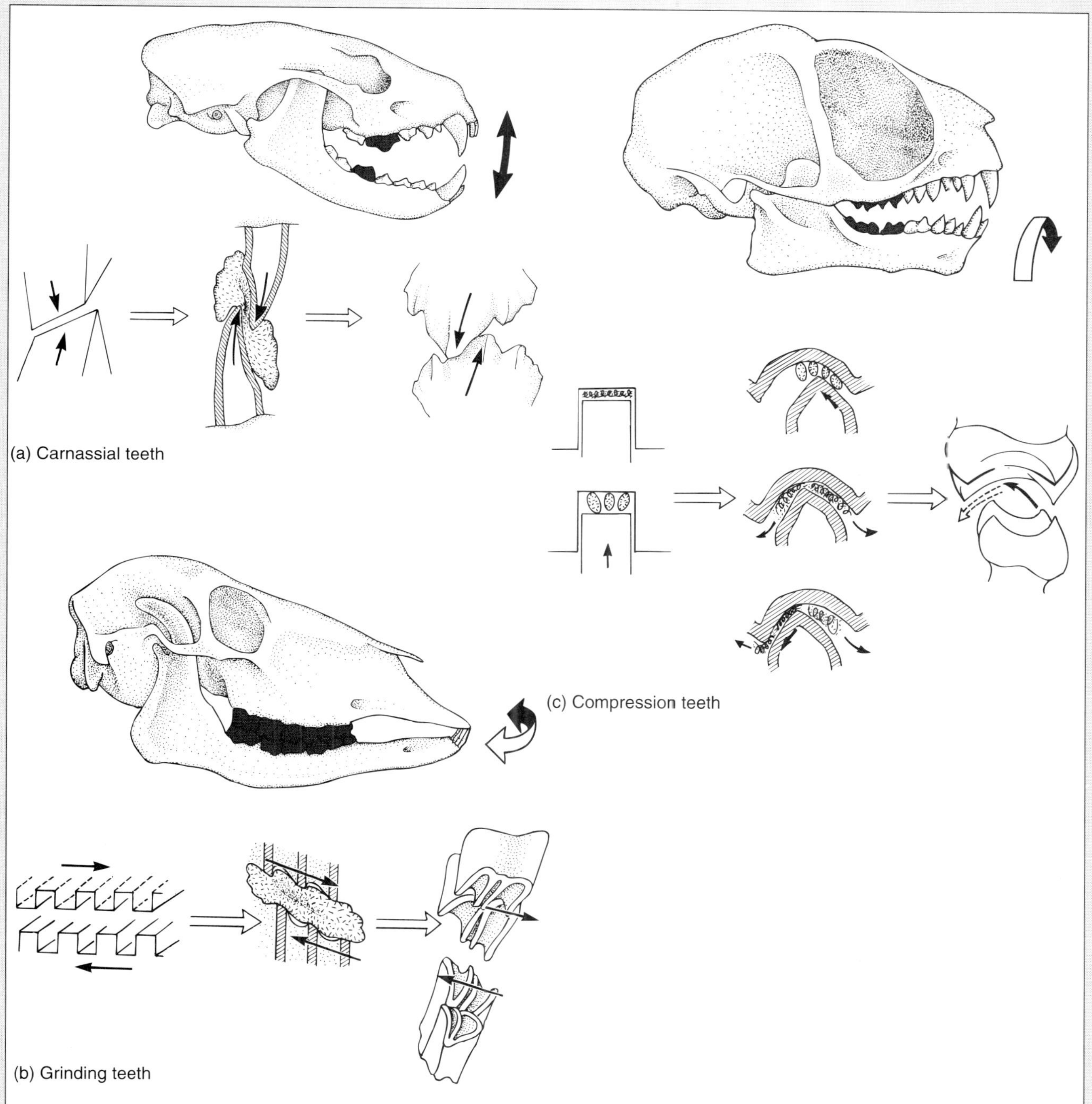

Box Figure 5.2
Mastication in mammals. (a) Carnivore skull showing position of carnassials (shaded). Carnassials function like scissors to slice through soft but resilient foods. (b) Artiodactyl skull showing position of grinding teeth (shaded). Corrugated occlusal surfaces of these teeth grind fibrous foods. (c) Primate skull showing position of compression teeth (shaded) that pulverize hard foods.
After M. Hildebrand, et al., eds., 1985, Functional vertebrate morphology, Belknap Press of Harvard University Press.

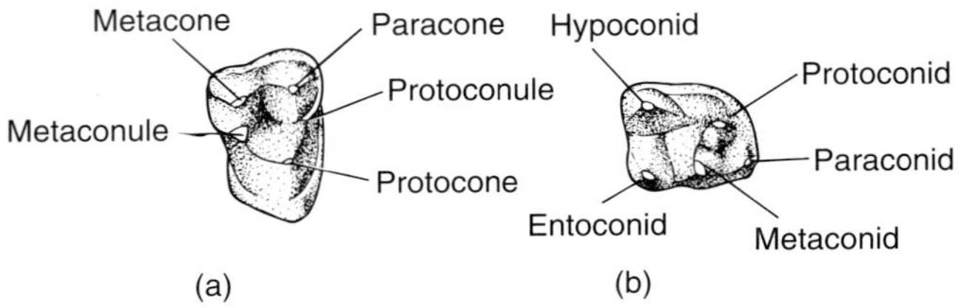

Figure 5.40
Molar anatomy in placental mammals. Doral view of crushing surface. (a) Upper right molar. (b) Lower left molar.
Modified from A.S. Romer and T.S. Parsons, 1985, The Vertebrate Body, Saunders College Publishing.

Form & Function

Mastication in mammals has been accompanied by very precise tooth occlusion to serve the mechanical breakdown of food. Precise, strong occlusion requires a firm skull, so mammals have lost the cranial kinesis of reptiles. In addition, the mammalian mandibular condyle often fits into a very restrictive articulation with the squamosal bone. When jaws close about this joint, upper and lower rows of teeth are placed in very precise alignment. This allows specialized teeth to function properly. To avoid disruption of the occlusial tooth row, teeth in most mammals are not continually replaced. Instead, most mammals exhibit diphyodonty, giving adults a "permanent" set of teeth and thereby a permanently functional occlusial tooth row.

Pattern of Evolutionary Change

Evolutionary change occurs in organisms by several mechanisms. The previous skeletal material and the examples in the boxed essays can illustrate a few of these ways. First, change can occur through the renovation of existing structures. As indicated in the boxed essays, anterior gill bars evolved into jaws, and several jawbones evolved into middle ear bones. These anatomical changes may also be accompanied by a change in function. For example, gill bars that support the respiratory apparatus shift to elements of the jaw associated with feeding. In turn, some of these jaw elements have shifted form and function from prey capture to the conduction of sound as part of the middle ear!

Second, change can proceed by the loss of some elements or by the enlargement of others. For example, the vertebral columns of amphibians and amniotes (reptiles, birds, and mammals) possess just a single centrum per body segment, resulting from the loss of either the intercentrum or pleurocentrum in the fish vertebral column, as illustrated in *Amia* (figure 5.8). By contrast, alligator snouts are elongated by drawing out ancestral facial and palatal bones and not by the addition of new elements.

Evolutionary changes in structure do not run under their own momentum. Instead, structural adaptations result from changes in a species' environment and lifestyle. For example, the major transitions from water-to-land or from land-to-air have required major modifications in the basic vertebrate body. As amphibians and amniotes became more terrestrial, the number of vertebral elements was reduced (the loss of either the intercentrum or pleurocentrum noted above). In addition, the demands for greater support for flight favored the fusion of vertebral elements in the sacrum of a bird and other parts of the skeleton. Lesser shifts in lifestyle have resulted in more subtle changes in anatomy.

In the laboratory you concentrated on identifying structures. This is valuable because a basic understanding of anatomy is an important foundation for many disciplines, including physiology, medicine, and even natural history. But if you step back from the task of identification, you will see that the various morphological forms are much more than a conglomeration of anatomical parts. Instead, the forms are adaptive responses to changes in function and environmental setting. Form, function, and environment must all be considered to better understand evolutionary change. Thus, in addition to finding the anatomy, take time to consider the comparative function of the parts and the likely environmental setting to which they are adapted.

Muscular System and External Anatomy 6

Introduction

Skeletal, smooth, and cardiac muscle help regulate physiological activities and permit movements of the skeleton. Skeletal muscles are the active companion of the skeleton, often considered together as a functional unit, the musculoskeletal system. Together, muscles, bones, and the connective tissues that join them permit body movements and locomotion vital to a vertebrate's existence. Smooth muscles affect the activity of the viscera. For example, muscles wrapping the tubular digestive tract constrict in peristaltic waves to mix and move the food within. Smooth muscles also form sphincters that control the passage of materials out of blood vessels and tubular portions of the digestive tract. Sheets of smooth muscle within the walls of the respiratory tract affect the flow of air to and from the lungs by changing the size of air passageways. Cardiac muscle generates rhythmic contractions to propel blood through an organism's vessels. Further, some unusual muscles have very specialized functions, including heat generation and the production of electrical fields useful in navigation, defense, and predation. The following chapter addresses only part of the muscular system, the skeletal muscles.

Terminology

A muscle is technically identified by where its ends attach—its origin and insertion (table 6.1). Its name may reflect a combination of the origin and insertion, in that order, into a compound word, such as pubofemoralis (pubis = "origin"; femur = "insertion"). Many muscle names in vertebrate anatomy are carried over from human anatomy. These names are often based not on attachments, homologies, or actions, but upon shape (clavotrapezius: -trapez = "table"), size (gluteus maximus: =-maxim = "largest"), or imaginary resemblances (gastrocnemius: gastro- = "stomach/belly").

Actions

Muscles can affect themselves and the skeletal system in many ways. For example, muscles can act in support of each other (synergists) or in opposition to each other (antagonists). The effect of muscles on the skeletal system has required a set of terminology (included in table 6.1) to specifically identify the consequences of muscle actions (figure 6.1)

Table 6.1

Mini-Glossary of Muscle Terminology

abduct—to move a limb away from the midline of the body
adduct—to move a limb toward the midline of the body
aponeurosis—flat, sheetlike tendon
depress—to move a part inferiorly, such as opening the jaws
extend—to straighten one part relative to another
flex—to bend one part relative to another
head—the part of the muscle that takes part in its origin
insertion—attachment site that moves relative to the origin
levate—to move a part superiorly, such as closing the jaws
ligament—connective tissue band or sheet joining bone to bone
origin—relative fixed site of muscle attachment
protract—to extend or protrude a part
retract—to draw back or pull in a part
tendon—connective tissue band joining muscle to bone

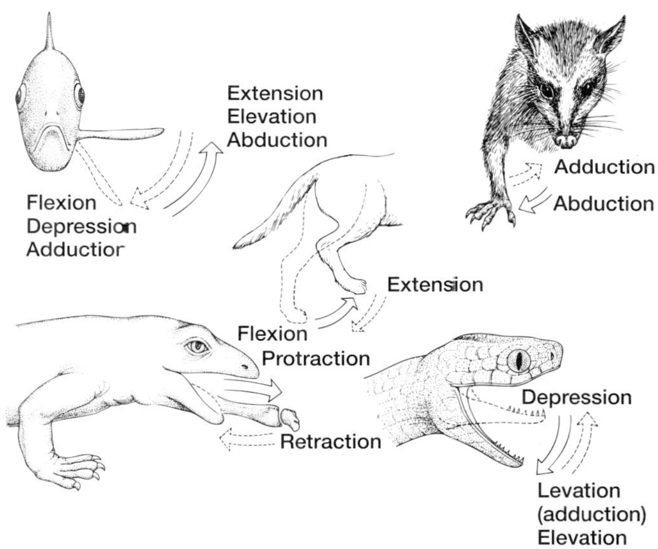

Figure 6.1
Muscle actions. Muscle adduction draws an appendage toward the ventral midline, and muscle abduction moves it away. Although these terms apply to tetrapod limbs and fish fins, the terms depression and flexion are sometimes used synonymously with adduction in fishes; extension and elevation are synonymous with abduction. In tetrapods, flexion means bending a part, extension means straightening it. Protraction sends a part outward from its base, retraction pulls it back. Opening the jaws is depression or abduction, and closing them is elevation, levation, or adduction.

General Muscle Groups

Parietal or Somatic Musculature

Derived from myotomes, these muscles are innervated by spinal nerves, and form voluntary muscles covering most of the body and appendages.

Hypobranchial Musculature

These muscles derive from myotomes that grow downward behind the head and then forward into the area beneath the gills. They are also innervated by spinal nerves.

Branchial or Branchiomeric Musculature

These derive from somitomeres and are innervated by cranial nerves.

Dissection

Dissection (Dis-sékt-shên) *n.* The careful exposure of anatomical parts, allowing students to discover and master the extraordinary morphological organization of an animal to understand the processes these parts perform and the remarkable evolutionary history from which they came.

The Process

The primary objective of a muscle dissection is to expose and study a muscle's origin, insertion, shape, and function by carefully separating it from the surrounding tissues. A good dissection does this while minimizing damage to the muscles and surrounding tissues, including blood vessels and nerves. Dissection requires patience, delicacy, and skill. Skeletal muscles are typically connected to each other and the skin by fascia, light colored and interwoven sheets of fibrous connective tissue that may look like spider webs. Fascia and fat typically obstruct the muscle boundaries and may render muscles difficult to see.

To dissect well, one needs little more than a good blunt probe (often a gloved finger is best) and a pair of grasping forces. Scissors are necessary only to cut skin and reflect muscles by cutting them perpendicularly along their midlength (reflecting a muscle in this way produces two long ends that can be found and brought together more easily than if the muscle is cut near its insertion or origin). Scalpels can cut quickly and easily along unnatural boundaries, permitting rapid and disastrous consequences in untrained hands. Their use is therefore discouraged.

Progress through your muscle dissection by (1) reading the detailed text description of the muscle's location, (2) searching for the reference points on the animal, (3) looking for the boundaries of the muscle (slight shifts in muscle fiber direction, identifying a distinct insertion or origin etc.), and (4) verifying the boundaries of the muscle through gentle probing. Once the boundaries are identified, use your probe and forceps to remove fascia and fat around the muscle from insertion to origin. Be careful not to break nerves and blood vessels that traverse these regions.

General Safety Precautions

Your instructor will inform you of risks and safety precautions associated with the chemicals used to fix and preserve animals that you dissect. Good progress has been made by most biological suppliers in reducing your exposure to noxious and potentially toxic chemicals. However, prudent behavior is still required.

When dissecting preserved tissues, care should be taken to avoid contacting or breathing the vapors of chemicals used in the preservation process. Your exposure can be reduced by wearing thin, flexible gloves and working in a well-ventilated area. In addition, wash with cold water the external and internal surfaces of your animal exposed through your dissection. Frequently wash away the fluids that seep from your animal into its dissection tray. Soft, gas permeable contact lenses can act as absorbent surfaces that hold the vapors against your eye. Avoid wearing such lenses when dissecting. Furthermore, protective eyewear should be worn by anyone <u>observing</u> (not just conducting) a dissection as small amounts of fluid are frequently splashed into the air (especially when turning an animal in its tray). Additional precautions and rules governing your dissection will be provided by your instructor.

External Anatomy

Take time to examine the external anatomy of your specimens before beginning your dissection of the muscular systems. Descriptions of the external anatomy precede the muscular dissection of the shark, *Necturus*, and cat.

Shark Dissection

External Anatomy

Major Anatomical Components

→ Find the <u>pectoral</u> and <u>pelvic</u> fins examined previously in the skeletal system. Note the sex of your specimen by determining if the pelvic fin bears claspers (see figure 5.16b and 6.7). The body is generally divided into three regions: the <u>head</u> extends from the tip of the snout to the cranial border of the pectoral fin; the <u>trunk</u> of the body is between the pectoral and pelvic fins; and the <u>tail</u> is caudal to the pelvic fin. Note that the dorsal side of the trunk is darker than the ventral side, a common pattern called <u>countershading</u>. Light from above casts a shadow on the ventral regions, producing a uniformly shaded shark more likely to blend into its environment.

Sharks have two unpaired <u>dorsal</u> <u>fins</u>, each with a prominent spine on the leading edge. (These spines, primarily used for defense, are often trimmed off be-

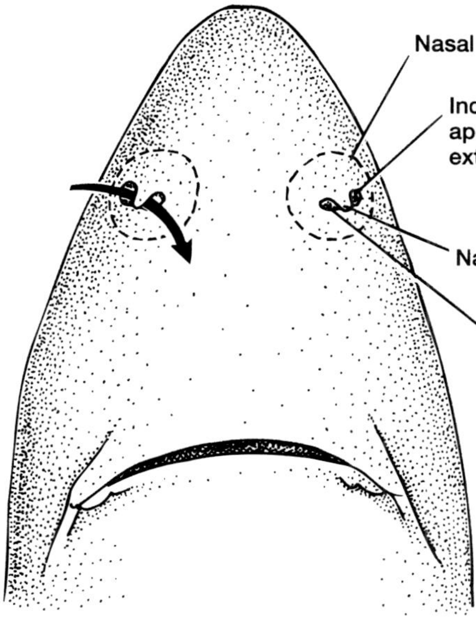

Figure 6.2
Nasal sacs of a shark. Ventral view of a shark head showing the direction water flows (solid arrow) through the nasal sac and across the olfactory epithelium.
After M.H. Wake (ed.), 1979 Hyman's Comparative Vertebrate Anatomy, University of Chicago Press.

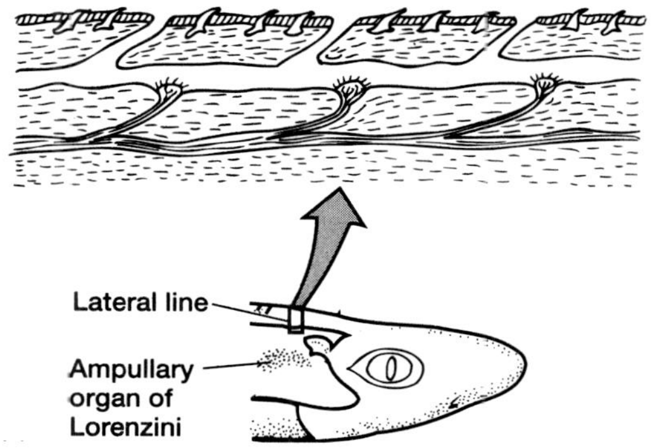

Figure 6.3
The lateral line system. Section through the skin of a shark showing the sunken lateral line canal opening to the surface via small pores.
After Goodrich.

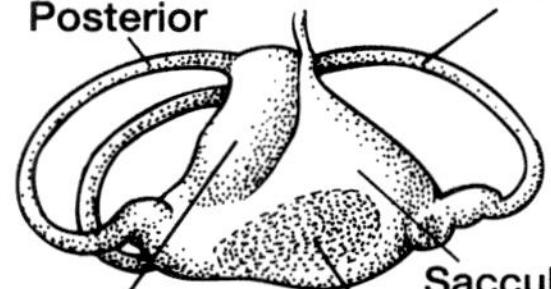

Figure 6.4
The vestibular apparatus of a shark.
After Goodrich.

fore you receive your animal to prevent damage to you and the storage bags.) The dorsal lobe of the tail fin is larger than the ventral lobe, forming a heterocercal tail configuration.

Sensory Systems

→ **Chemoreception** Just cranial and ventral to the eyes are paired nares, which open into blind-ended olfactory sacs for chemoreception (smelling) (figure 6.2).

Lateral Line System Sharks and many other aquatic vertebrates (but not aquatic birds or mammals) sense vibrations and movement in the water by a lateral line system composed of bundles of sensory cells and canals located within the skin (figure 6.3). These canals are concentrated along the head and extend along the sides of the body onto the tail. The largest canal, the lateral line, is a light thin stripe extending just above the most lateral side of the trunk.

Electroreceptors Gently squeeze the shark's snout and note the thick fluid exuding from pores on the dorsal and lateral sides of the head. These pores are openings into the ampullae of Lorenzini, a modified portion of the lateral line system sensitive to electrical fields.

Vestibular Apparatus The vestibular apparatus (figure 6.4) is a paired balancing organ that arises phylogenetically from the lateral line system. It is suspended in the otic capsule identified previously when studying the skeletal system (see figure 5.29). A pair of endolymphatic ducts connect the vestibular apparatus to the outside of the body through endolymphatic pores. Although difficult to see without a dissection microscope, these pores are located along the dorsal midline just caudal to the eyes. The three semicircular canals, part of the vestibular apparatus, are positioned in different planes to better sense specific movements. As the body moves, fluid in these canals shifts, triggering sensory signals conveying information about the nature of the movement.

Form & Function

Position yourself in a location that allows you to safely close your eyes. Now slowly tilt your head back and forth from side to side. How can you tell that your head is tilted? Like sharks and all other vertebrates, we have a vestibular apparatus that provides similar information about the position and movement of our heads. In addition, stretch receptors in your neck muscles let you know that the muscles on one side of your head are stretched more than other neck muscles, providing additional positional information.

Respiratory System

→ Sharks respire by ventilating their gills with water pulled into the mouth as the floor of the mouth is depressed and the pharynx is expanded. As the floor of the mouth is elevated, the water is forced through the pharynx into separate gill chambers containing gills richly supplied with blood. The water next moves away from the gills and out of the body through separate gill slits. Identify the five gill slits located on the posterior and lateral sides of the head. An additional opening, the spiracle (seen in figure 6.6), is technically the first gill slit in dogfish sharks (thus six gill slits in total). It is a rounded opening, lacking gills, and located just behind the eye on the dorsal side of the head. In bottom-dwelling skates and rays, the ventral mouth may be partially buried. The dorsally placed spiracle is thus in an unobstructed position to allow water to be drawn in for respiration. However, for most sharks, the function of the spiracle is unknown.

Cloaca

→ Examine the ventral side of the body between the cranial edges of the pelvic fins to find the medial depression, the cloaca (cloaca = "sewer"). The urinary and digestive tracts of both sexes empty into the cloaca, giving significance to its literal meaning of "sewer." The urinary system enters the middle of the cloaca through the tip of the nipplelike urinary papilla. The uterus of the female empties separately into the cloaca, and the reproductive tract of a male sends sperm along the clasper.

Skinning the Shark

→ Position your shark in the dissection tray so that the ventral side of the trunk is directed toward you. To begin skinning the shark, you may need to use scissors to make a small incision midway between pelvic and pectoral fins. (A slit in the ventral musculature may have already been made to permit the injection of latex into the blood vessels of the shark.) At the cut edge, use a probe and forceps to separate and pull a flap of skin away from the shark. Then when a good grip can be had, pull the skin away from the underlying muscle. Pull while using a probe or finger to hold the musculature down at the junction where the skin is still attached. Caution should be used to ensure that muscle is not torn out as the skin is removed. The skinning process can be continued by gently pulling on the skin and pushing down on the muscle with finger or probe.

Try to leave large flaps of skin still attached near the dorsal midline. These flaps can be repositioned to protect the underlying musculature during storage. Continue to reflect skin to expose the muscles on the ventral and right sides of the head and trunk up to the dorsal midline. Completely remove skin from the dorsal side of the head forward to the tip of the snout. The most fragile musculature will be encountered on the ventral and lateral sides of the head behind the jaws and around the gill slits. Extra care should be taken to minimize damage to these delicate muscles. Work slowly and carefully. The entire skinning of the shark may take two to four hours. Thin but quite tough fascia (connective tissue) covering the muscles should also be cleared, but just sufficiently to reveal underlying muscles. A thorough job of removing fascia can wait until the underlying musculature is identified in the dissection below.

Musculature

Insertion, Origin, and Function

Knowing the origin and insertion of a muscle makes identification easier, especially if the muscle's fibers closely parallel one or more of its neighbors. Use table 6.2 to assist your identification of the following muscles and to help you relate their position and shape to their functions.

Axial Musculature

→ The axial musculature arises from myotomes that differentiate from somites (figure 6.5). These myotomes grow and expand along the sides of the body, forming the musculature associated with the vertebral column, ribs, and lateral body wall. On the shark, the slanted blocks of muscle are myomeres separated from each other by myosepta (figure 6.6). The long horizontal line is the outer margin of the horizontal septum running from tail to head. This divides the body musculature into epaxial (above) and hypaxial (below) musculature within which specifically named muscle groups are recognized. The epaxial muscles form the dorsal longitudinal bundles and appear to be a single mass externally, but in cross section are two or three distinct bundles. Follow these bundles forward to their insertion at the back of the chondrocranium (see figure 6.9b).

The hypaxial musculature divides into the more dorsal lateral longitudinal bundle (just below the horizontal skeletogenous septum) and ventral longitudinal bundle forming most of the ventral surface of the trunk (figure 6.6). Follow the lateral longitudinal bundle forward to its attachment on the scapular process (figure 6.6). A thin line, the linea alba, occurs where the paired ventral longitudinal bundles meet along the ventral midline (see figure 6.9a).

Fin Musculature

→ If you have not already done so, clear skin from both sides of the pelvic and pectoral fins. The dorsal surface of the fins is covered by extensors of the fins, the ventral surface by flexors of the fins (figures 6.6, 6.7, and 6.9). In males, the siphon covers the ventral side of the pelvic fin (figure 6.7). Peel it away on one pelvic fin and note that the flexor is divided into proximal and distal halves near the edge of the metapterygium.

Table 6.2

The Origin, Insertion, and Function of Shark Musculature

Muscle	Origin	Insertion	Function
Fin Musculature			
pectoral extensors	scapular process and adjacent fascia	radials and ceratotrichia	elevates pectoral fin
pectoral flexors	coracoid bar	radials and ceratotrichia	depresses pectoral fin
pelvic extensors			
superficial part	iliac process of puboischiac bar	radials and ceratotrichia	elevates pelvic fin
deep part	metapterygium	radials and ceratotrichia	elevates pelvic fin
pelvic flexors			
proximal part	linea alba and puboischiac bar	metapterygium	depresses pelvic fin
distal part	metapterygium	radials and ceratotrichia	depresses pelvic fin
Superficial Branchial Constrictor Muscles			
adductor mandibulae	posterior part of the palatoquadrate cartilage	Meckel's cartilage	closes the mouth
dorsal and ventral constrictors 3–6	vertical raphe	vertical raphe	compresses gill pouches
interhyoideus	midventral raphe	ceratohyal	compresses gill pouches
intermandibularis	Meckel's cartilage and fascia of the adductor mandibulae	midventral raphe	raises the floor of the mouth
levator hyomanidibulae	otic capsule	hyomandibular	compresses gill pouches
preorbitalis	midventral surface of the chondrocranium	adductor mandibulae	closes the mouth
spiracularis	otic capsule	palatoquadrate cartilage	elevates the palatoquadrate
Levator Series			
cucullaris	dorsal longitudinal bundle fascia	epibranchial cartilage and scapular process	elevates the gill arches and pectoral girdle
interarcuals	pharyngobranchials	epibranchials	expands the pharynx
levator palatoquadrati	otic capsule	palatoquadrate cartilage	raises the upper jaw
Hypobranchial Musculature			
common coracoarcuals	coracoid bar	coracohyoid	opens the mouth
coracomandibularis	fascia of common coracoarcuals	Meckel's cartilage	opens the mouth
coracohyoid	common coracoarcuals	basihyal	opens the mouth
Extrinsic Eye Muscles			
inferior oblique	anterior end of orbit	ventral side of eyeball	rotates eye ventrally
inferior rectus	posterior end of orbit	ventral side of eyeball	rotates eye ventrally
lateral rectus	posterior end of orbit	lateral side of eyeball	rotates eye posteriorly
medial rectus	posterior end of orbit	medial side of eyeball	rotates eye cranially
superior oblique	anterior end of orbit	dorsal side of eyeball	rotates eye dorsally
superior rectus	posterior end of orbit	dorsal side of eyeball	rotates eye dorsally

Branchial Musculature

→ **Constrictor Series** The superficial constrictor musculature covers the gill region and consists of six dorsal and six ventral constrictors, above and below the external gill slits, respectively (figure 6.8). The first constrictor muscle in this series is highly modified, as is the first gill arch it moves.

The first dorsal constrictor consists of three muscles (see figure 6.6). The adductor mandibulae (quadratomandibularis) is the massive hemisphere of muscle behind the eye. The spiracularis runs tightly along the anterior margin of the spiracle and inserts on the palatoquadrate. Clear fascia to find it and another muscle that closely parallels it anteriorly, the levator palatoquadrati, actually part of the levator series traced later. The third component of the first dorsal constrictor is the preorbitalis, adjacent to the deep ventral surface of the eyeball.

The first ventral constrictor consists of the paired intermandibularis (figure 6.9a) on the ventral surface of the throat running between Meckel's cartilages and divided by a median raphe.

The second dorsal constrictor consists of a broad sheet of muscle between the spiracle and raphe of the

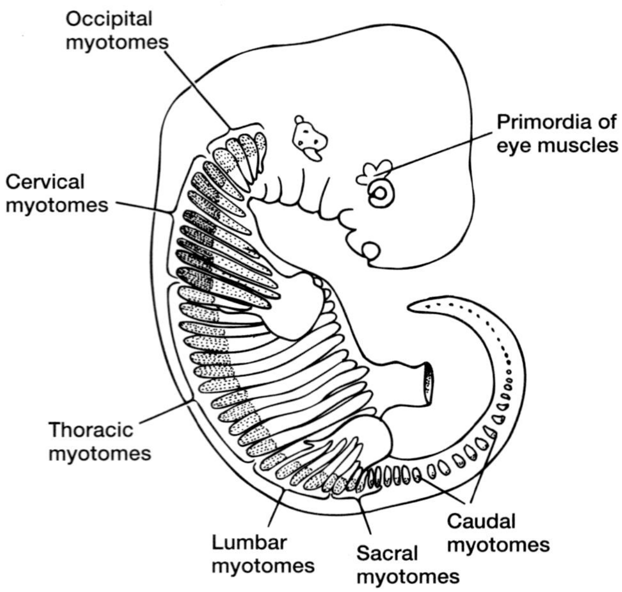

Figure 6.5
Muscles derived from embryonic myotomes in a reptile (lizard). During embryonic development, the myotomes expand into respective areas of the body.

first complete external gill slit (see figure 6.6). The anterior half of this muscle is the levator hyomandibulae (epihyoideus), which extends caudally to the posterior margin of the adductor mandibulae. Fibers of the posterior half of the second dorsal constrictor sweep downward past the rear of the adductor mandibulae to below the gills. This posterior half constitutes the second dorsal gill constrictor, proper.

The second ventral constrictor, the interhyoideus, is a sheet of muscle just dorsal to the intermandibularis. To find it, cut a window in the belly of the thin intermandibularis to reveal the parallel fibers of the interhyoideus running at a slightly different angle (figure 6.9a)

Dorsal and ventral constrictors 3–6 follow, separated from each other by a raphe (figure 6.6).

Levator Series The first levator, levator palatoquadrati, is closely applied to the anterior face of the spiracularis (figure 6.6). The second levator, levator hyoideus, is impractical to distinguish because it fuses with the medial face of the levator hyomandibulae.

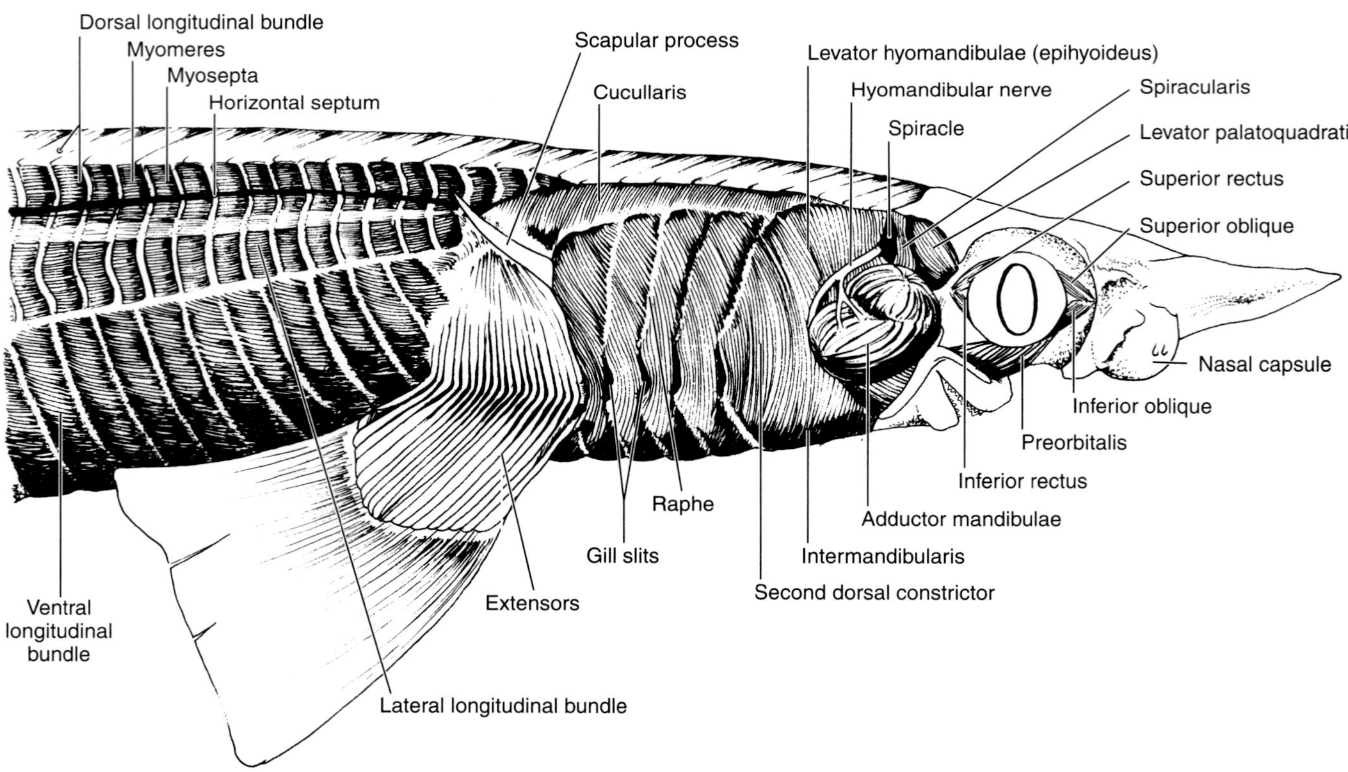

Figure 6.6
Lateral view of the trunk, branchial, and cephalic musculature of a shark.

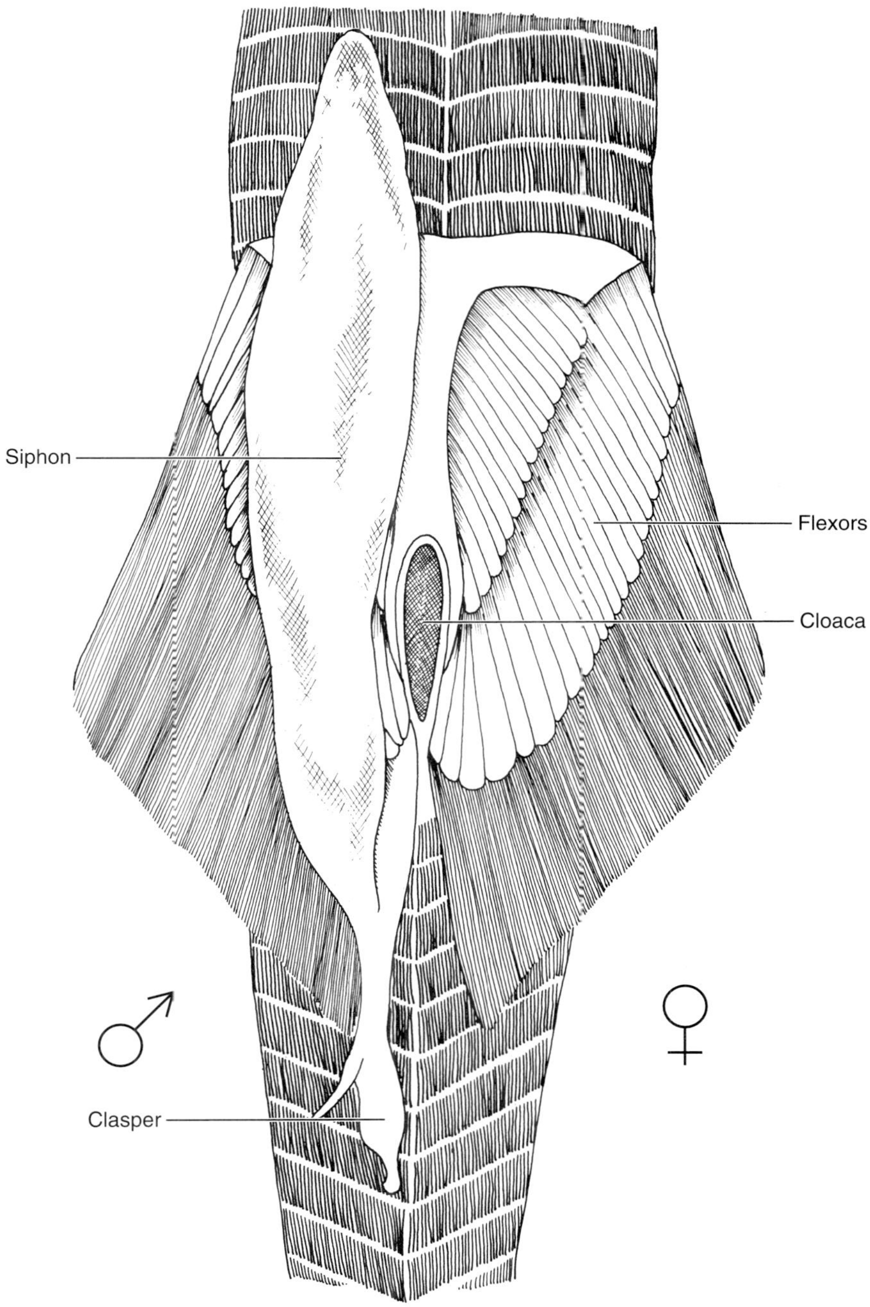

Figure 6.7
Ventral view of the pelvic fins of a male (viewer's left) and female (right).

Levators 3–8 fuse into a single long, thin wedge of muscle, the cucullaris, lying in front of the scapular process between the top of the gill pouches and the dorsal longitudinal bundles (figures 6.6 and 6.9b).

Interarcual Series Separate the upper margin of the gill pouches from the lower edge of the cucullaris (figure 6.9b). Spread the two regions until the deep pharyngobranchial and epibranchial gill arch elements come into view. Between these cartilages lie the series of small interarcuals, which move the gill arch cartilages.

Hypobranchial Musculature These muscles run on the ventral side of the shark from the pectoral girdle to

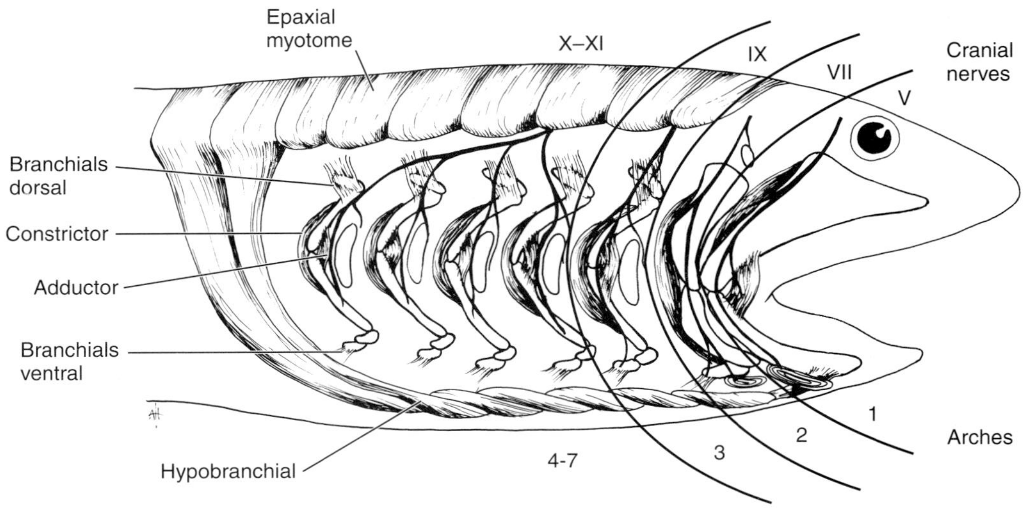

1st	V	<u>Levators</u> Levator palatoquadrati <u>Constrictors</u> (Dorsal) Adductor mandibulae Preorbitalis Spiracularis (Ventral) Intermandibularis	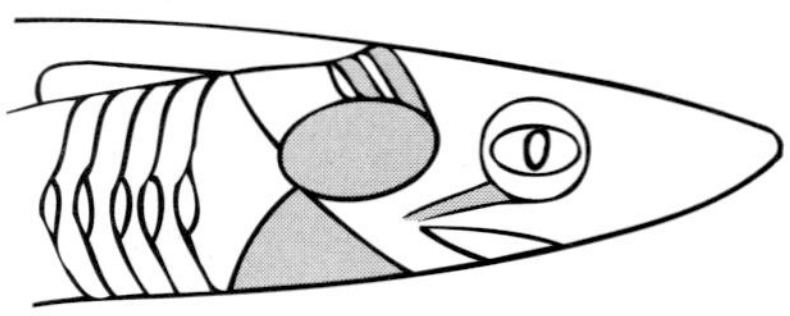
2d	VII	<u>Levators</u> Levator hyoideus <u>Constrictors</u> (Dorsal) Second dorsal constrictor levator hyomandibulae (Ventral) Interhyoideus	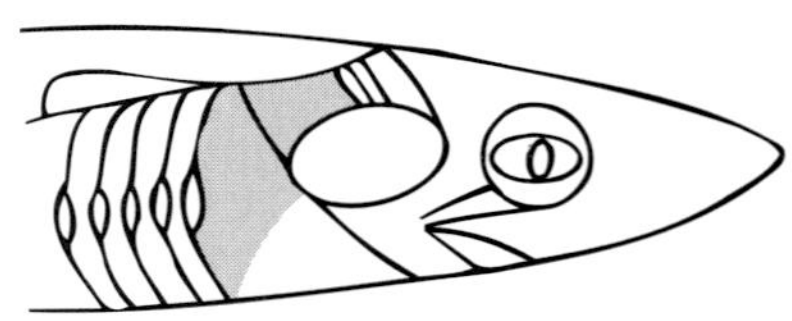
3d	IX	<u>Levators</u> Cucullaris <u>Constrictors</u> (Dorsal) Superficial constrictors, dorsal (Ventral) Superficial constrictors, ventral	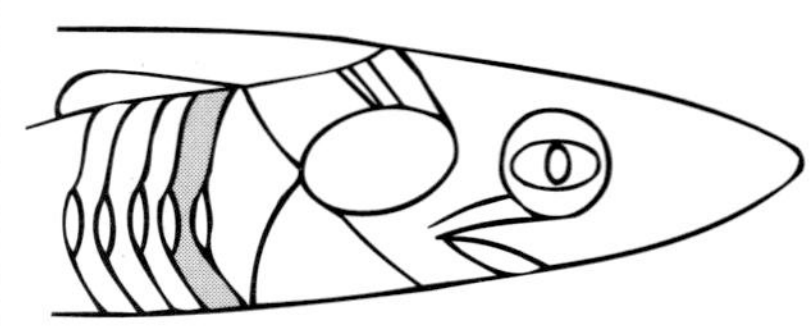
4–7th	X	<u>Levators</u> Cucullaris Interarcuals <u>Constrictors</u> (Dorsal) Superficial constrictors, dorsal (Ventral) Superficial constrictors, ventral	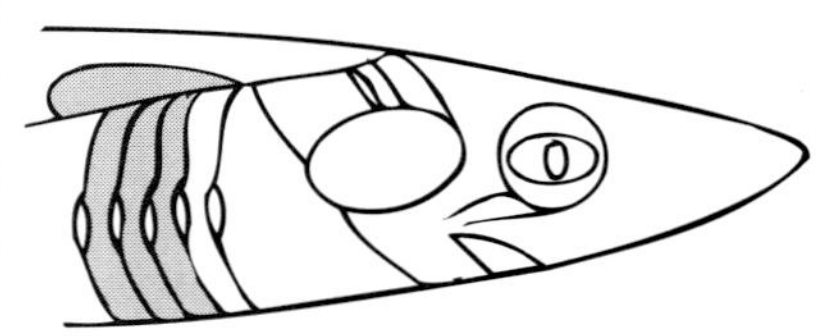

Figure 6.8
Branchiomeric musculature. Jaw muscles and their cranial nerve supply tend to stay with their respective branchial arch during the course of subsequent evolution. Each arch has levator and constrictor muscles that, respectively, elevate and close the articulated elements. Cranial nerves V, VII, IX, and X–XI supply muscles of arches 1, 2, 3, and 4–7, respectively. Fidelity of muscles, nerves, and arches generally was maintained as the branchial arches evolved and subsequently became modified into components of the jaws. The specific derivatives of the levator and constrictor series of muscles in each arch of a shark are shown together with their cranial nerve supply.

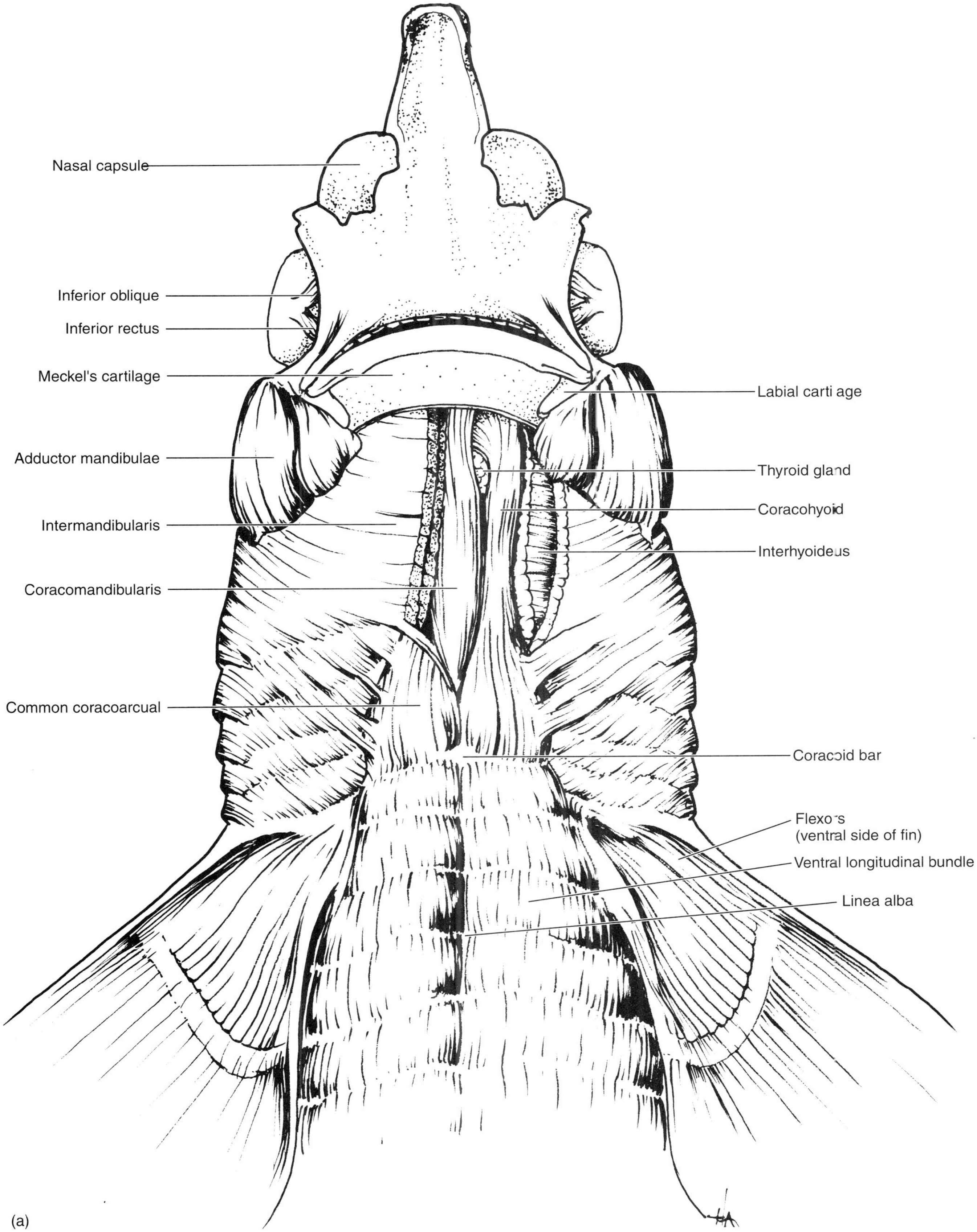

Figure 6.9
Trunk, branchial, and cephalic musculature of a shark. (a) Ventral view and (b) dorsal view.

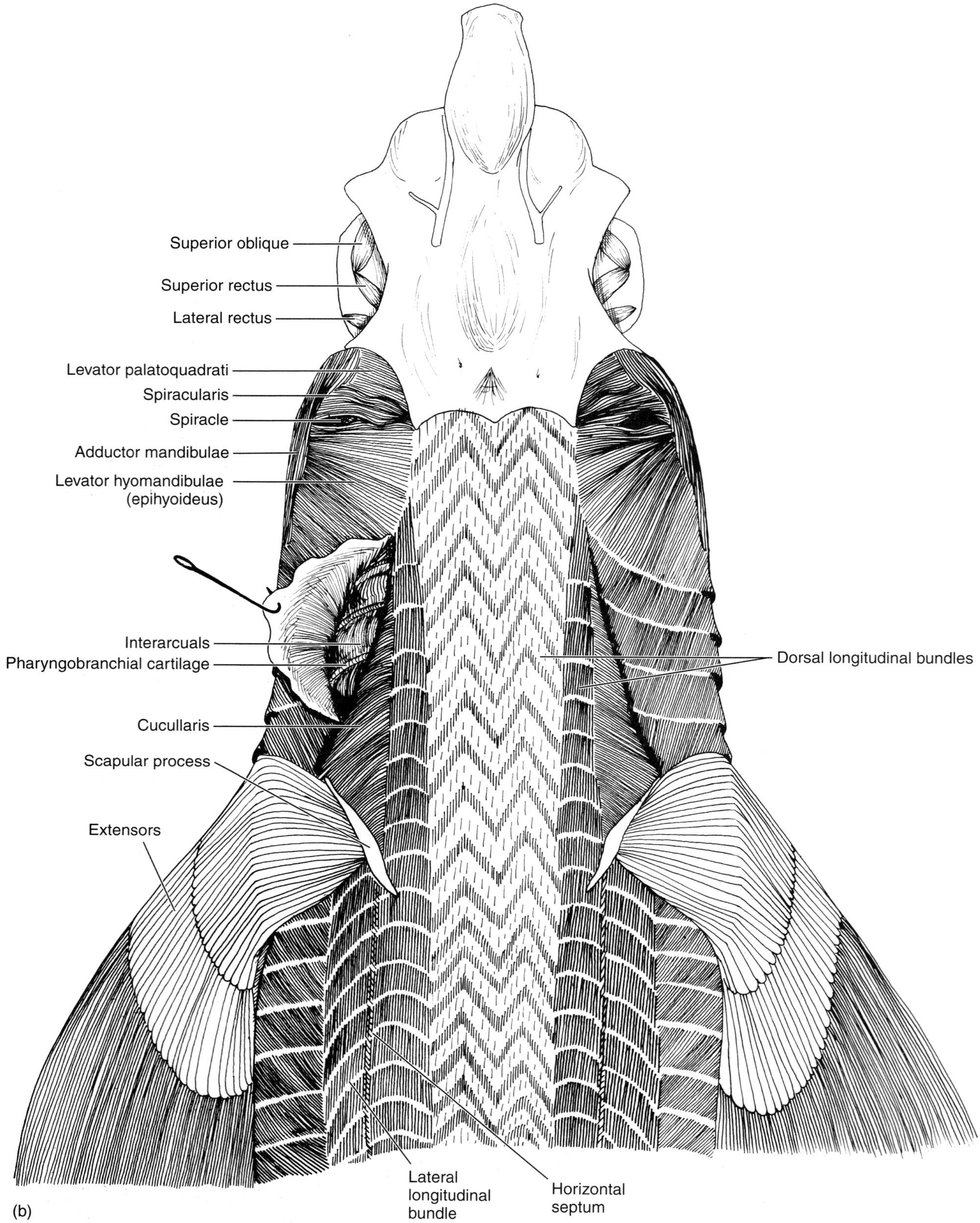

Figure 6.9
Continued.

the mandible (figure 6.9a). They lie deep to the ventral constrictors identified previously. In the ventral midline, just anterior to the coracoid bar, lies a paired muscle triangular in shape, the common coracoarcuals. Cut along the lateral edge of the left (apparent right) common coracoarcual just deep enough to clear the intermandibular and interhyoideus (figure 6.9a). Carefully fold back the flaps to expose the muscles but do not cut deeper than directed as this endangers delicate blood vessels traced later.

The thin muscle (actually a paired muscle tightly fused together) directly above (dorsal to) the median raphe is the coracomandibularis, on its way to the mandible. Above (dorsal) to it are the paired coracohyoids, continuous caudally with the common coracoarcuals. The thyroid gland appears as a squashed dark mass slightly to the left of center, between the coracohyoids and coracomandibularis.

Extrinsic Eye Muscles

→ Expose the left eyeball by removing the supraorbital crest over the orbit. Clean out debris, taking care not to break any delicate nerves (white threadlike structures) or muscles. Viewed from above, four thick ribbons of muscle leave from a common site on the posterior medial wall of the orbit (figure 6.10). These four rectus muscles of the eye diverge and attach to the top, bottom, medial, and lateral sides of the eyeball. These are the superior (top), inferior (bottom), medial, and lateral rectus muscles. In addition, two oblique muscles extend from the anterior medial wall of the orbit. (Thus six eye muscles in total.) The inferior oblique inserts below the eyeball and the superior oblique above the eye near the insertion of the superior rectus. The larger white cords are major nerves: the superior ophthalmic coursing along the back of the orbit, the optic nerve projecting medially from the back of the eye between medial and inferior oblique muscles, and the infraorbital nerve running across the preorbitalis muscle (figure 6.10). (Preserve the integrity of the other nerves in this region for easier identification in the later section on the nervous system.)

Between the diverging rectus muscles lies the cartilaginous optic pedicel, shaped somewhat like a golf tee. Finally, expose the nasal capsule just anterior to the eye for a firsthand view (figure 6.6).

Necturus

External Anatomy

The mudpuppy, *Necturus*, is a common inhabitant of clear streams in the midwestern and eastern United States. Unlike most salamanders, it retains many larval features into the adult stage, a condition called **paedomorphosis**. For example, most salamander larvae are aquatic, hatching from eggs laid in water. Usually after several months the larvae **transform** (undergo **metamorphosis**) into terrestrial juveniles, which grow and develop on land into adults. However, *Necturus* do not transform. Instead, they remain aquatic throughout their lives.

→ The large, bushy, and dark external gills located at the rear of the head are a good example of a larval characteristic retained into adulthood. Small, dark eyes reside on the front dorsal side of the head. At the tip of the snout are the paired external nares through which air can be drawn into the mouth. Air moving through the external nares passes a chemosensory organ before entering the buccal cavity (mouth) through the internal nares. The sharp teeth, identified previously when studying the skull, are bound laterally by thick lips.

The expansive trunk occupies most of the body between the pectoral and pelvic girdles. A distinct gular fold (figure 6.11) occurs on the ventral side of the neck where the trunk merges with the head. The cloaca should next be found along the ventral midline at the caudal end of the trunk. As in the shark, this chamber is the location where the digestive, urinary, and reproductive tracts deposit their products. Surrounding the cloacal chamber is a large hump, the cloacal gland, which releases products used during reproduction (see figure 6.14).

Dissection

→ Begin separating the skin from underlying muscles at the edge of a cut along the ventral trunk. (As in the shark, a slit may already have been made in this region to inject blood vessels.) Completely remove the skin from the ventral surface and right side of the body to the dorsal midline. Cut a circle around the external gills and leave them in place. While pulling the skin off the external gills, care should be taken not to damage the delicate subarcual muscles. Pull the skin over each limb, leaving it inside out, like removing a sweater.

To minimize damage to the muscles, continue to practice the precautions used for the shark. Muscles running from body to limbs and around the gills are especially delicate, usually fan-shaped, and often adhere to the integument. Thus, in removing the skin, be especially careful not to tear these muscles away.

Musculature

Insertion, Origin, and Function

→ As noted previously with the shark, knowing the origin, insertion, and function of a muscle makes identification easier. Use table 6.3 to assist your identification of the *Necturus* muscles and to help you relate the position and shape of the muscles to their functions.

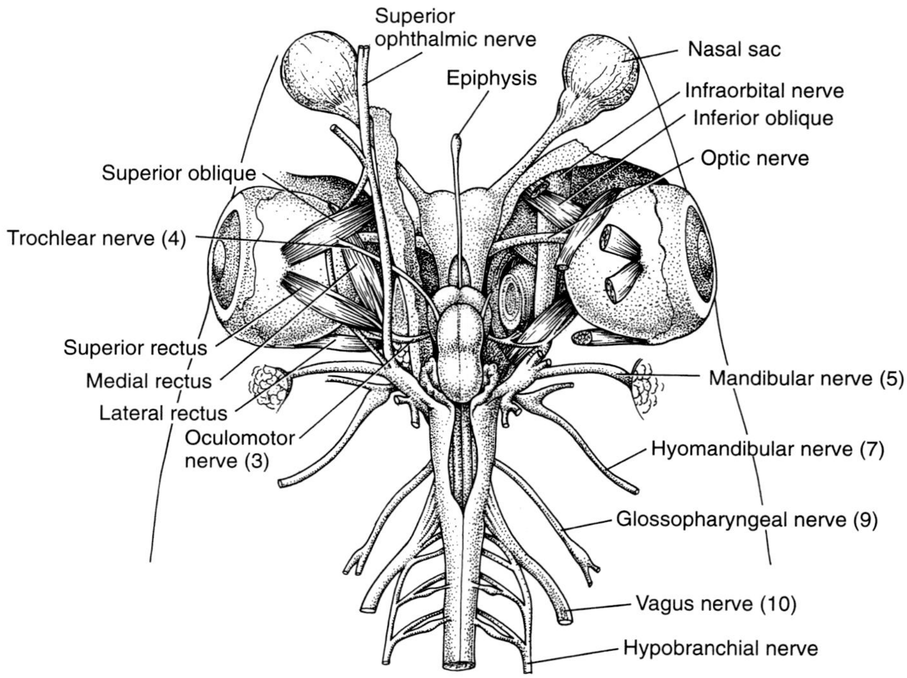

Figure 6.10
Extrinsic eye musculature of a shark (dorsal view). The extrinsic eye muscles are derived from somitomeres and rotate the eyeball within the orbit in order to direct the gaze. The roof of the chondrocranium has been removed to expose several extrinsic muscles (left). The superior oblique, superior rectus, medial rectus, and lateral rectus muscles have been cut to expose the deeper extrinsic muscles (right).

Ventral Pectoral Girdle, Forelimb, and Throat

→ Begin by finding the fan-shaped pectoralis muscles ventrally, between the forelimbs (figure 6.11). Just caudal are the rectus abdominis muscles that originate on the linea alba and converge laterally to the humerus. Anterior to the pectoralis lies the supracoracoideus, riding on the coracoid cartilage. To help distinguish it from the pectoralis, grasp the most superficial coracoid cartilage with tweezers and slide it forward and back. The line between muscles can then be better established. The elongate procoracoid cartilage runs forward on the lateral side of the neck and is covered by the procoracohumeralis muscle passing backward to insert on the humerus. Probe this muscle to uncover the procoracoid cartilage. Between these paired cartilages lies the ventral neck muscles, the rectus cervicis, segmented by inscriptions, but lacking a linea alba. A small, short muscle, the omoarcual, inserts on the procoracoid and lies between the supracoracoideus and procoracohumeralis. It is apparently derived from the rectus cervicis.

Across the ventral surface of the upper arm lies two small muscles, the humeroantebrachialis (anterior) and coracobrachialis (posterior). The ventral surface of the forearm contains several flexors but these will not be followed individually.

Ventrally the throat is covered by transverse fibers between the mandibles and the median raphe (figure 6.11). The anterior part of this sheet is the intermandibularis, which indistinguishably merges into the interhyoideus, forming the posterior half. Make a slit along the median raphe and open the flaps. Beneath is the paired, slender geniohyoid passing from the chin posteriorly to a basibranchial cartilage. Lateral to the geniohyoids are the large paired branchiohyoideus muscles arching forward to the chin. Retract the branchiohyoideus muscles laterally to expose the cartilaginous gill arches. Deep between the gill arches are the subarcuals, short wedges of longitudinal muscles running between successive gill arches. Note the attachment of the rectus cervicis medially to the gill arches. Dorsal to the rectus cervicis and between it and the last gill cartilage runs the transversis ventrales (not in a figure).

Lateral Pectoral Girdle and Forelimb

→ Turn the specimen so its exposed right side faces you and pull forward the external gills to reveal the several muscles above the limb (figure 6.12a). Above the external gills are two muscles that may at first appear as one. The anterior, levatores arcuum, inserts on the arches. The posterior, dilatator laryngis (partially cov-

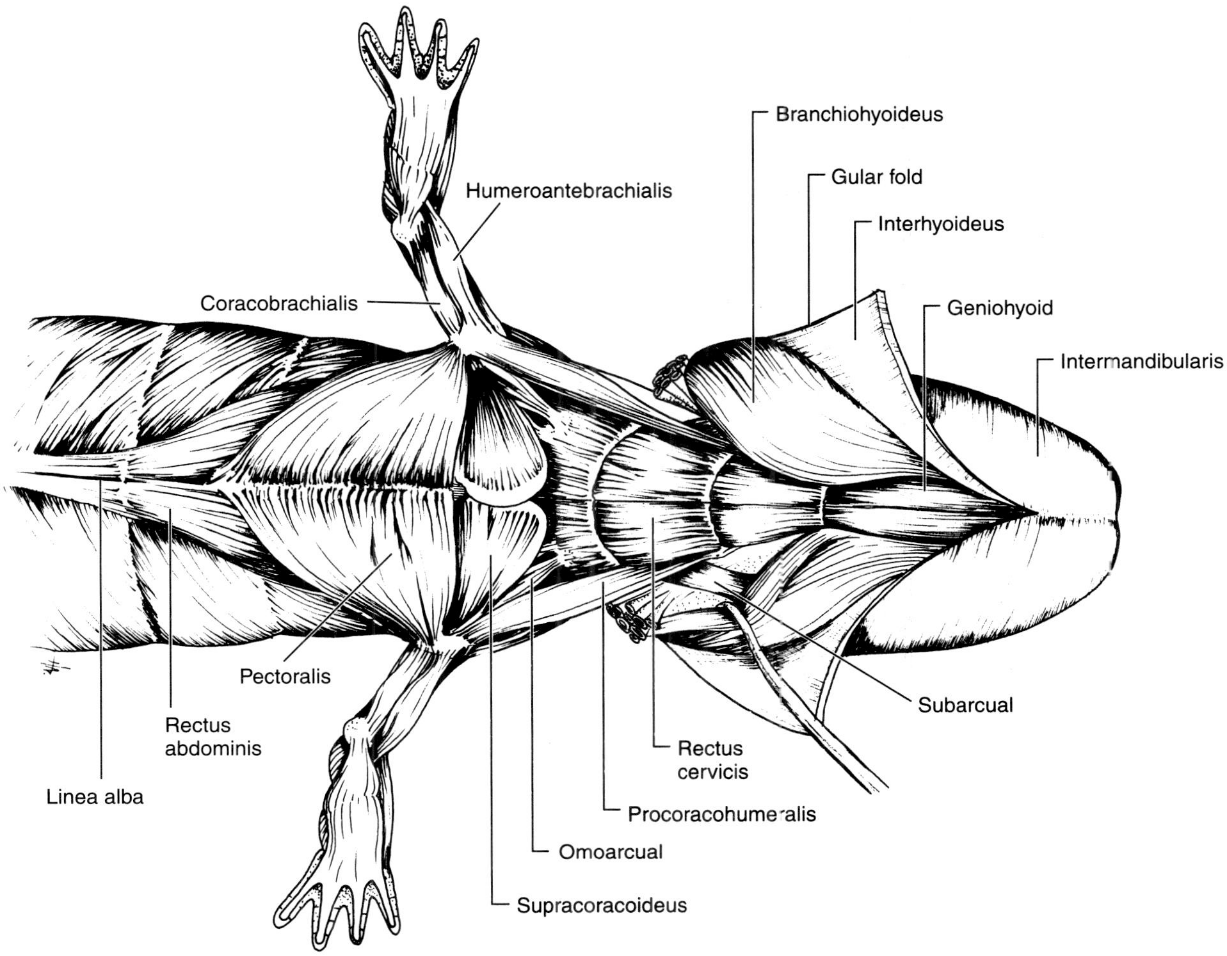

Figure 6.11
Ventral trunk, branchial, and cephalic musculature of *Necturus*.

ering the origin of cucullaris) inserts deeper (figure 6.12a,b). Beginning along the ventral edge of the dilatator laryngis and above the procoracohumeralis are four muscles in the following order (figure 6.12a). Most anterior is the thin pectoriscapularis followed by the cucullaris, the broader dorsalis scapulae, and the latissimus dorsi. Each of these four muscles converges on the shoulder near the junction of the scapula and humerus.

Across the dorsal surface of the upper arm lies the large triceps brachii (= anconeus). The dorsal surface of the forearm contains several extensors that will not be identified individually.

Lateral Head

→ Again identify the branchiohyoideus, which forms a large mass immediately in front of the gills (figure 6.12a). Along the dorsal and anterior margin of the branchiohyoideus is the lateral edge of the depressor mandibulae. Both muscles originate on gill bars, but the depressor mandibulae passes forward to the rear of the mandible, while the branchiohyoideus passes ventrally and forward to the chin. To separate them, trace the anterior edge of the branchiohyoideus dorsally. Note how its fibers arch posteriorly, and part the two muscles along this line.

Dorsally, between the eye and dorsalis trunci, resides a large muscle mass, the adductor mandibulae (figure 6.12b). It is divisible into a medial levator mandibulae anterior and a larger and more lateral levator mandibulae externus.

Trunk and Tail Musculature

→ Epaxial and hypaxial musculature along the trunk are separated by a horizontal line, the outer margin of the horizontal septum (figure 6.13). The superficial part

Table 6.3

The Origin, Insertion, and Function of *Necturus* Musculature

Muscle	Origin	Insertion	Function
Ventral Pectoral Muscles			
omoarcual	rectus cervicis	humerus	draws humerus craniad
pectoralis	linea alba and coracoid cartilage	humerus	adducts forelimb
procoracohumeralis	procoracoid cartilage	humerus	draws humerus craniad
supracoracoideus	coracoid cartilage	humerus	adducts forelimb
Ventral Throat Muscles			
branchiohyoideus	first branchial arch	ceratohyal cartilage	moves branchial arches and hyoid
geniohyoid	medial mandible	second basibranchial cartilage	depresses lower jaw and draws hyoid cranially
interhyoideus	fascia covering the branchiohyoideus	median raphe	elevates floor of throat
intermandibularis	rami of mandible	median raphe	elevates floor of throat
rectus cervicis	rectus abdominis	branchial cartilages	retracts tongue and lowers head
subarcual	gill cartilage	gill cartilage	depresses gills
transversis ventralis	rectus cervicis	last gill cartilage	depresses gills
Arm and Forearm Muscles			
coracobrachialis	coracoid cartilage	humerus	adducts forelimb
humeroantebrachialis	humerus	radius	flexes forearm
triceps brachii	glenoid fossa, humerus, and coracoid cartilage	ulna	extends forearm
Dorsal and Lateral Muscles of the Head and Gills			
depressor mandibulae	third branchial arch	posterior end of mandible	depresses lower jaw
dilatator laryngis	fascia near dorsal midline	lateral cartilages of larynx	dilates larynx
levator mandibulae anterior	frontal and parietal bones	lateral dentary	elevates lower jaw
levator mandibulae posterior	frontal and parietal bones	lateral dentary	elevates lower jaw
levatores arcuum	fascia of epibranchial muscles	epibranchial cartilages	elevates external gills
Lateral Pectoral Muscles			
cucullaris	fascia of epibranchial muscles	scapula	elevates the pectoral girdle
dorsalis scapulae	suprascapular cartilage	humerus	elevates the forelimb
latissimus dorsi	fascia of epibranchial muscles	shoulder and proximal end of humerus	draws humerus caudad
pectoriscapularis	epibranchial cartilage	scapula and procoracoid cartilage	draws shoulder and forelimb craniad
Dorsal–Epaxial Trunk Muscles			
dorsalis trunci	myosepta and transverse processes of vertebrae	myosepta and transverse processes of vertebrae	flexes the vertebral column laterally
Ventral–Hypaxial Trunk Muscles			
external oblique	transverse processes of trunk vertebrae	connective tissue of rectus abdominis	compresses the coelom and flexes and rotates the trunk
internal oblique	transverse processes of trunk vertebrae	connective tissue of rectus abdominis	compresses the coelom and flexes and rotates the trunk
rectus abdominis	anterior margin of the puboischiac plate	rectus cervicis	flexes the trunk ventrally
transversus	transverse processes of trunk vertebrae	connective tissue of rectus abdominis	compresses the coelom and flexes and rotates the trunk
Pelvic Girdle and Hindlimb Muscles			
caudocruralis	posterior border of the puboischiotibialis	haemal arches of caudal vertebrae 3–6	flexes tail
caudofemoralis	haemal arches of caudal vertebrae 3–6	femur	flexes tail and draws hindlimb caudad
ilioextensorius	ilium	tibia and fascia covering the knee	extends the shank
iliofibularis	ilium	fibula	extends the shank and draws the hindlimb caudad
iliotibialis	ilium	tibia	abducts the hindlimb
ischiocaudalis	ischium	haemal arches of caudal vertebrae 3–6	flexes tail

(Continued)

Table 6.3 *continued*

The Origin, Insertion, and Function of *Necturus* Musculature

Muscle	Origin	Insertion	Function
Pelvic Girdle and Hindlimb Muscles—continued			
ischioflexorius	ischium	fascia of foot	flexes shank and toes
pubo-ischio-femoralis externus	puboischiac plate	femur	adducts the hindlimb
pubo-ischio-femoralis internus	internal surface of the puboischiac plate	femur	draws the femur craniad
pubo-ischio-tibialis	posterior puboischiac plate	tibia	adducts the hindlimb
pubotibialis	anterior puboischiac plate	tibia	adducts the hindlimb

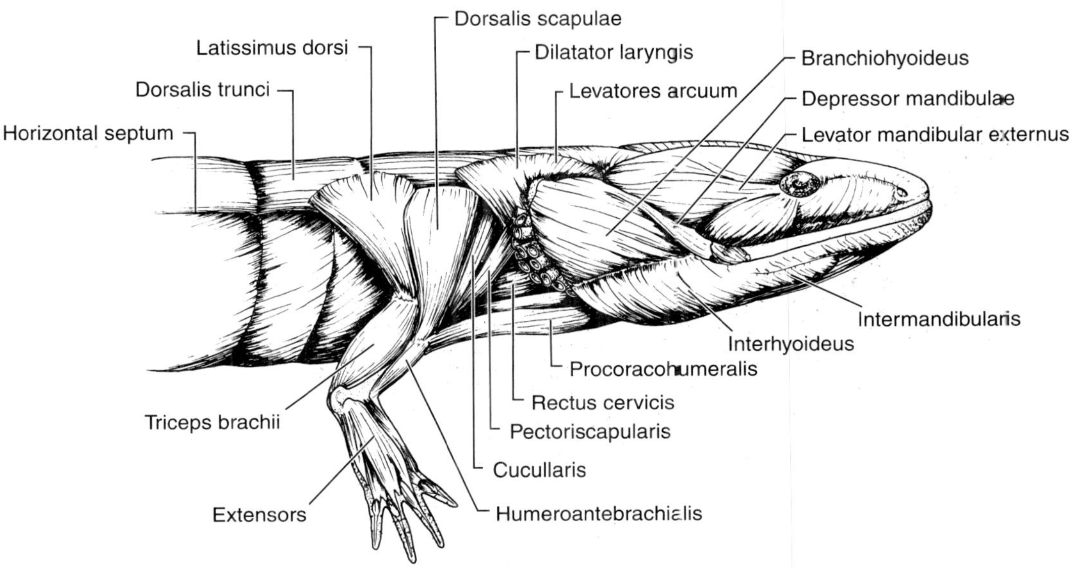

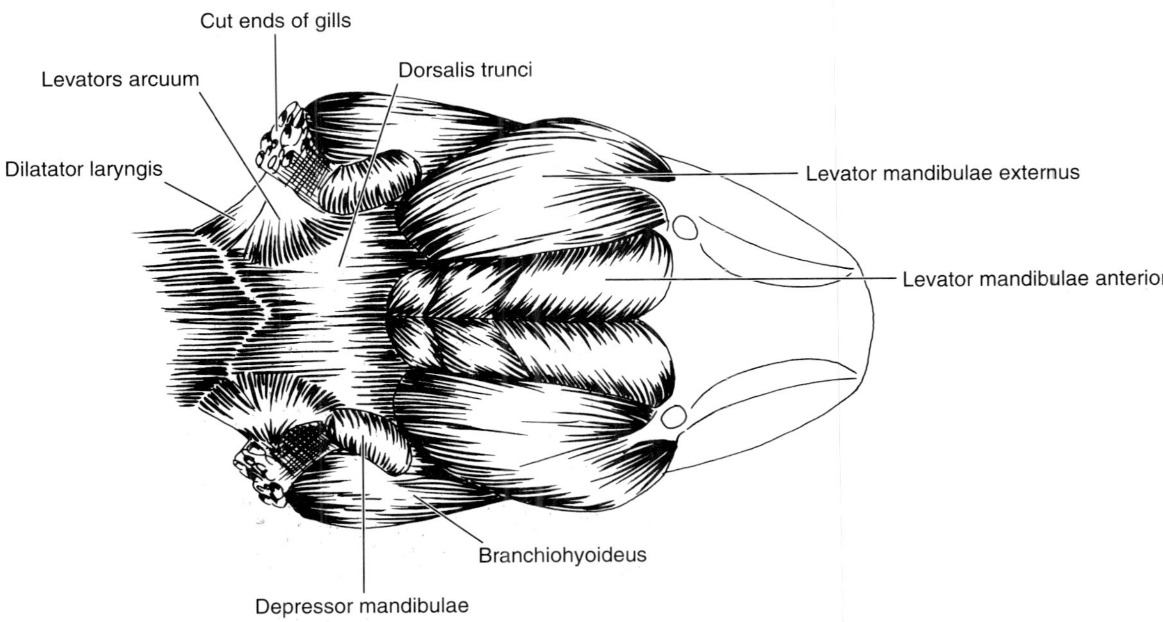

Figure 6.12
Trunk, branchial, and cephalic musculature of *Necturus*. (a) Lateral and (b) dorsal view.

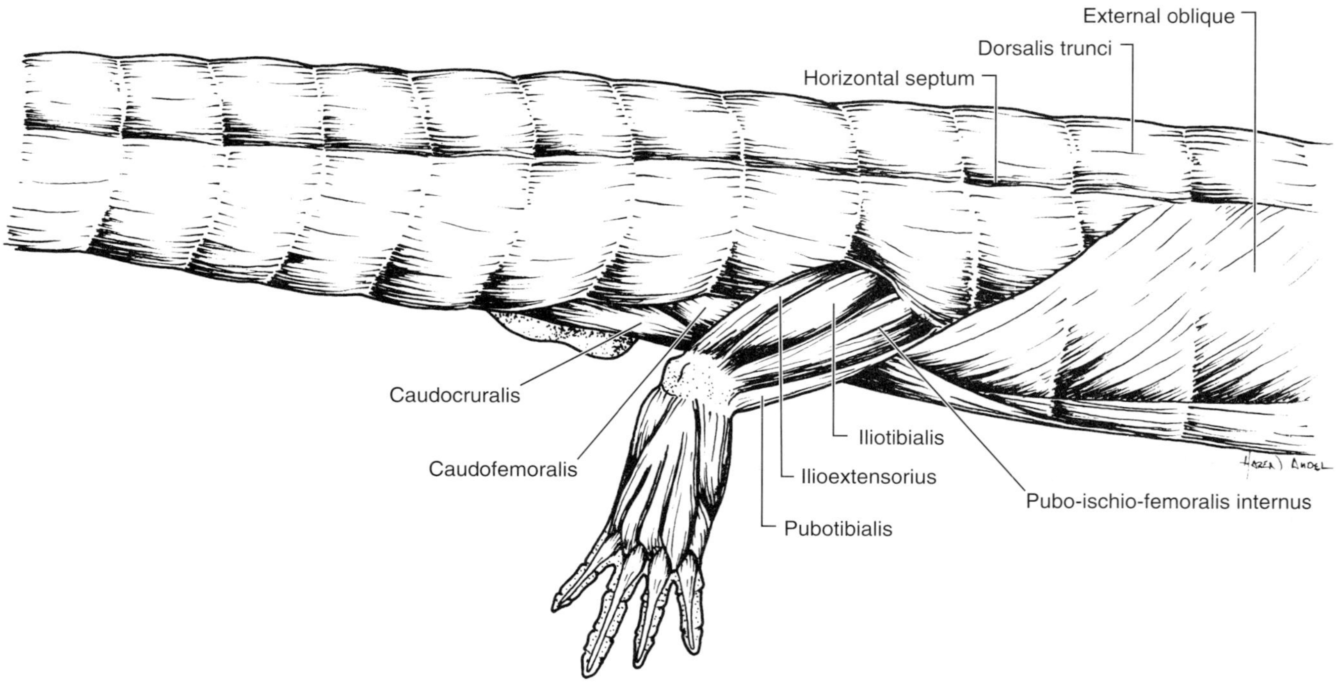

Figure 6.13
Lateral trunk and hindlimb musculature of *Necturus.*

of the epaxial muscles form the dorsalis trunci, whose fibers run horizontally between myosepta and deep to transverse processes of the vertebrae (figures 6.12a and 6.13).

The hypaxial muscles form three major lateral layers (figure 6.13). Superficially lies the external oblique, with fibers slanting caudad and ventrad. Cut a window in this thin sheet to reveal the intermediate internal oblique, slanting in the opposite direction. Deeper still is the transversus, a very thin layer reached just before penetrating into the body cavity and identified by its slightly different fiber slant, somewhat transversely around the body. The ventral most hypaxial muscle is the paired rectus abdominis, extending from the anterior edge of the pubic cartilage to the base of the rectus cervicis near the pectoral girdle (figure 6.14). Along the ventral midline between the rectus abdominis is a thin connective strip, the linea alba (figure 6.14).

Pelvic Girdle and Hindlimb

→ Rubbery cloacal glands surround the cloaca. Carefully probe between these and the more dorsal musculature to remove and discard one side of the gland (figure 6.14). Laterally, on either side of the cloaca is a downward projecting shelf of hypaxial myotomes. Between these hypaxial muscles and the cloacal glands are three pairs of muscles: two ventral and one dorsal. The most lateral of the ventral pair is the caudocruralis, which runs between the tail and the pubo-ischio-tibialis muscle, which itself covers the posterior half of the puboischiac plate (figure 6.14). The medial member of the pair is the ischiocaudalis, which runs between the tail and the ischium. Deflect (pull aside) the bellies of the ischiocaudalis and caudocruralis medially. This should slightly expose dorsally yet a third caudal muscle, the caudofemoralis, also seen by looking laterally (figure 6.13). The insertion of this third muscle should be confirmed by following it forward to the proximal end of the femur.

The pubo-ischio-tibialis, just identified, narrows laterally, and as the name suggests, inserts on the tibia (figure 6.14). Lying anterior to and occupying the front part of the pubic cartilage is the pubo-ischio-femoralis externus, whose fibers converge at their insertion on the femur. Note how these last two muscles partially overlap at their origins. Next identify the muscles of the thigh. On the ventral surface, from posterior to anterior, is the flat ischioflexorius, pubo-ischio-tibialis, already described, and the round pubotibialis. Separate these three muscles from each other near their insertions and confirm origins and insertions implied by their names.

Continuing forward from the anterior edge of the pubotibialis and across the dorsal side of the thigh are four muscles (figure 6.13). Lying along the anterior

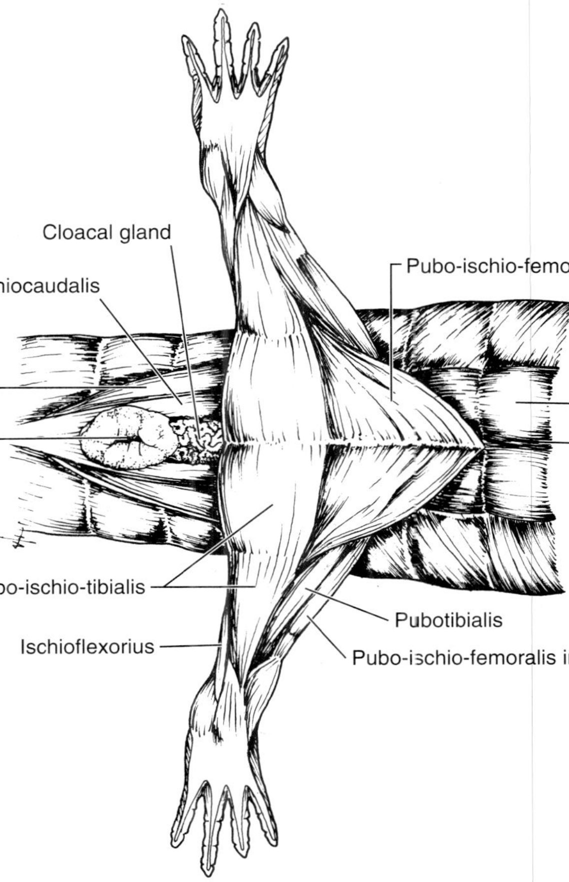

Figure 6.14
Ventral trunk and hindlimb musculature of *Necturus*.

edge of the pubotibialis is the pubo-ischio-femoralis internus, then the more dorsal iliotibialis, the more posterior ilioextensorius, and finally behind it the iliofibularis (not pictured as it occurs on the most posterior edge hidden in dorsal and lateral views). Following around the posterior (postaxial) side of the thigh returns you to the ischioflexorius identified earlier. Spend time now delineating with blunt probe or forceps each muscle and confirm attachments suggested by the name of each muscle.

Cat

External Anatomy

→ Most students have a general familiarity with cat anatomy. However, some features, their names, and their specific functions are not well known. The pinna of the ears are the external, flexible portions that initially collect sound. On the snout, long hairs, or vibrissae, provide tactile information. Numerous tori form thick pads that cushion the feet.

Sexing Your Cat

→ At the caudal end of the body, identify the anus, directly below the tail. Look along the midventral line just anterior to the anus. Male cats have a single fur-covered scrotum forming a noticeable lumpy projection (see figure 6.21). Just cranial to the scrotum is the penis, likely withdrawn and difficult to detect. Female cats will instead have a second opening, the urogenital sinus, in the position of a male's scrotum. (This is no accident, as the genitals of male and female mammals develop from common embryonic structures.) Further details of the reproductive systems will be addressed later in the chapter devoted to the reproductive system.

Approach

The cat has the same basic muscle regions as the shark and *Necturus*, except that there is a greater number of named muscles, probably reflecting the greater complexity of motion and refinement of action. The right side of the cat will be saved for later study of the circulatory system. Thus, confine your muscle dissection to the cat's **left side**. Even here, however, be sure to avoid unnecessarily destroying nerves or injected blood vessels.

Reflecting Muscles

Remember that when you need to expose a deeper structure covered by a muscle, **reflect** the muscle by cutting it through its belly (middle) so you can later reposition severed ends of the muscle for study and review.

Skinning the Cat

→ The skinning should be done so that, at the end of each laboratory, the skin can be refolded to its original position. This helps retain moisture and prevents drying of the tissues between dissections.

Most instructors have specific preferences for the removal of the skin. (After all, "there is more than one way to skin a cat.") In general, use a pair of scissors to make a shallow incision along the ventral midline from the chin to the groin. (As noted in the shark and *Necturus*, an incision may have already been made, often laterally, to inject a portion of the circulatory system. If so, begin with this incision instead of making another.) On the left side, extend the midventral incision laterally along the fore- and hindlimbs to the base of the paws. At the terminus of these incisions into the limbs, extend the cut circularly all the way around the ends of the paw. Now the skin can and should be freed from underlying tissue all the way to the dorsal midline—taking care not to pull attached muscle with it. Those with female cats will encounter the thick, glandular mammary glands along the belly. These should accompany the skin as it is worked free from the abdominal region.

Between the skin and muscle lies fascia, which includes the cobweblike superficial fascia and the deep fascia, often impregnated with yellow fat deposits, that clings to the surface of the muscles. This should all be removed to reveal muscular details. As you remove skin, you may notice a very thin layer of muscle sometimes attached to the skin, or clinging to the underlying musculature. This is the dermal muscle. In the head and neck it is called the platysma, but elsewhere the cutaneous maximus. It acts to move the skin and in the head contributes to changes in facial expression.

General Regions

→ The white streak down the ventral midline along the abdomen is the linea alba. Along the back, middorsally over the lumbar region, is an extensive area covered with white sheets of connective tissue. The sheets are collectively the lumbodorsal fascia. The axilla (or armpit) is the angle between the base of the forelimb and body. The angle between the medial base of the hindlimb and abdominal wall is the inguinal region.

Musculature

Abdomen, Ribs

→ (1) Rectus abdominis
The rectus abdominis runs along the abdomen adjacent to and parallel with the linea alba (figure 6.15). It originates on the pubic symphysis and inserts on the sternum and costal cartilages.
(2) External oblique
The superficial layer of muscle lateral to the rectus abdominis that forms the side of the abdomen is the external oblique. Its fibers slant downward in a dorsocranial to a ventrocaudal direction.
(3) Internal oblique
Separate a few fibers of the external oblique with a dissecting needle until fibers passing in another direction are encountered. This is the internal oblique, which also connects to the ventral midline by a broad but more lateral sheet of fascia (figure 6.15).
(4) Transversus
Again pick away a few fibers from the internal oblique until fibers running in yet a third direction are found. This is the deepest of the three muscle layers of the abdomen, the transversus.
(5–6) External and Internal Intercostals
Cut an opening in the thin external oblique covering the posterior and ventral part of the rib cage (figure 6.15). This exposes the ribs and the short fibers that run between adjacent ribs. The superficial muscular layer between ribs is the external intercostal. The deeper layer, reached through the external intercostals and running at a different slant, is the internal intercostal muscle.

To confirm your identification, notice that the layer of internal intercostals persists, all the way to the ventral tips of the ribs, whereas the external intercostals stop short of the rib tips.

Superficial Chest Musculature

→ Spread the forelimbs, tying them back if necessary, to bring the exposed muscles of the chest into better view.

(1–2) Clavotrapezius and Clavobrachialis
Clearly expose and follow the long slab of muscle that runs from the forearm, over the shoulder, into the neck, and to the head (figure 6.15). Run a blunt probe along this muscle as it passes over the shoulder and into the neck. You will discover here a hard bone within the tissue of the muscle, the free moving clavicle. The portion of this muscle from clavicle to the head is the clavotrapezius and from clavicle to forearm, the clavobrachialis.

The muscle bands running from the chest out to the arm constitute the pectoralis group of muscles (figure 6.15). In the cat, the pectoralis group is differentiated into four named muscles, discussed next. This

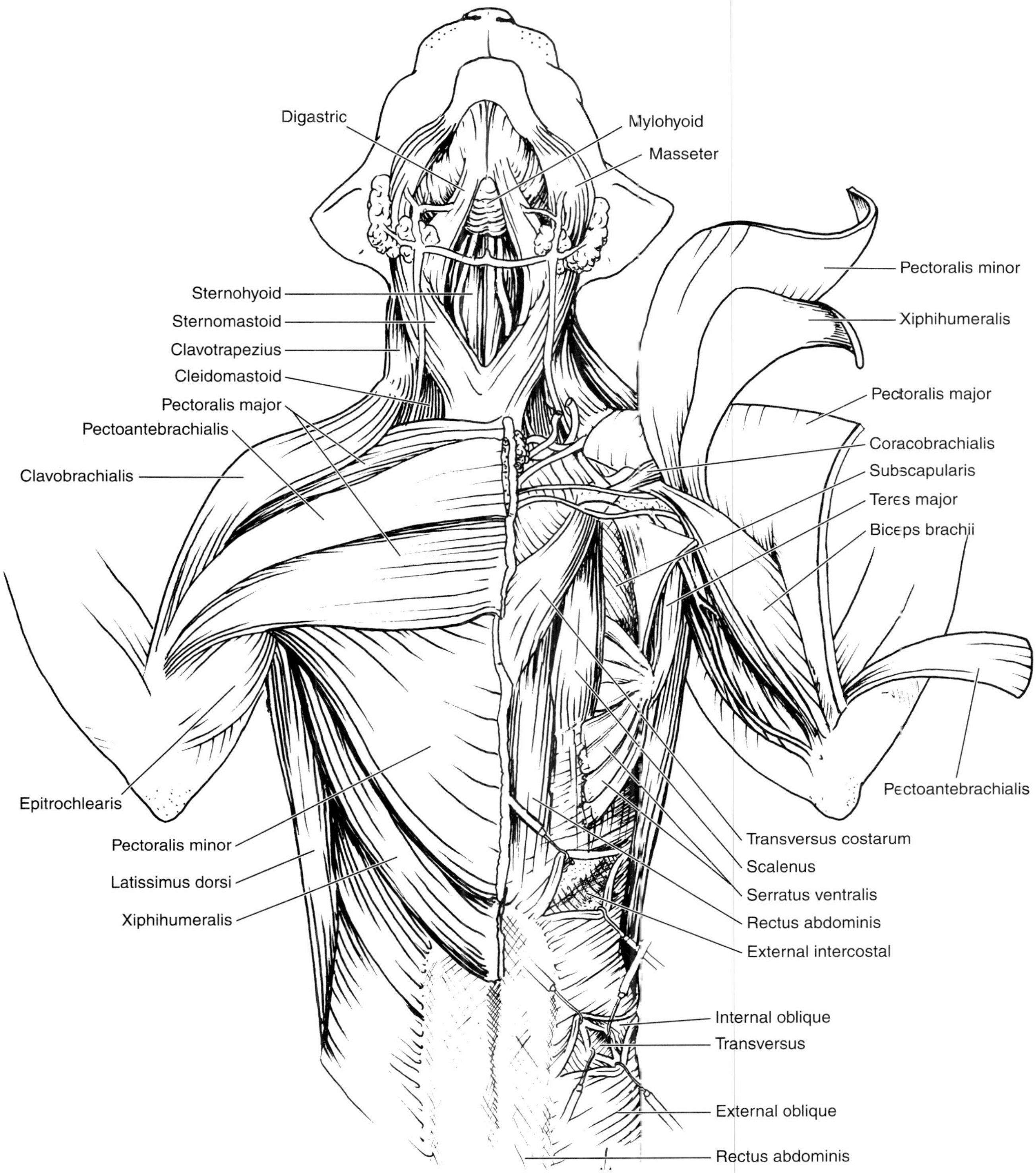

Figure 6.15
Superficial (viewer's left) and deep (viewer's right) ventral trunk, branchial, and cephalic musculature of a cat.

differentiation is not at first easy to see (all originate upon different parts of the sternum). It is best to rely upon insertions to help in the separation and upon slightly different orientation of fibers within each muscle.

(3) Pectoantebrachialis (= Pectoantibrachialis) This muscle inserts via a tendon on the upper part of the forearm. It is also the most superficial of the four pectoralis muscles, applied to the middle surface of the pectoralis major.

(4) Pectoralis major
Slip a blunt probe under the pectoantebrachialis to separate it along its length from the underlying pectoralis major. Part of the broad pectoralis major also lies covered by the clavobrachialis. Follow the pectoralis major to its long insertion along the anterolateral side of the humerus.
(5) Pectoralis minor
This muscle follows next posteriorly and is in turn partially covered by the more anterior and superficial pectoralis major. Follow it to its long insertion on the humerus next to the insertion of pectoralis major.
(6) Xiphihumeralis
This is the last of the pectoralis group and one of the most delicate and difficult to separate. It originates along the xiphoid process of the sternum. It slants into the limb passing dorsally above the pectoralis minor to insert on the humerus. The posterior edge of the xiphihumeralis is often ill defined because connective tissue joins it with a quite extensive and unrelated muscle, the latissimus dorsi.
The latissimus dorsi originates in the area of the lumbodorsal fascia and inserts on the humerus. This muscle will be addressed again when discussing the forelimb.

Deep Chest and Shoulder Musculature

→ Cut through the bellies of the pectoralis muscles so the ventral end of the scapula can be pulled away from the body wall—be careful not to damage the abundant and prominent blood vessels and nerves (figure 6.15).

(1) Serratus ventralis
The fan-shaped muscle on the wall of the rib cage facing the scapula is the serratus ventralis (figure 6.15). It converges dorsally to insert on the vertebral border of the scapula.
The smaller serratus dorsalis lies in the body wall, runs only over the ribs and does not pass to the limb. It will not be followed.
(2) Scalenus
Running across the origin of the fan-shaped serratus ventralis is the scalenus (figure 6.15). It is actually composed of three separate muscles that originate on separate ribs and converge as they pass forward to insert on the cervical vertebrae.
(3) Transversus costarum
This single strap of muscle lies deep to the pectoralis minor (figure 6.15). It originates from the sternum via an aponeurosis and inserts on the first rib adjacent to the point of entrance and exit of major blood vessels and nerves from the thoracic cavity. Notice that the aponeurosis of the rectus abdominis lies beneath (deep to) the origin of transversus costarum on the sternum.
(4) Subscapularis
The subscapular fossa on the medial side of the scapula is occupied by the subscapularis muscle (figure 6.15). It is a pinnate muscle whose fibers converge ventrally on a tendon to the lesser tuberosity of the humerus.
(5) Coracobrachialis
At about the point where the converging fibers of the subscapularis pass across the edge of the glenoid fossa on their way to the humerus, the tendon is crossed by a tiny muscle, the coracobrachialis (figure 6.15).
(6) Teres major
The teres major occupies the ventrolateral border of the scapula. Separate it clearly from the subscapularis and confirm its origin from the axillary border and insertion (with the latissimus dorsi) on the medial surface of the humerus (figure 6.15).

Neck and Throat Musculature

→ To expose these muscles you may need to further free the overlying skin (figure 6.16a). Be sure to preserve blood vessels, nerves, and glands as best you can for later dissection.

(1) Sternomastoid
This muscle originates on the manubrium of the sternum and on the median raphe (figure 6.16a). It sweeps forward along the neck to insert on the lambdoidal ridge and mastoid process of the skull. Before following it to its insertion, note: (a) it is crossed superficially by the blue latex injected external jugular vein, and (b) two salivary glands are also in contact with its superficial surface. The largest gland beneath the pinna of the ear is the parotid gland, and the smaller one ventral to it is the submaxillary gland.
(2) Cleidomastoid
Lateral to the sternomastoid and beneath the clavotrapezius lies the slender cleidomastoid (figure 6.16a). As the name suggests, the cleidomastoid originates on the clavicle (cleido = "clavicle") and inserts on the mastoid process. The clavicle was previously identified between the clavotrapezius and clavobrachialis muscles.
(3) Sternohyoid
This paired muscle runs between the two sternomastoids directly forward from the sternum along the ventral midline to the body of the hyoid (figure 6.16a). At about this insertion it is crossed by the transverse jugular vein. Free the sternohyoid along its length to confirm its origin and insertion.
(4) Geniohyoid
Parallel fibers of the geniohyoid continue on from the body of the hyoid forward to the chin, where they take origin (figure 6.16a). A transversely crossing sheet of fibers often hides this origin. This obscuring sheet is the mylohyoid.
(5) Mylohyoid
This transverse sheet of muscle joins the two tips of the mandibular rami (figure 6.16a). A faint median raphe is often evident. Cut this muscle along its raphe to better expose the geniohyoid.

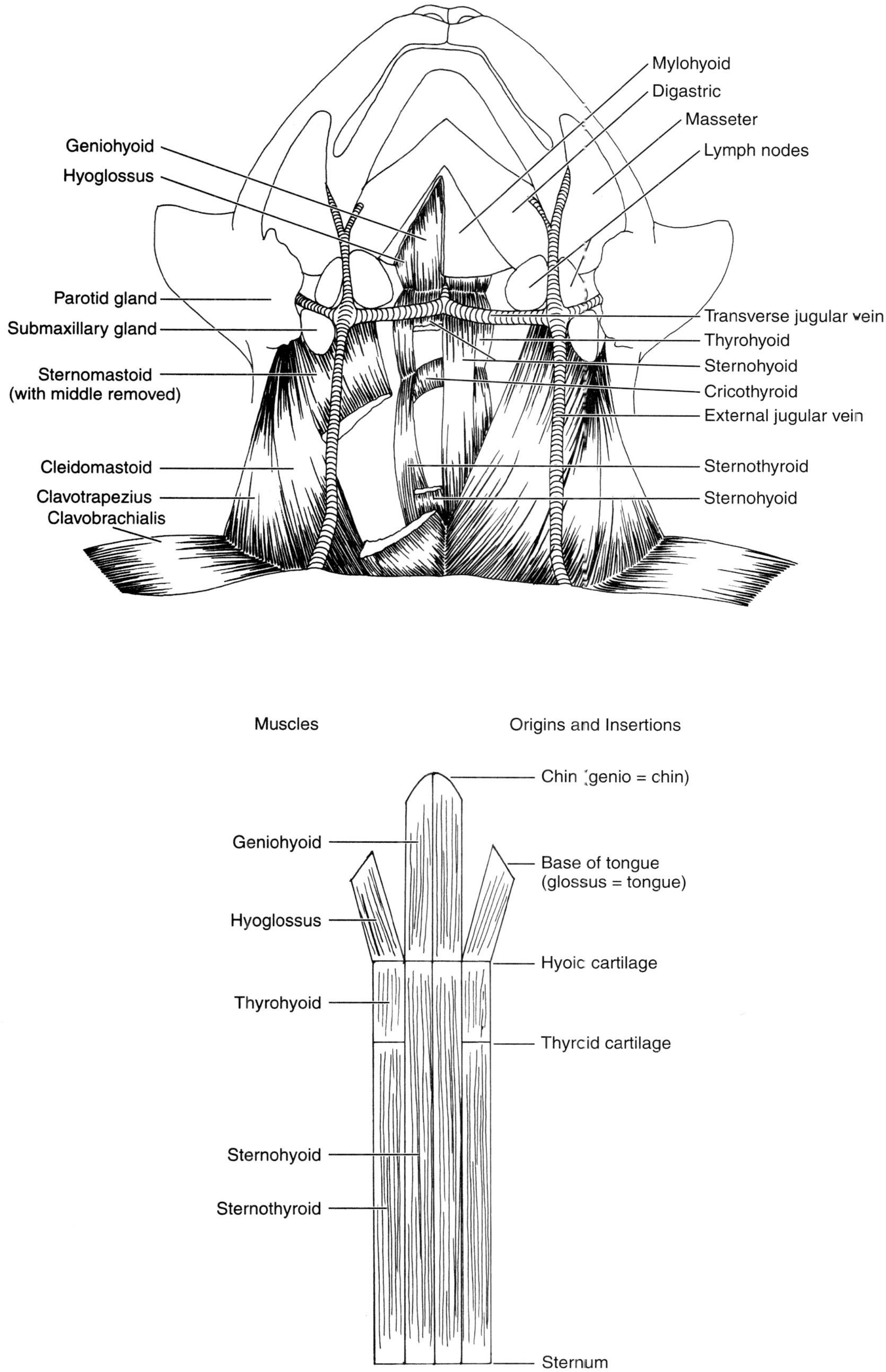

Figure 6.16
Superficial (viewer's right) and deep (viewer's left) head and neck musculature of a cat. (a) Ventral view. (b) Diagram indicating the origination of the names of some neck muscles based upon their origins and insertions.

(6) Digastric
Running parallel with and along the ventral edge of the mandible is the digastric (figure 6.16a). Along one side, follow it deep to confirm its origin from the jugular and mastoid processes of the occipital bone of the skull.
(7–8) Hyoglossus and Thyrohyoid
Lateral to and parallel with the geniohyoid is the broader hyoglossus (figure 6.16a). It will at first look like a continuous muscle sheet from the larynx up into the tongue, but investigation with a blunt probe will confirm the presence of the hyoid in the middle of this muscle mass. The stretch from the hyoid forward into the base of the tongue is the hyoglossus. The part from the hyoid posteriorly to the thyroid cartilage of the larynx is the thyrohyoid muscle. Separate the hyoglossus from surrounding muscles, then grasp it with tweezers, and give a gentle pull. The tongue, into which it inserts, should move.
(9–10) Sternothyroid and Cricothyroid
The expanded larynx, onto which the thyrohyoid inserts, begins the windpipe or trachea, a ringed, flexible tube. Just posterior to the larynx, on the sides of the trachea, lies the paired thyroid gland.

The sternothyroid is a slender muscle running just lateral to the sternohyoid from the sternum to the base of the larynx. The cricothyroid is a small patch of muscle on the larynx itself near the insertion of the sternothyroid.

As you work through the neck muscles, remember that their names are often taken from the structures they attach to—origin, then insertion (table 6.4 and figure 6.16b)

Table 6.4

The Origin, Insertion, and Function of Cat Musculature

Muscle	Origin	Insertion	Function
Abdomen and Ribs			
external intercostal	rib	rib	draws ribs craniad
external oblique	lumbodorsal fascia and ribs	aponeurosis on the linea alba	compresses abdominal viscera
internal intercostal	rib	rib	draws ribs caudad
internal oblique	pelvis and lumbodorsal fascia	aponeurosis on the linea alba	compresses abdominal viscera
rectus abdominis	pubic symphysis	sternum and costal cartilages	flexes trunk and compress abdominal viscera
transversus	ilium, lumbar vertebrae, and posterior ribs	linea alba	compresses abdominal viscera
Chest Musculature			
clavobrachialis	clavicle	proximal end of ulna	draws humerus craniad
pectoantebrachialis	manubrium	fascia of forelimb near the elbow	adducts and rotates the forelimb
pectoralis major	cranial sternebrae	proximal half of humerus	adducts and rotates the forelimb
pectoralis minor	caudal sternebrae	proximal half of humerus	adducts and rotates the forelimb
xiphihumeralis	xiphoid process	proximal end of humerus	adducts and rotates the forelimb
Neck and Throat			
cleidomastoid	clavicle	mastoid process	rotates head
cricothyroid	cricoid cartilage	thyroid cartilage	moves the larynx
digastric	mastoid and jugular processes of occipital	mandible	depresses the lower jaw
geniohyoid	mandible	hyoid	draws hyoid craniad
hyoglossus	hyoid	tongue	retracts the tongue
mylohyoid	mandible	median raphe	elevates floor of mouth
sternohyoid	manubrium of sternum	hyoid	draws hyoid caudad
sternomastoid	manubrium of sternum	lambdoidal ridge and mastoid process	rotates head
sternothyroid	manubrium of sternum	thyroid cartilage	draws larynx caudad
thyrohyoid	thyroid cartilage	hyoid	draws larynx craniad
Lateral Jaw			
masseter	zygomatic arch	mandible	elevates lower jaw
temporalis	lambdoidal crest and zygomatic process of frontal bone	coronoid process of mandible	elevates lower jaw

Table 6.4 *continued*

The Origin, Insertion, and Function of Cat Musculature

Muscle	Origin	Insertion	Function
Dorsal Shoulder–*continued*			
acromiodeltoid	acromion process of scapula	humerus	flexes and rotates humerus
acromiotrapezius	neural spines of cervical and first thoracic vertebrae	metacromion process and spine of scapula	draws scapula dorsally and medially
clavotrapezius	nuchal line of skull and dorsal midline of neck	clavicle	draws humerus craniad
infraspinatus	infraspinous fossa of scapula	greater tuberosity of humerus	rotates the humerus
latissimus dorsi	lumbodorsal fascia and neural spines of the last thoracic and lumbar vertebrae	medial surface of humerus	elevates the arm and draws it caudad
levator scapulae ventralis	occipital bone and transverse process of atlas	metacromion process	draws scapula craniad
rhomboideus	neural spines of thoracic vertebrae	ventral border of scapula	draws scapula dorsally
rhomboideus capitis	superior nuchal line	angle of scapula	draws scapula craniad
spinodeltoid	spine of the scapula	deltoid ridge of humerus	elevates and rotates the humerus
spinotrapezius	neural spines of the thoracic vertebrae	fascia around scapula	draws scapula dorsally and caudad
supraspinatus	supraspinous fossa	greater tuberosity of humerus	extends the humerus
teres minor	axillary border of scapula	greater tuberosity of humerus	rotates humerus
Ventral Shoulder			
coracobrachialis	coracoid process	proximal end of humerus	adducts humerus
scalenus	ribs	transverse processes of cervical vertebrae	flexes neck and draws ribs craniad
serratus ventralis	first ten ribs	vertebral border of scapula	draws scapula ventrally and craniad
subscapularis	subscapular fossa	lesser tuberosity of humerus	draws humerus medially
teres major	axillary border of scapula	medial surface of humerus	draws humerus medially
transversus costarum	sternum	first rib	draws sternum craniad
Upper Arm–Brachium			
biceps brachii	border of glenoid fossa	tuberosity of radius	flexes forearm
brachialis	lateral surface of humerus	below semilunar notch of ulna	flexes forearm
epitrochlearis	lateral surface of latissimus dorsi	olecranon process of ulna	rotates ulna
triceps brachii	shaft of humerus	olecranon process of ulna	extends forearm
Forearm–Antebrachium Dorsal Side			
abductor pollicis longus	lateral surface of radius and ulna	first metacarpal	abducts the forelimb
brachioradialis	middle of humerus	distal end of radius	supinator of the manus
extensor carpi radialis brevis	distal portion of the humerus	third metacarpal	extends the manus
extensor carpi radialis longus	distal end of humerus	second metacarpal	extends the manus
extensor carpi ulnaris	lateral epicondyle of the humerus	fifth metacarpal	extends the manus
extensor digiti secundi	proximal end of ulna	middle phalanx of second digit	extends second digit
extensor digitorum communis	distal portion of the humerus	digits II–V	extends the manus
extensor digitorum lateralis	distal portion of the humerus	digits II–V	extends the digits of the manus
extensor pollicis	proximal end of ulna	middle phalanx of first digit	extends first digit
supinator	lateral epicondyle of humerus	radius	supinator of the manus
Forearm–Antebrachium Ventral Side			
flexor carpi radialis	medial epicondyle of humerus	digits II and III	flexes the manus
flexor carpi ulnaris	medial epicondyle of humerus and ulna	carpal pisiform bone	flexes the manus
flexor digitorum profundus	five heads: 1 head ulna, 3 heads humerus, and 1 head radius	digits I–V	flexes the manus
flexor digitorum superficialis	medial epicondyle of humerus and surface of flexor digitorum profundus	digits II–V	flexes the manus
pronator teres	medial epicondyle of humerus	medial border of radius	pronates the manus

Table 6.4 ***continued***

The Origin, Insertion, and Function of Cat Musculature

Muscle	Origin	Insertion	Function
Muscles of the Lateral Thigh			
biceps femoris	ischial tuberosity	proximal third of tibia	flexes the forearm
caudofemoralis	proximal caudal vertebrae	patella	abducts thigh
gluteus maximus	last sacral and first caudal vertebra	fascia lata	abducts thigh
gluteus medius	iliac crest, sacral, and first caudal vertebrae	greater trochanter of femur	abducts thigh
tensor fasciae latae	ilium	fascia lata	extends the fascia lata
Muscles of the Medial Thigh			
adductor femoris	pubic symphysis	ventral surface of femur	extends thigh
adductor longus	pubic symphysis	femur	adducts thigh
gracilis	pubic symphysis	fascia of lower thigh	adducts thigh
rectus femoris	ilium	patella	extends lower leg
sartorius	iliac crest	patella and proximal tibia	adducts and rotates thigh
semimembranosus	ischium	medial epicondyle of femur and proximal tibia	extends thigh
semitendinosus	ischium	tibia	flexes lower leg
vastus intermedius	femur	patella	extends lower leg
vastus lateralis	femur	patella	extends lower leg
vastus medialis	femur	patella	extends lower leg
Muscles of the Shank			
extensor digitorum longus	lateral epicondyle of femur	digits II–V	extends digits
flexor digitorum longus	middle tibia and head of fibula	digits I–V	flexes digits
flexor hallucis longus	tibia and fibula	tendon of flexor digitorum longus	flexes digits
gastrocnemius	femur and tendon of plantaris	calcaneum	extends foot
peroneus group	fibula	metatarsals	some flex foot some extend foot
plantaris	patella and femur	calcaneum	flexes digits
soleus	proximal fibula	calcaneum	extends foot
tibialis anterior	proximal tibia and fibula	first metatarsal	flexes foot
tibialis posterior	tibia and fibula	tarsals	extends foot

Lateral Jaw Musculature

→ (1) Masseter

On one side, pull the skin away from the head. Cut a circle around the base of the pinna (= external ear) if need be to free the skin. Be careful of blood vessels, nerves, and glands.

The large, domed masseter muscle lies below the pinna in the cheek region (figure 6.16a).

(2) Temporalis

The temporalis muscle primarily lies above the base of the pinna and is covered by a shiny tendon (figure 6.17).

Forelimb Musculature

→ **Dorsal/Lateral Shoulder—Superficial Musculature**

Lay your cat on its side with the exposed muscles of the left side facing up toward you.

(1) Latissimus dorsi

The broad exposed curtain of muscle covering the side of the chest and abdomen is the latissimus dorsi (figure 6.17). It originates from the lumbodorsal fascia in the lumbar region plus adjacent neural spines and passes forward converging upon the axilla. It inserts on the medial surface of the humerus.

(2) Clavotrapezius

Refamiliarize yourself with this muscle identified earlier, but here seen from a different position (figure 6.17). The clavotrapezius is the first of three muscles of the trapezius group. Note its broad origin from the superior nuchal line and median dorsal line of the neck.

(3) Acromiotrapezius

The origin of this muscle lies immediately behind the clavotrapezius along neural spines of the cervical and first thoracic vertebrae (figure 6.17). Its fibers converge to insert on the metacromion process and spine of the scapula.

(4) Spinotrapezius

The spinotrapezius is the smallest and most posterior of the three trapezius muscles (figure 6.17). It lies on the posterior edge of the acromiotrapezius. Arising from

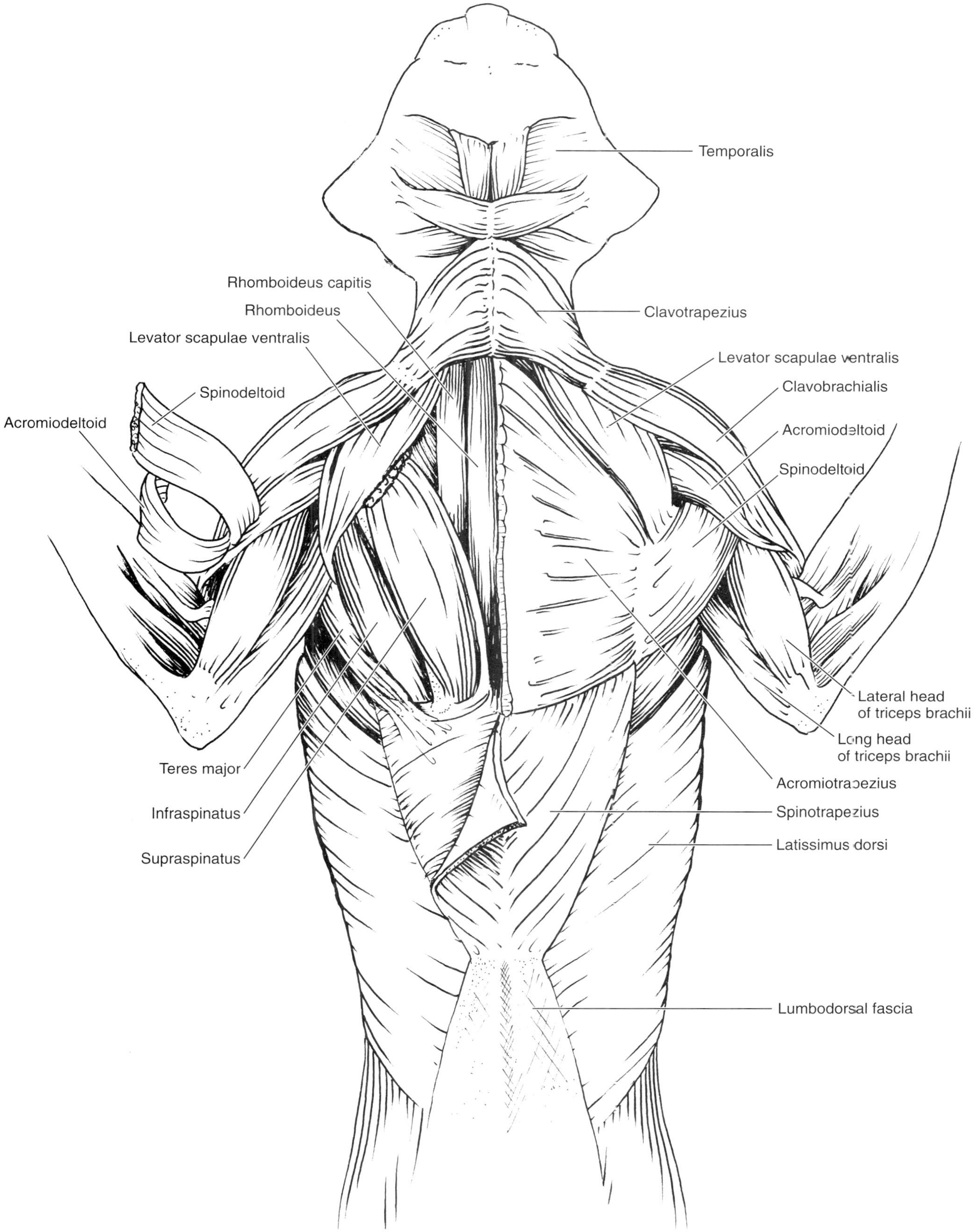

Figure 6.17
Superficial (viewer's right) and deep (viewer's left) dorsal trunk, brachial, and cephalic musculature of a cat.

Patterns & Connections

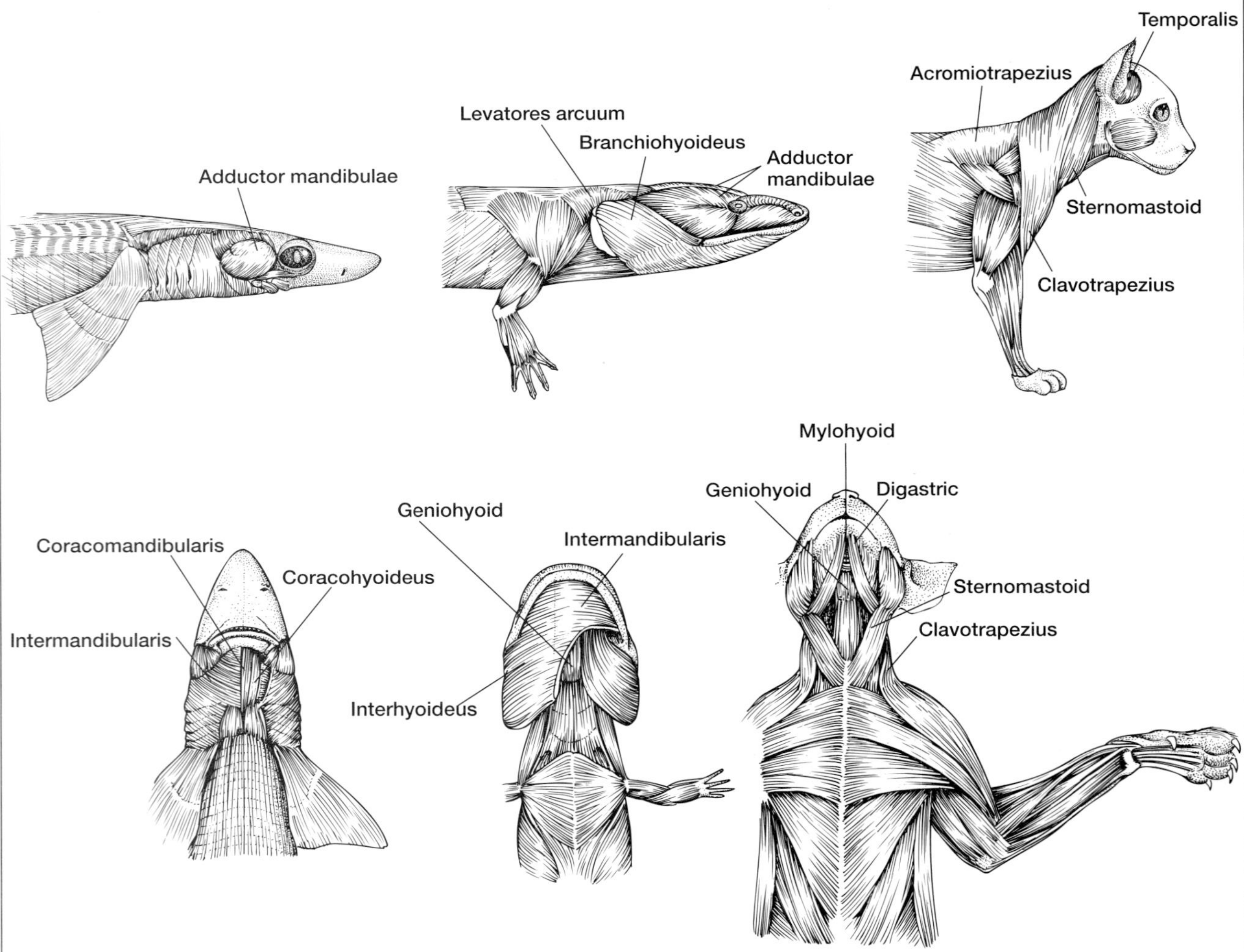

Homologous muscles of cranial and shoulder musculature in a shark, salamander, and cat. Lateral views (a–c) and ventral views (d–f).

The ventral musculature of the chest and throat of these three vertebrates reveals a common pattern. Notice, for example, that each of these animals has a set of muscles that extends along the midline from the throat to the chin (coracomandibularis and coracohyoideus in the shark and the geniohyoid muscles of *Necturus* and the cat). In addition, each animal has another set of muscles running perpendicularly across the throat (intermandibularis and interhyoideus in the shark and *Necturus* and the mylohyoid in a cat). Although these three animals are quite different, they share morphological patterns based upon a common ancestral design. Thus, these muscles are homologous. Table 6.5 includes these and other homologous muscles of the cranial musculature.

neural spines of the thoracic vertebrae, it passes beneath the acromiotrapezius, and inserts on the spine of the scapula and adjacent fascia.

(5) Levator scapulae ventralis

This is a small wedge of muscle that lies between the clavotrapezius and acromiotrapezius (figure 6.17). Confirm that the levator scapulae ventralis originates from the transverse process of the atlas and occipital bone. It inserts on the metacromion process and adjacent fascia.

In the cat, two[1] members of the deltoid group are present.

(6) Acromiodeltoid

The short, acromiodeltoid lies just ventral to the insertion of the levator scapulae ventralis and dorsal to the clavotrapezius. Confirm that it arises on the acromion and inserts on the humerus.

[1]Some consider the clavobrachialis to be a third deltoid present in the cat and so prefer the name *clavodeltoid* for this muscle.

Table 6.5

Homologies of Cranial Musculature

Arch	Cranial Nerve Supply	Shark	*Necturus*	Cat or Mink
Branchiomeric Musculature				
I	V	Levator palatoquadrati	Adductor mandibulae	Masseter
		Spiracularis	(Levator mandibulae)	Temporalis
		Adductor mandibulae		Pterygoids
		Preorbitalis		Tensor veli palati
				Tensor tympani
		Intermandibularis	Intermandibularis	Mylohyoid
				Anterior digastric
2	VII	Levator hyomandibulae	Depressor mandibulae	Stapedius
		(Epihyoidean)	Interhyoideus	Platysma and facial muscles (part)
		Interhyoideus	Constrictor colli	Posterior digastric Stylohyoid
3	IX	Superficial constrictor	Branchiohyoideus	Intrinsic muscles of the larynx and certain muscles of the pharynx
4	XI	Cucullaris	Cucullaris	Trapezius complex
	X		Levatores arcuum	Sternocleido-mastoid complex
		Interarcuals		
		Superficial constrictors	Dilatator laryngis	
		and interbranchials	Subarcuals	
			Transversi ventrales	
			Depressors arcuum	
Hypobranchial Musculature				
		Coracoarcuals	Rectus cervicis	Geniohyoid, others of the tongue and larynx
		Coracohyoideus	Geniohyoid	
		Coracoman-dibularis	Pectoriscapularis	Rectus cervicis, sternohyoid, omohyoid, thyrohyoid
		Coracobrachialis		

(7) Spinodeltoid
This muscle lies just posterior to the acromiodeltoid, which partially obscures its insertion on the deltoid ridge of the humerus (figure 6.17). It originates on the spine of the scapula next to the insertion of the acromiotrapezius.

Dorsal/Lateral Shoulder—Deep Musculature Turn your cat onto its belly so the exposed muscles of the upper part of the shoulder face you. Reflect the acromiotrapezius across its middle (figure 6.17).
(1–2) Infraspinatus and Supraspinatus
The infraspinatus and supraspinatus occupy their respective fossae on the lateral side of the scapula, namely the infraspinous fossa and supraspinous fossa. You may need to spread the spinotrapezius and latissimus dorsi to expose the infraspinatus.
(3) Teres Minor
Reflect the spinodeltoid to reveal the deep muscles below. Fibers of the infraspinatus converge and pass to their insertion on the humerus. Between the infraspinatus and two heads of the triceps lies the teres minor. This is a small muscle usually clinging to the edge of the infraspinatus. Separate it and locate its approximate origin and insertion.
(4) Rhomboideus
Slightly pull the scapula away from the body wall. The extensive fan-shaped muscle emanating from an insertion on the vertebral border of the scapula and taking origin extensively along the neural spines of the neck is collectively the rhomboideus (figure 6.17). The first (anterior) strap of this muscle is somewhat separate from the rest and takes origin from the superior nuchal line. This single strap of muscle is termed the rhomboideus capitis.

Upper Arm (Brachium) Musculature Return the cat again to its side with exposed muscles of the forelimb facing you.
(1) Triceps brachii
There are three heads that occupy the back of the upper arm and collectively insert on the olecranon (figure 6.18a). The lateral head of the triceps lies along the lateral side of the upper arm. The equally prominent long head of the triceps occupies the posterior side of the upper arm. Spread the long and lateral heads to find the smaller medial head of the triceps deep to the prominent radial nerve and lying against the humerus.
(2) Brachialis
The brachialis is parallel with, but anterior to, the lateral head of the triceps along the lateral side of the humerus (figure 6.18a). Its origin on the humerus lies hidden beneath the acromiodeltoid. Follow the brachialis, under the lateral head of the triceps, to its insertion on the forearm next to the insertion of the biceps brachii described below.
(3) Epitrochlearis
Again lay your cat on its back to examine the medial surface of the arm. The medial face of the long head of the triceps is covered by a broad, flat muscle, the epitrochlearis (figure 6.19a). It originates from the latissimus dorsi and triceps. Insertion is by an aponeurosis to the olecranon. Cut through the belly of this muscle to discover the shiny medial surface of the long head of the triceps.
(4) Biceps brachii
Parallel with the long head of the triceps and the epitrochlearis, but along the anterior margin of the humerus, is the biceps brachii (figure 6.15a). Follow this muscle to its insertion on the forearm via a narrow, but strong tendon.

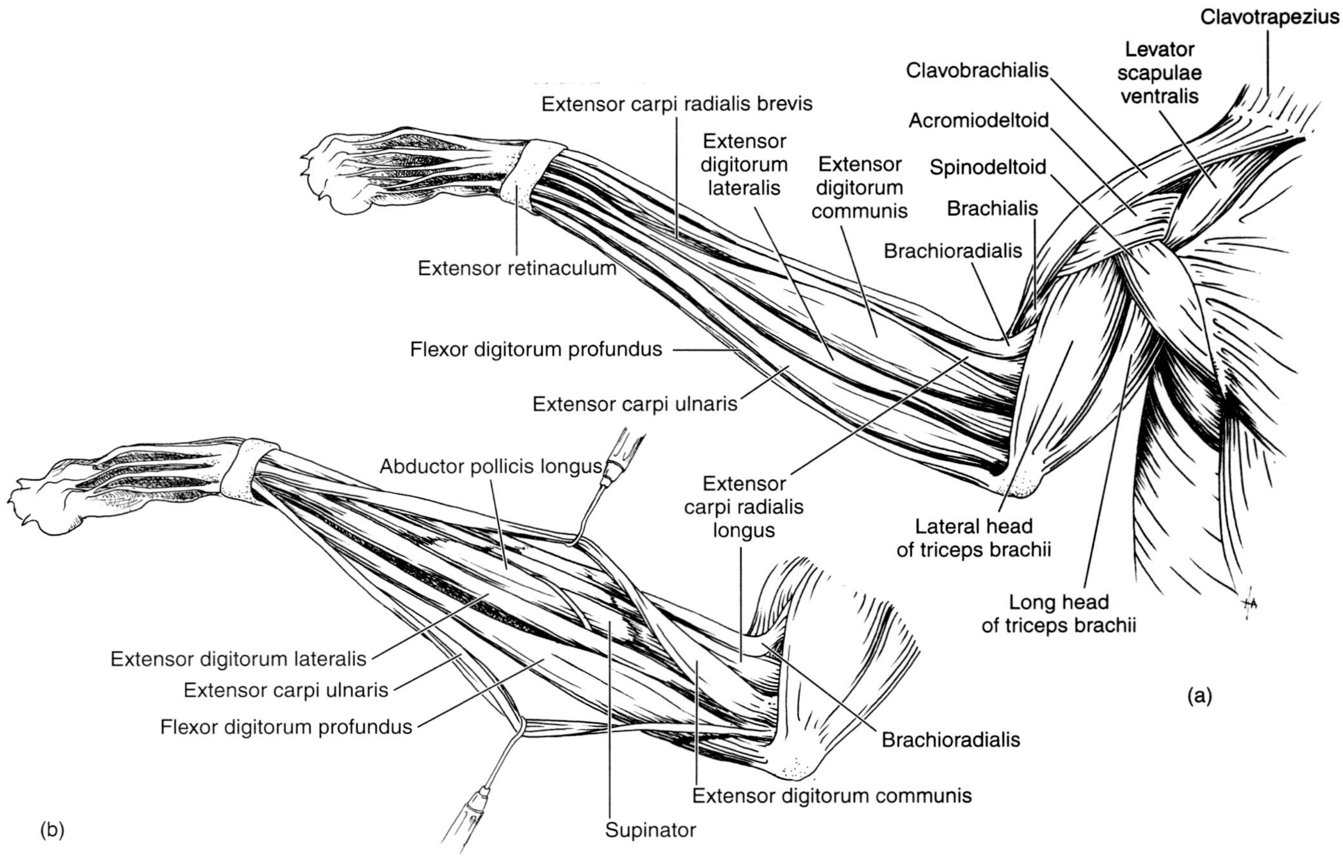

Figure 6.18
Musculature of the dorsal left forelimb of a cat. (a) Superficial and (b) deep.

➞ Forearm (Antebrachium) Musculature—Dorsal Side

Ten muscles form the dorsal surface of a cat's forearm (figure 6.18a). Remove skin on the dorsal and ventral surfaces of the foot (= manus) leaving the tendons intact. Near the wrist, a connective tissue band, a retinaculum (pl. retinacula) encircles many of the tendons. On the dorsal side, it is the extensor retinaculum, and along the ventral side, the flexor retinaculum. Use caution to keep these retinacula intact when tracing tendons.

Most of the muscles on the dorsal side extend elements of the foot or wrist. After tracing a tendon onto its insertion, firmly grab the tendon and give it a short tug to appreciate the muscle's function (table 6.4).

(1) Brachioradialis

This long muscle travels down the upper arm to emerge along the anterior margin of the lateral head of the triceps brachii. It continues along the medial surface of the elbow joint to insert on the distal end of the radius. When dissected free of other muscles, it hangs loosely away from the forearm.

(2) Extensor Carpi Radialis Longus

This muscle originates near the distal end of the humerus and forms the most medial member of the tightly grouped forearm muscles. (Medial means toward the side of the first digit.) Its single and thin tendon forms near the distal third of the radius and distinguishes it from adjacent muscles.

(3) Extensor Digitorum Communis

Next, moving laterally near the elbow joint, is the extensor digitorum communis. Near the distal third of the radius it sends four long, tightly packed, and thin tendons under the extensor retinaculum. The tendons then diverge to separately attach to digits II–V.

(4) Extensor Carpi Radialis Brevis

Separate the "longus" and "communis" muscles just identified. Between them lies the deeper extensor carpi radialis brevis (figure 6.18a), a thicker muscle whose single tendon courses distally with the "longus" to insert upon the third metacarpal bone.

(5) Extensor Digitorum Lateralis

This muscle is about the same size as the "communis" and parallels it along its lateral side (figure 6.18a). The "lateralis" also produces three or four very thin tendons that run collectively to the same digits, but under the tendons of the "communis" (the number varies within the species of domestic cats).

(6–7) Abductor Pollicis Longus and Supinator

Deep between the "communis" and "lateralis" (the two muscles with three to four tendons each), are two more muscles (figure 6.18b). The abductor pollicis longus (ex-

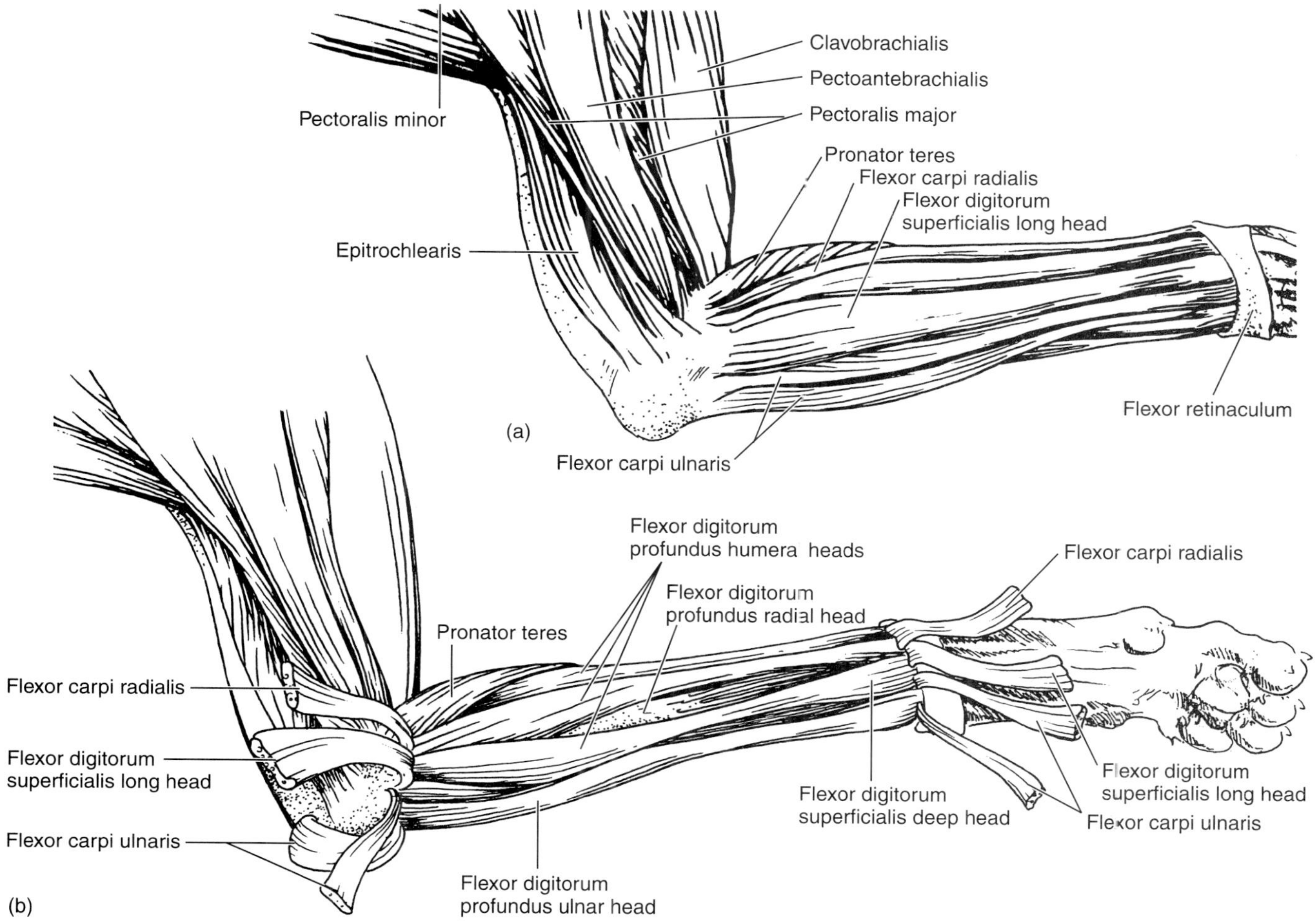

Figure 6.19
Musculature of the ventral left forelimb of a cat. (a) Superficial and (b) deep.

tensor pollicis brevis) is a wide muscle wrapped tightly and transversely around the distal half of the ulna and radius. Just proximal to the "longus," in the same position between the "communis" and "lateralis," is the supinator.
(8) Extensor Carpi Ulnaris
Just lateral to the proximal end of the "lateralis" is the extensor carpi ulnaris, approximately the same diameter as the "lateralis" but ending in a single, broad, tendon that inserts upon the base of the fifth metacarpal bone (figure 6.18a). Just lateral to the "ulnaris," forming the most lateral margin of the ventral forearm, is the ulnar head of the flexor digitorum profundus, described more extensively below in the section on the dorsal flexor muscles of the forearm.
(9–10) Extensor Pollicis and Extensor Digiti Secundi
Directly below the "ulnaris" is a pair of muscles that are collectively smaller in diameter than any of the muscles previously examined. The more medial extensor pollicis (extensor digiti I) has a longer and larger tendon that passes to the first digit. The tendon of the very thin and more lateral extensor digiti secundi (extensor digiti II) merges with the tendon of the "longus."

→ **Forearm (Antebrachium) Musculature—Ventral Side**
A considerable sheet of superficial fascia overlies much of the ventral surface and must be reflected carefully to spare the musculature below.
(1) Flexor Digitorum Superficialis
This is the most prominent, superficial muscle, forming a broad thin sheet along the medial portion of the forearm (figure 6.19a). Its distal tendon travels under the flexor retinaculum and subdivides into four tendons that attach to digits II–V. A deeper, second head will be described below in the section on the flexor digitorum profundus.
(2) Flexor Carpi Ulnaris
This muscle, with two heads, is positioned lateral to the "superficialis" previously identified (figure 6.19a). The smaller, more lateral head overlaps a deeper, larger head. The tendons converge and run together to the carpal pisiform bone.
(3) Flexor Carpi Radialis
This muscle is found on the opposite (medial) side of the "superficialis" (figure 6.19a). Sometimes a head of the flexor digitorum profundus (not yet described) extends between the flexor carpi radialis and the flexor

Form & Function

After identifying all of the forearm muscles, you may wonder why so many small, discrete muscles occur at these positions instead of a few larger muscles like those of the upper arm or shoulder. The forearms of humans are similarly complex. Imagine trying to play the piano or take notes in lecture if most of your digits moved simultaneously. In the forearm, one large action becomes divided between many separate muscles, permitting a wider range of manipulative skills. The digits can act independently of one another because each is moved by its own set of muscles. In cats, many discrete forearm muscles permit their paws to quickly adjust to irregularities in the terrain when running.

digitorum superficialis. The long tendon of the spindle-shaped flexor carpi radialis passes under the flexor retinaculum and attaches to digits II and III. To further confirm the flexor carpi radialis, identify the pronator teres described directly below.

(4) Pronator Teres

This wedge-shaped muscle primarily located along the proximal half of the forearm resides on the medial border of the flexor carpi radialis (figure 6.19a). The muscle fibers of the pronator teres initially run diagonally from their origin on the humerus.

(5) Flexor Digitorum Profundus

Pull apart the "superficialis" and "radialis" from the "ulnaris" (figure 6.19b). (Alternately, you may cut and reflect these muscles along their midlengths.) Most of what you see are the five heads of the flexor digitorum profundus: one ulnar head, one radial head, and the three humeral heads arising from the medial epicondyle of the humerus. The tendons of these five heads converge and collectively run below the flexor retinaculum to insert on separate digits. One of the humeral heads narrows along the distal third of the ulna and then expands to form a deep head of the "superficialis." The most medial of the humeral heads was noted previously, as it may sometimes appear at the surface between the "superficialis" and the "radialis." The ulnar head is the largest head located along the lateral margin of the forearm. It was noted previously lateral to the extensor carpi ulnaris when the extensors were examined (figure 6. 18a). The radial head of the flexor digitorum profundus is tightly applied to the distal third of the ulna.

(6) Lumbricales

These small muscles are found only on the hand and will not be separately identified (figure 6.19a).

Hindlimb Musculature

→ If you have not already done so, extend the midventral incision through the cat's skin posteriorly to the anus. Then, beginning at this incision, continue the cut outward along the inner (medial) side of one hindlimb. Note that one of the limbs was used as the site of latex injection into the blood vessels. You may wish to select the opposite for this dissection of the hindlimb musculature, or perhaps perform a superficial dissection on one side and a deep dissection on the other. Be careful now, as throughout the dissection, not to destroy any of the major blood vessels or nerves coursing through the limb. In addition, extra caution will be needed along the ventral midline to preserve the spermatic cords in male cats.

Once the proper incisions have been made, use your fingers or a blunt probe to loosen the skin and gradually work it loose from the underlying muscles all the way to the tip of the foot. Free the skin completely from the limb, cutting it near the claws if necessary, but leave the flap connected at the dorsal midline on the back so that it might be replaced over the limb when you finish each day's dissection.

With the skinning of the limb completed, place the cat on its belly in the tray before you. The first part of the dissection will begin on the lateral, outer surface of the thigh. Consult figure 6.20 for general orientation, then return to this point in the description.

Muscles of the Lateral Thigh The thigh is covered by a superficial fascia, a connective tissue that covers the limb muscles. Use your fingers to carefully remove this fascia, being certain that any clinging muscle fibers are not removed along with it. Large and very tough connective tissue sheets, such as the large lumbodorsal fascia along the dorsal midline and the fascia lata on the lateral sides of the legs should not be removed.

(1) Sartorius

This thin strap of a muscle is found along the anterior margin of the thigh (figure 6.20). Most of the sartorius is actually found behind, on the medial side, away from you as you now view it. However, it will make a useful landmark during the dissection. The next four muscles are to be found in this region.

(2) Tensor Fasciae Latae

This muscle is just below (ventral to) and behind the preceding. It is a triangular-shaped muscle encased within a tough, tendinous sheet, the fascia lata, that spreads ventrally across the limb toward the knee (figure 6.20). The tensor fasciae latae and fascia lata wrap the anterior margin of the thigh. They will be identified again when the medial side of the thigh is examined. Just dorsal to the tensor fasciae latae is a tough connective sheet that extends over the next muscle and inserts on the ilium. Make an incision in the middle of this dorsal fascia parallel to the axis of the body and reflect the fascia back.

(3) Gluteus Medius

This muscle is to be found as the large, triangular muscle at the most anterior and dorsal part of the thigh just exposed by the reflection of the dorsal fascia (figure 6.20). In cats, the gluteus medius is larger than the gluteus maximus that follows caudally.

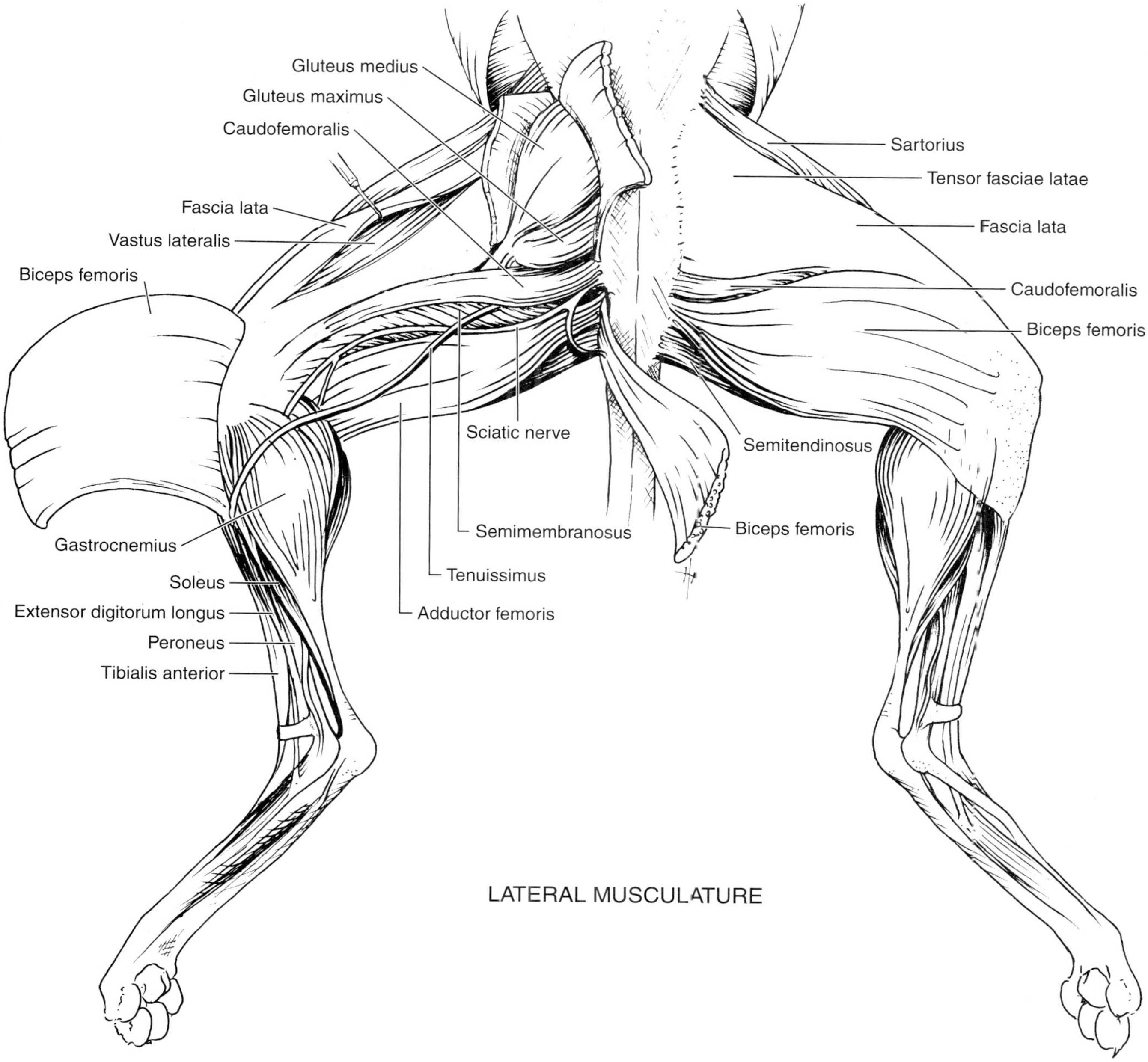

Figure 6.20
Lateral musculature of the hindlimb of a cat. Superficial (viewer's right) and deep (viewer's left).

(4) <u>Gluteus maximus</u>
This smaller muscle is found immediately caudal to the gluteus medius along the dorsal midline (figure 6.20).

(5) <u>Caudofemoralis</u>
This is a long thin muscle, about the diameter of the gluteus maximus. It is found between the gluteus maximus and biceps femoris, found next (figure 6.20).

(6–7) <u>Biceps Femoris</u> <u>and</u> <u>Tenuissimus</u>
This is the largest of the muscles so far described and forms the posterior and lateral muscle mass of the thigh (figure 6.20). By pulling the biceps femoris laterally, you can find the sciatic nerve (and its branches) that pass deep to the biceps femoris. This is a good opportunity to locate the <u>tenuissimus</u>. It is a thin muscle, about the size of the nerve, but can be distinguished from the nerve by its browner color and that it shows no distal branching. Once you find these underlying structures, you may wish to reflect the biceps femoris to better appreciate their shape and position.

(8) <u>Vastus Lateralis</u>
Lift up the thin tensor fasciae latae near its junction with the fascia lata to reveal the vastus lateralis immediately below (figure 6.20).

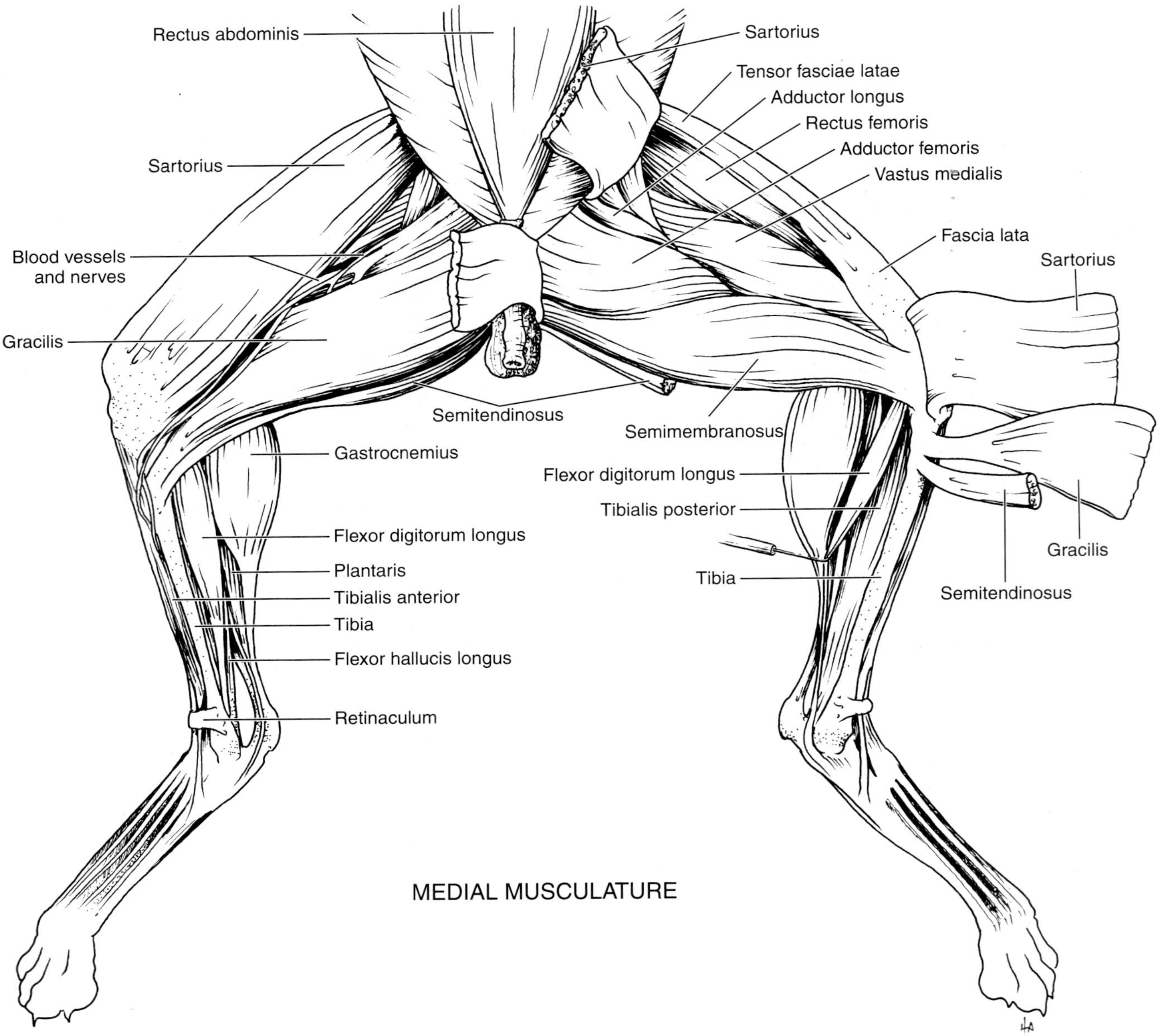

Figure 6.21
Medial musculature of the hindlimb of a cat. Superficial (viewer's left) and deep (viewer's right).

→ Muscles of the Medial Thigh

(1–2) Gracilis and Sartorius

Turn the cat over on to its back. The inner (medial) side of the thigh can now be seen and any fascia should be cleared away. Take care not to destroy the major nerves and blood vessels that run along the middle of the thigh, as they are useful landmarks. Two thin straps of muscles lie on either side of these blood vessels and nerves: the gracilis, posterior, and the sartorius, anterior (figure 6.21). Reflect the sartorius at its middle to better expose the muscles beneath it. Note the posterior margins of the tensor fasciae latae and fascia lata that wrap the anterior margin of the thigh just anterior to the sartorius.

(3–4) Rectus Femoris and Vastus Medialis

Two muscles can now be seen below the sartorius and anterior to the blood vessels and nerves. The rectus femoris lies near the ventral midline and is overlapped along its most lateral half by the more caudal vastus medialis (figure 6.21).

(5) Vastus Lateralis

Just anterior to the rectus femoris, and under the medial portions of the tensor fasciae latae, is the medial margin of the vastus lateralis (identified previously on the lateral side of the thigh).

(6) Vastus Intermedius

By widely spreading the rectus femoris from the vastus medialis, this deep muscle wrapped partly around the shaft of the femur can be seen.

To expose the next four muscles, cut through the middle of the gracilis, separating it from underlying muscles with your fingers, and fold back the two flaps (figure 6.21). Beginning at the groin and working caudally along the thigh, the muscles in order are as follows:

Patterns & Connections

You have just identified the rectus femoris and three "vastus" muscles. Collectively, these form the quadriceps muscles commonly referred to as "the quads" in athletes (unfortunately, all too often in regards to an injury). All four insert onto the patella (and adjacent ligaments) and act to extend the shank. Thus they are important in powerful lifting (pumping) of the leg when running. When an athlete "pulls a quad," one or more of these muscles or their tendons has been ripped. The injury may take weeks for a full recovery. Properly warming up and stretching reduces the likelihood of such an injury.

(7) Adductor longus

It is quite small and the most anterior of the four taking origin from the pubis.

(8) Adductor Femoris

This follows posterior to the preceding and is nearly three times longer. Its origin is near the origin of the gracilis.

(9) Semimembranosus

This is next posteriorly, and is the largest of the four. Its insertion is covered by the gracilis.

(10) Semitendinosus

It is the most posterior of the thigh muscles and usually must be separated from the preceding. It inserts on the tibia below the knee, where its tendon may be located and followed back to the belly of the muscle.

Patterns & Connections

Athletes often refer to the hamstring muscles or "hams," the collective term for posterior thigh muscles that include the biceps femoris, semimembranosus, and semitendinosus. They originate on the ischium and collectively insert on the knee (patella, femur, and tibia, respectively). The name "hamstring" comes from the terminology of butchers who used this strong tendon to hang hams during the smoking process.

→ **Muscles of the Shank** Some of the best aids in identification of shank muscles are the sites of insertion of tendons. Thus, clean away fascia from the shank so that the cordlike tendons to the heel, ankle, and digits can be seen. The insertions of the gracilis and biceps femoris should not be removed, but it will be necessary to pull them back as far as possible to reveal underlying muscles. Do this as needed as the dissection progresses.

(1-2) Tibialis Anterior and Extensor Digitorum Longus

Find the tibia. Its anterior *medial* side will be free of muscle, but on its anterior *lateral* side is a slab of muscle closely applied to and taking origin from it, the tibialis anterior (figure 6.21). It is very likely that you, at first, have also located the extensor digitorum longus. To be certain, check the sites of insertion of the long tendons emanating from each. The tibialis anterior inserts by a single tendon that passes to the medial side of the foot. The extensor digitorum longus forms a stout tendon that if followed into the foot divides into four tendons that pass along the top of the foot; each inserts on a digit (figure 6.20). (*Note:* Often a ligamentous band is preserved in the dissection that passes transversely across the tendons of both muscles near the end of the tibia. This is another retinaculum, and serves to prevent the tendons from bowing away from the ankle during foot flexion [figure 6.21].)

(3) Peroneus Muscles

Return to the lateral side of the shank (figure 6.20). These muscles are partly covered by the extensor digitorum longus just identified. Although it is impractical to individually separate each of the peroneus muscles, their presence is suggested by the several slender tendons that pass over the lateral side of the ankle to insert on the bottom of the foot.

(4) Gastrocnemius

This is the large muscle forming the posterior muscle mass of the shank (figure 6.20). It constitutes most of the so-called calf. By a strong tendon (Achilles) it inserts on the calcaneum. It is more-or-less divisible into two large portions, lateral and medial.

(5) Soleus

If orientation is changed, return again to the lateral side of the shank (figure 6.20). This muscle runs internal to the lateral gastrocnemius and is in contact with the peroneus muscles ventrally. The soleus tapers to a tendon that joins the tendon of the gastrocnemius.

Turn the cat onto its back to next examine muscles of the medial side of the shank. Clean away surface fascia to reveal slender muscles and tendons between the gastrocnemius and tibia.

(6) Plantaris

On the medial side of the shank, locate the medial portion of the gastrocnemius (figure 6.21). Cut the middle of the semitendinosus, fold it back, and clean away fascia. Pull the medial portion of the gastrocnemius away from the tibia, peer inside the opened space, and note that the plantaris is enclosed. Its identification may be

Form & Function

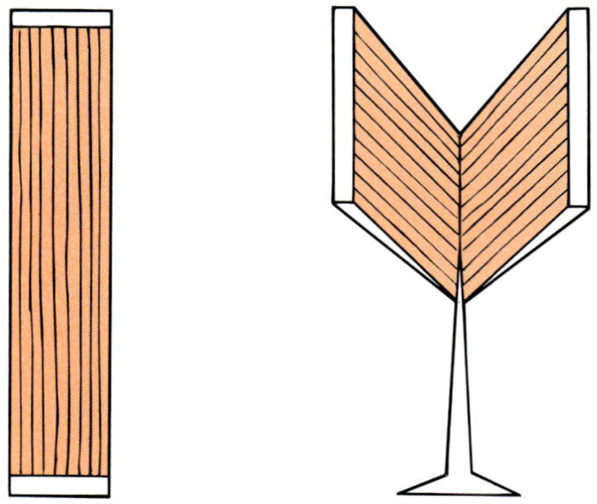

Parallel and pinnate muscles. The pinnate orientation permits the packing of more fibers in the same volume than the parallel arrangement does. Thus pinnate muscles are generally better suited for moving heavy loads short distances.

Compare the orientation of the muscle fibers in the gastrocnemius and sternohyoid muscles. Which muscle has a parallel and which has a pinnate fiber arrangement? Notice how the converging fibers of the gastrocnemius insert onto the calcaneum. This strong muscle is used to extend the foot and lift the body. The sternohyoid fibers are not arranged for strength. Instead, they contract to hold the hyoid in position or to move the hyoid caudally. From the study of these two muscles, we would predict that the forces needed to move the hyoid are much less than the forces needed to extend the foot.

confirmed by lifting it from the gastrocnemius and noting its shiny aponeurosis. Confirm its partial fusion with the lateral portion of the gastrocnemius.

(7–8) Flexor Digitorum Longus and Flexor Hallucis Longus

Find the tibialis anterior (previously located); next to this is the tibia; next comes the flexor digitorum longus, and finally the fleshy flexor hallucis longus, tightly fused to the exposed medial surface of the flexor digitorum longus muscle (figure 6.21). Follow the flexor hallucis longus distally to the heal where it becomes a tendon passing to the foot. Pull on this stout tendon to confirm its action on the foot.

(9) Tibialis Posterior

This muscle lies between the flexor digitorum longus and the tibia (figure 6.21). To find it, use a blunt probe to split the seemingly single tendon of the flexor digitorum longus. A long tendon is separated that can be found to pass upward (proximally) between the flexor digitorum longus and the tibia to the belly of tibialis posterior.

Form & Function

Limb tendons of a giraffe. Tendons distribute the forces of muscle contractions to sites distant from the muscle itself. The limb muscles of a giraffe are located close to the body, but tendons of these muscles extend outward along the leg bones and deliver their forces at the giraffe's hooves.

Most of the active muscle mass (the muscle fibers) that moves a giraffe's legs is located proximally, while tendons running out along the leg apply the force distally, at the hoof. By moving most of the muscle mass closer to the body, overall limb inertia is reduced. In cursorial animals, this allows the legs to move faster, because the limbs have overall less mass (inertia) to accelerate/decelerate back and forth, during limb oscillations. The muscles of a cat's legs show this same feature. Examine the overall shape of the flexor digitorum longus. Notice that the greatest bulk of the muscle is proximally positioned, just below the knee. The long tendon transmits the contractile force distally to insert on the ventral side of the digits.

7 Digestive Systems

Introduction

The business of turning a meal into usable energy begins with the digestive system. Food is broken down mechanically by chewing, churning, or grinding, and chemically by the action of digestive enzymes released into the digestive tract as food passes. The form of the digestive system can be viewed in terms of these functions. If teeth and grinding jaws are present, mechanical breakdown of food, or **digesta**, begins in the buccal cavity. In birds and alligators, an especially muscularized part of the stomach, the **gizzard,** holds ingested grit (figure 7.1). The **gizzard** mechanically grinds food swallowed whole into small pieces, increasing the surface area for the action of chemicals. Digestive enzymes are added directly from microscopic glands within the walls of the digestive tract. Larger glands, outside the tract, also pour their chemical products into the digestive tract through ducts.

One way to increase the effectiveness of digestion is to prolong the time food is exposed to mechanical and chemical breakdown. For example, **spiral valves** in many types of fish (figure 7.2) turn digesta along a lengthened course, **ceca** temporarily sequester food, and longer intestines extend the time within the tract.

As food is broken down into fundamental components—carbohydrates, fatty acids, protein—it is absorbed through the walls of the **alimentary canal**, picked up by the circulatory system, and delivered to active parts of the organism or into sites of storage. The circulatory system is covered in the next chapter, but here we should notice the general position of part of the digestive system within the blood vascular system of supply (**arteries**) and drainage (**veins**). The amount of blood supply is generally an indication of the level of activity of the particular part of the digestive system and the route by which breakdown of products of digestion are carried away.

Shark

Pleuroperitoneal Cavity

Viscera

→ Being careful that the deep blade of your scissors does not catch viscera beneath the belly wall, make an incision from cloaca forward through the pelvic girdle up to the pectoral girdle, keeping just to the left of the ventral midline. You may wish to make short lateral cuts at each end of this incision so that the flaps of body wall open like doors to reveal the viscera packed in the exposed <u>pleuroperitoneal cavity</u>. Posteriorly, this cavity communicates with the exterior through a pair of channels, the <u>abdominal pores</u>, that open at the bases of the pelvic fins on either side of the cloaca. Pass a blunt probe along the inside of the caudal end of the pleuroperitoneal cavity to confirm the position and course of an abdominal pore.

> **Form & Function**
>
> The liver of sharks contains large amounts of oil (in some sharks, the oil alone can constitute 16–24% of the body weight). Oils are less dense than water, so they make the shark more buoyant, saving the shark energy that would otherwise be spent keeping it from sinking.

Down the middle of the pleuroperitoneal cavity runs the gut, with an S-shaped curve halfway along its length (figure 7.3). To either side lie the long <u>left</u> and <u>right lobes of the liver</u>, and anteriorly where they meet, a short <u>median lobe</u> of the liver, thus three lobes in all. Along the right lateral wall of the median lobe is a collapsed, dark green <u>gallbladder</u>. From the gallbladder runs a cord, the <u>common bile duct</u>, accompanied by blood vessels. The bile duct runs to the gut, specifically the <u>duodenum</u> at one end of the S-shaped curve. Lying ventrally on the duodenum is the rounded <u>ventral lobe of the pancreas</u>. Slightly spread the bend of the duodenum to discover the long <u>dorsal lobe of the pancreas</u> passing posteriorly; the short connection between dorsal and ventral lobes is the <u>isthmus</u>. Formally, the duodenum begins just following a constriction in the gut, the <u>pyloric sphincter</u>, the last part of the stomach. The stomach region just before this sphincter is the <u>pylorus</u>, which leads to a great bend in the organ. This bend and the more anterior part of the stomach is the <u>body</u>. The right and left sides of the stomach are respectively termed the <u>lesser curvature</u> (right) and <u>greater curvature</u> (left). The wedge-shaped <u>spleen</u> rides posteriorly along the greater curvature of the stomach. No external demarcation marks the boundary between stomach and

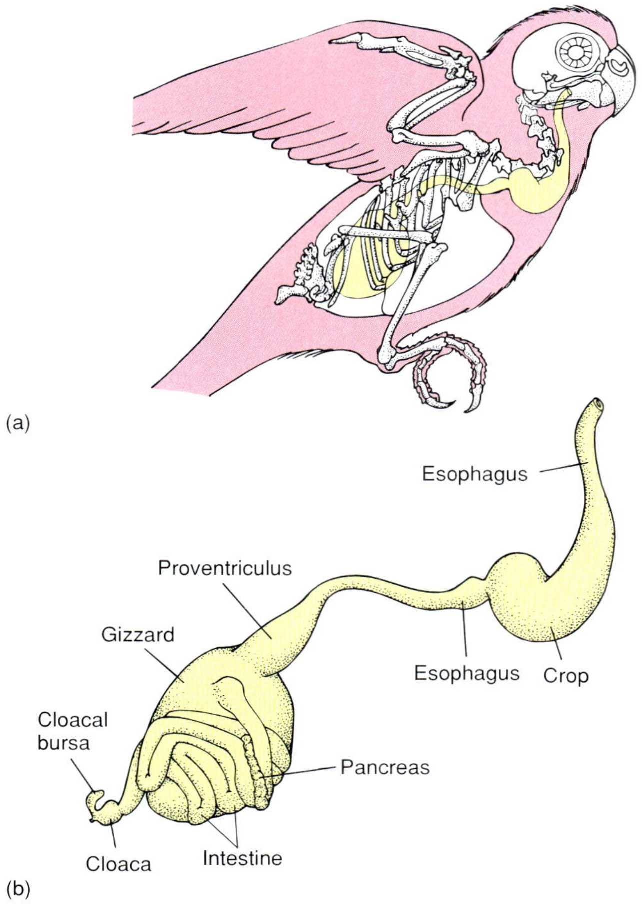

Figure 7.1
Alimentary canal of a parakeet. (a) Approximate position of the alimentary canal within the bird. (b) Alimentary canal enlarged.
After H.E. Evans, 1982, "Anatomy of the Budgerigar" in Diseases of cage and aviary birds, ed. Margaret L. Petrak, 2d ed., Lea & Febiger, Philadelphia.

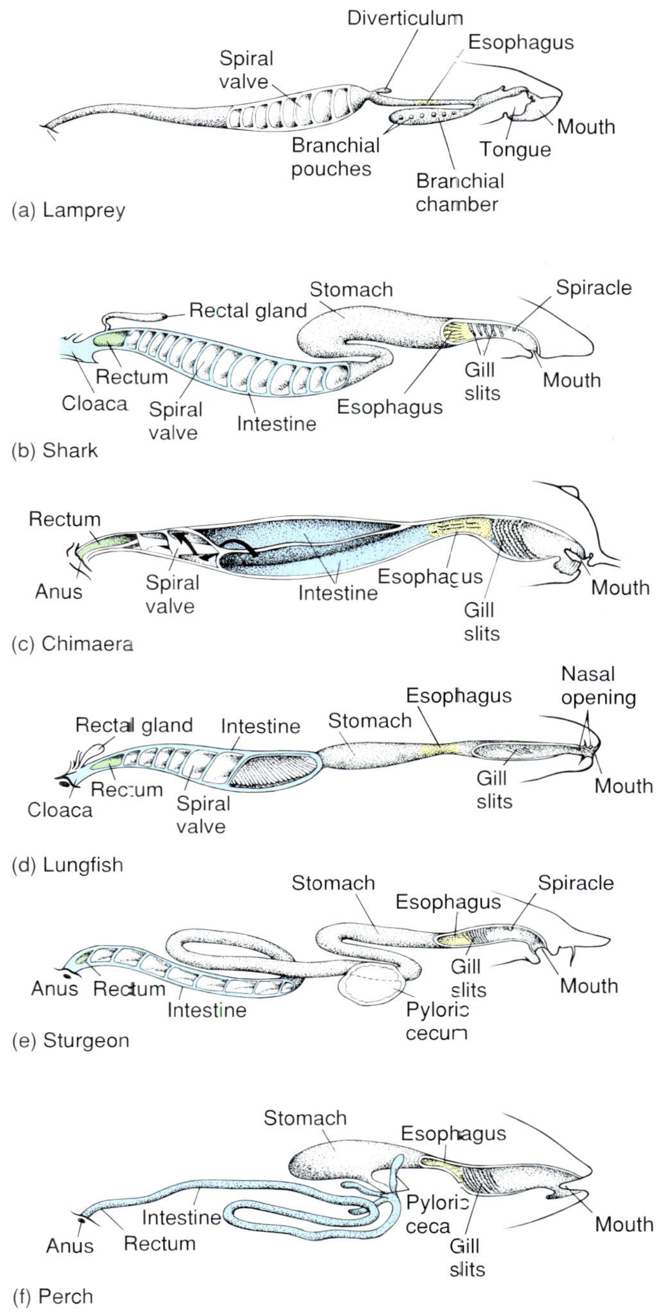

Figure 7.2
Digestive tracts of selected fishes. (a) Lamprey. (b) Shark. (c) Chimaera. (d) Lungfish. (e) Sturgeon. (f) Perch.
From Dean.

esophagus. However, by making a lengthwise incision in the body of the stomach forward to the still intact pectoral girdle, the distinctive internal structure permits certain identification of both as follows: the wall of the stomach forms longitudinal folds, rugae; the wall of the esophagus possesses stumpy, fingerlike projections, papillae.

Return now to the duodenum (figure 7.3). Passing posteriorly, the duodenum enlarges, continuing into the valvular intestine. The outer walls of the intestine are circularly creased by an internal subdivision, the spiral valve (figure 7.2b). Cut open, lengthwise, the wall of the intestine to identify the spiral valve, the defining feature of the valvular intestine. The duodenum lacks a spiral valve; thus its specific boundaries, between spiral valve and pyloric sphincter, can be settled.

Posterior to the valvular intestine, the gut narrows into the colon, which lacks a spiral valve and ends in the slitlike cloaca (figure 7.3). Dorsal to the colon, and joining it near the entrance to the cloaca, is the fusiform rectal gland.

Form & Function

A spiral valve within the lumen of the intestine (figure 7.2) lengthens the route traveled by food and thereby increases time for digestion. Simply making the alimentary canal longer also prolongs the digestive processes. Carnivores, with a nutrient dense food supply, typically have shorter intestines than those of herbivores, which must extract nutrients from resistant plant cells.

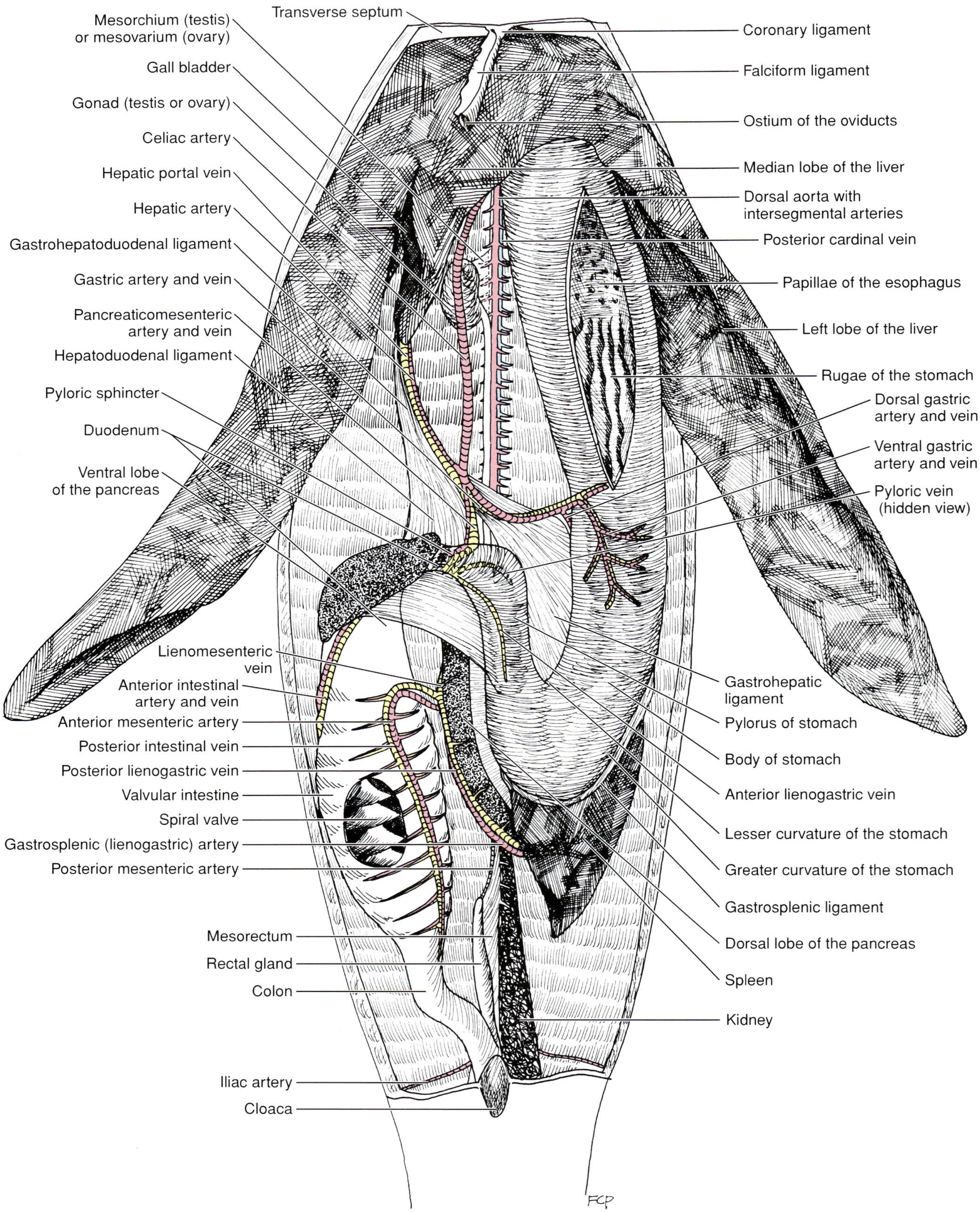

Figure 7.3
Ventral view of the pleuroperitoneal cavity of a shark.

Form & Function

Although it opens into the rectum, the rectal gland of sharks is not directly involved in digestion. Instead, it collects excess salts in the circulation and then excretes them via the digestive tract.

Anteriorly, above the liver lobes are the paired gonads (testes or ovaries). Check the sex of your shark (note pelvic fins), then examine gonads of a shark of the opposite sex.

Mesenteries

The entire pleuroperitoneal cavity is lined by the shiny **parietal peritoneum** (figure 7.4d). Poke a blunt probe slowly into the liver, noting that a thin surface membrane must break before the probe enters the liver tissue. This lining membrane, covering all viscera, is the **visceral peritoneum** (= serosa; figure 7.4d). Probe the parietal peritoneum similarly. The sheets of membranes suspending viscera and joining them at places with each other are the **mesenteries**. Vertebrate embryos commonly possess a broad **dorsal mesentery** joining the viscera to the dorsal body wall (figure 7.4d), and a similarly long **ventral mesentery** joining viscera to the ventral body wall (figure 7.4d). Only portions of the ventral mesentery persist into the adult stages of most vertebrates. However, broader regions of the dorsal mesentery occur frequently in adult vertebrates.

→ **Dorsal Mesentery** Pull the viscera to one side, noting that most mesenteries attach to the dorsal midline of the pleuroperitoneal cavity by a prominent, thin, and translucent sheet, the dorsal mesentery. Specific regions of the dorsal mesentery can be recognized. That part of the dorsal mesentery supporting the esophagus and stomach is the mesogaster (= greater omentum). The part to the anterior half of the valvular intestine is the mesentery[1] proper. Pull the spleen from its nestled position on the stomach to reveal the gastrosplenic ligament joining the two organs (figure 7.3). The short, isolated, expanse of dorsal mesentery suspending the rectal gland is the mesorectum.

→ **Ventral Mesentery** The ventral mesentery is represented at a few specific sites (figures 7.3 and 7.4d). Return to the bile duct near its departure from the gall bladder. The gastrohepatoduodenal ligament (= lesser omentum) runs along with blood vessels in a thin, ribbonlike mesentery extending to the liver. Carefully spread the organs to notice that this ligament actually branches upon its approach to the gut. One fork, the hepatoduodenal ligament, goes to the body of the duodenum. The other fork, the gastrohepatic ligament, goes to the body of the stomach along its lesser curvature.

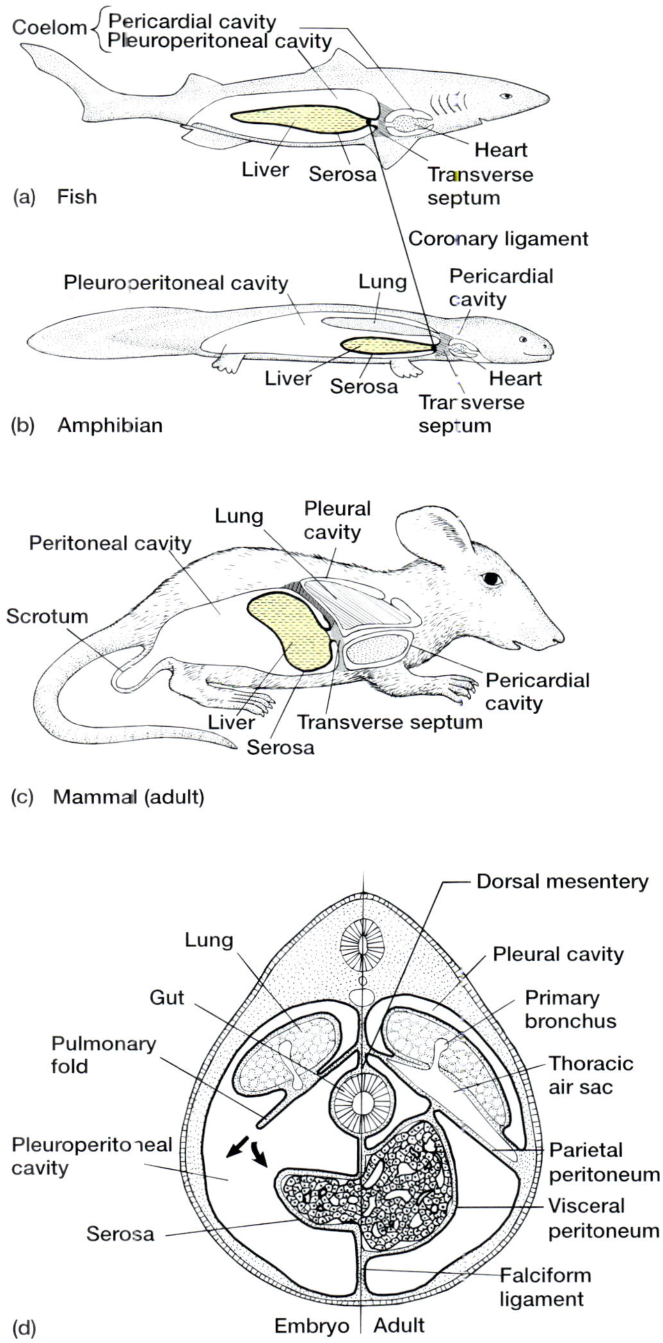

Figure 7.4
Body cavities. The coelom, arising in somitomeres, becomes divided by a fibrous transverse septum into pericardial and pleuroperitoneal cavities in fishes (a), amphibians (b), and mammals (c). In all three animals, the small coronary ligament attaches the cranial end of the liver to either the transverse septum or diaphragm. (d) Cross section of a bird illustrating the embryonic (left) and adult (right) cavities. In the embryo, the pulmonary fold grows obliquely to establish contact with the liver and the body wall. This confines the lung to its pleural cavity.

[1] An unfortunate term, since here it is used in a limited sense, while elsewhere mesentery, in a broad sense, means the whole class of such membranes supporting viscera.

Form & Function

Why do vertebrates have mesenteries? Mesenteries, like those just examined, keep organs in position and define spaces in which active organs can operate more freely. In addition, mesenteries restrict the amount of tensile force experienced by the many blood and lymphatic vessels that extend between the body wall and internal organs. Without mesenteries, shifting organs could wrap these vessels or pull them apart.

Return to the anterior blunt end of the liver. A transverse septum (figure 7.3) separates the pleuroperitoneal cavity from the pericardial cavity. The anterior end of the liver attaches to the septum by the coronary ligament, which, strictly speaking, is a derivative of and thus part of the transverse septum (figure 7.4a). This very short mesentery may be seen by gently pulling the liver caudally away from the transverse septum. The falciform (= suspensory) ligament is a short mesentery that attaches the cranial end of the liver to the ventral body wall (figures 7.3 and 7.4d). It appears as a ventral continuation of the coronary ligament. However, the falciform ligament may be partially torn by the initial midventral cut made when opening the pleuroperitoneal cavity. The common opening or ostium of the oviducts lies within the falciform ligament.

→ **Mesenteries of the Reproductive Tract** The mesorchium suspends the testis, the mesovarium the ovary, and in mature females, the mesotubarium holds the oviduct.

Pericardial Cavity

→ The pericardial cavity lies in front of the coracoid bar, dorsally over the common coracoarcual muscles (figure 7.4a). Carefully remove these muscles to expose the parietal pericardium, a white membrane lining the pericardial cavity. Cut through the parietal pericardium to reveal the heart within. The heart is covered with a very thin, visceral pericardium. The chambers of the heart will be identified with the circulatory system.

Necturus

Pleuroperitoneal Cavity

Viscera

→ To inject blood vessels with colored latex, the supply house preparing *Necturus* specimens must make a midventral incision to reach vessels in the pleuroperitoneal cavity (figure 7.4b). Usually the dark brown to black liver protrudes through this incision (figure 7.5). Extend this incision forward to the pectoral girdle and posteriorly through the pelvic girdle, cloacal gland, and into the cloaca. Take care not to damage viscera with the deep blade of your scissors. Keep slightly to the left of the ventral midline, noting now, and attempting to keep intact, the delicate falciform ligament from liver to ventral body wall. (Recall that this same ligament was likely damaged when examining the shark.)

Lift the posterior end of the liver to reveal the pea-sized and green gallbladder (figure 7.5). Gently push the liver to the right side of pleuroperitoneal cavity to better reveal the more dorsal tan stomach, and the dark, red spleen on the left lateral side of the stomach. The slender, saclike, and paired lungs are nestled along the dorsolateral sides of the pleuroperitoneal cavity.

Next trace the parts of the digestive tract. The cranial end of the stomach is nestled in the most anterior end of the pleuroperitoneal cavity (figure 7.5). It extends caudally and laterally until it abruptly narrows and ends in a constriction, the pyloric sphincter. Taking care not to damage mesenteries, cut open the stomach to expose the internal, lengthwise folds or rugae. Beyond the pyloric sphincter, the digestive tract continues as the long, but highly folded small intestine. Cut into the lumen of the small intestine at a convenient point to confirm the absence of a spiral valve, but presence of small, wavy folds, called plicae. The duodenum is the first short section of the small intestine. The light, pink pancreas is found cradled within the first bend of the duodenum. The short cystic duct from the gallbladder joins hepatic ducts draining the liver to form the common bile duct that finally empties into the duodenum. However, tracing these small ducts in a *Necturus* usually requires patience and a dissection microscope. Near the end of the small intestine the gut straightens and widens to form the large intestine. A single, shriveled urinary bladder is found along the ventral midline ventral to the large intestine. The urinary bladder and large intestine empty separately into the cloaca, identified previously during the muscle dissection.

The urinary kidney lies above the large intestine, tightly applied to the dorsal midline at the caudal end of the pleuroperitoneal cavity (figure 7.5). Solid elongate testes or much larger, granular ovaries are found lateral to the kidneys of sexually mature specimens. The kidneys, gonads, and their ducts constitute the urogenital system and will be examined in greater detail in the chapter on reproductive systems.

Mesenteries

→ As in the shark, the pleuroperitoneal cavity is lined by the parietal peritoneum, and the organs are covered by the visceral peritoneum. Specific regions of the dorsal mesentery are recognized (figure 7.5). The mesogaster supports the anterior half of the stomach. The gastrosplenic ligament joins the spleen to the stomach. On the left side of the body is a very narrow but long mesentery, the pulmonary ligament, running along most of the length of the lung, and attaching the lung to the mesogaster. However, the pulmonary ligament of the right

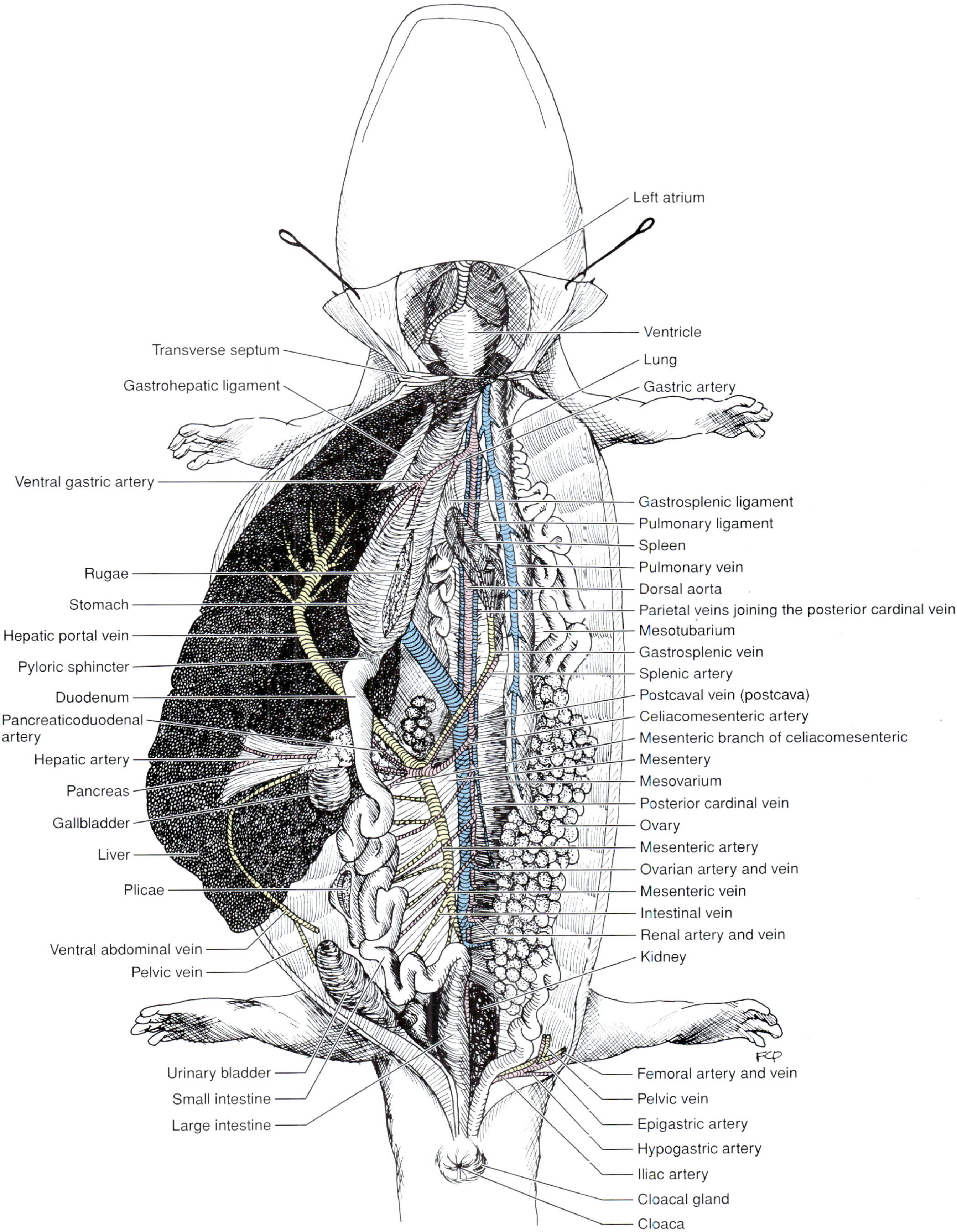

Figure 7.5
Ventral view of the pleuroperitoneal cavity of *Necturus*.

lung joins not the mesogaster, but a separate mesentery, the hepatocavopulmonary ligament. As the name suggests, this mesentery supports the liver (hence hepato), a large blood vessel called the postcaval vein (hence cavo), and receives attachment of the mesentery from the right lung (hence pulmonary). The small intestine is supported by the mesentery proper, which is continuous with the mesorectum supporting the large intestine.

As in the shark, only parts of the ventral mesentery are present (figure 7.5). The falciform ligament, identified earlier, runs from the liver to the ventral body wall. The gastrohepatic ligament joins the anterior end of the stomach to the dorsal surface of the liver. It is best seen from the left side by spreading the two organs. The hepatoduodenal ligament joins the duodenum and caudal part of the liver, but wraps the middle portion of the pancreas, thus making identification challenging. The urinary bladder attaches by the median ligament of the bladder to the ventral body wall. Finally, the mesorchium suspends the testis, a mesovarium suspends the ovary, and a mesotubarium suspends the oviduct.

As in the shark, a transverse septum forms the anterior end of the pleuroperitoneal cavity (figure 7.4b). The short coronary ligament, arising from the surface of the liver, attaches to this septum (figure 7.4b).

Pericardial Cavity

→ Pick away the triangular block of hypobranchial muscles between the procoracohumeralis muscles to expose the deep parietal pericardium, a membrane enclosing the pericardial cavity, which contains the heart. The very thin covering over the heart is the visceral pericardium.

The transverse septum divides the pericardial and pleuroperitoneal cavities. To see it better, extend forward the midventral incision used to open the pleuroperitoneal cavity. Spread the pectoral girdle at the caudal end of the pericardial cavity. Be careful not to split or damage the large blood vessels coming from the liver. Snip connective tissues to relieve tension and to better expose the septum. In *Necturus*, the transverse septum is a composite of peritoneum and muscle. Check again the attachment of the coronary ligament to the septum.

Cat

Salivary Glands

→ Five pairs of salivary glands are present in the cat (figure 7.6). The largest gland is the parotid gland that lies just beneath the skin at the base of the pinna of the ear. (If the parotid is absent, check the reflected skin to see if it pulled away during skinning.) Free skin up to the ear to expose this gland. Its duct, the white cordlike parotid duct, passes across the middle of the masseter muscle to empty into the mouth opposite the last premolar. Not to be confused with this duct are thin branches of the facial nerve that run superficially at the top and bottom of this muscle. The round submaxillary gland (= mandibular gland) touches the ventral corner of the parotid gland and is crossed by the posterior facial vein, the most dorsal of three tributaries to the external jugular vein running along the side of the neck. The long middle tributary is the transverse jugular vein, passing under the throat and joining external jugulars of both sides. The submaxillary duct leaves the anterior margin of its gland, courses forward, and disappears beneath the lower edge of the masseter. A small glandular wedge, the sublingual gland, rides along the submaxillary duct. The sublingual and submaxillary ducts run together to the mouth. Spend some time confirming the boundaries of these three glands, their ducts, and distinguishing them from lymph nodes and branches of nerves.

The molar gland lies on the external surface of the mandible between the corner of the mouth and anterior facial vein. Its several short ducts, impractical to see, empty on the inside of the cheek.

The infraorbital gland lies against the anterior edge of the zygomatic arch, at the bottom of the orbit, but will not be examined at present.

Oral Cavity

→ The buccal cavity, or mouth, extends from the lips (labia) to the end of the hard palate (figure 7.7a). The bony roof of the buccal cavity is roughened by washboardlike ridges, palatal rugae. The pharynx is the space continuing behind the level of the hard palate to the level of the opening into the larynx. Its roof is the soft palate. The part of the buccal cavity between lips and teeth is the vestibule. Below the nose, pull the lips away from the teeth to find a vertical, medial flap, the labial frenulum, attaching lips to the gingiva (= gums), and the median lingual frenulum attaching the anterior tongue to the floor of the buccal cavity. At the base of this frenulum, near its free edge, locate the small pimplelike swelling, the sublingual papilla, that in the cat bears the openings for the ducts from both sublingual and submaxillary glands.

Patterns & Connections

Many of these buccal structures can be found in your own mouth. Probe your tongue upward between your top front teeth and lips. There you should feel a thin membrane, the labial frenulum. A lingual frenulum attaches the anterior base of your tongue to the floor of your mouth; however, a mirror will likely be needed to see this connection. In humans as in the cat, the submandibular glands empty near the lingual frenulum. But unlike the cat, the sublingual glands in humans empty separately through many openings in the floor of the buccal cavity.

The roof of our buccal cavity is similarly composed of a hard and soft palate. However, the junction of these two regions is difficult to find without deep probing with a clean finger (don't try this in lab).

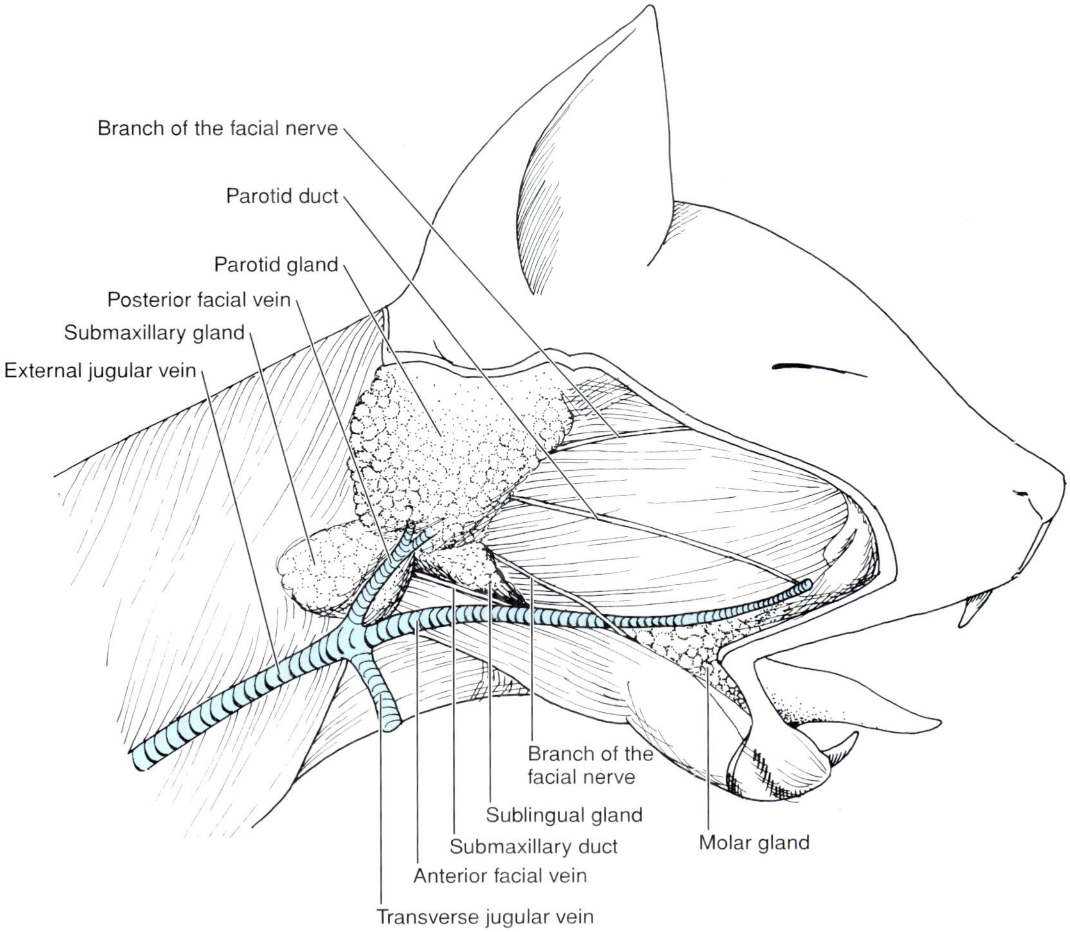

Figure 7.6
Ventral view of the salivary glands of a cat.

Preparing the Pharyngeal Region

→ To examine features near the larynx, you must prepare the right side of the oral cavity. This is done by rotating the right mandible, making a major cut along the right side, and laying open the right side of the face and neck. Proceed as follows: With bone or heavy scissors, cut through the mandibular symphysis to divide the lower jaw. Next cut muscles, skin, and the soft tissues lining the floor of the buccal cavity, close to their attachments to the right mandible. This should free the mandible so it can then be rotated laterally, out of the way. Finally, with one blade of the scissors inside, cut posteriorly, following the oral cavity. Continue this cut about one inch or until the tip of the trachea can be easily seen (figure 7.7a). Spread the margins of the cut to identify the following structures.

Buccal and Pharyngeal Regions

→ The tongue is covered by four types of papillae (figure 7.7b). The filiform papillae bear tiny spinelike projections. These papillae are small at the margins and enlarge toward the middle of the tongue's surface. Interspersed among them are the fungiform papillae that lack the spinelike projections. At the posterior sides of the tongue are the large, soft foliate papillae. Finally, along the posterior dorsum of the tongue are several oblique rows of columnar vallate papillae set in pits.

At the anterior end of the hard palate, behind the incisor teeth, lies a pair of small openings for the nasopalatine ducts (= incisive ducts) coming from the vomeronasal organ, small in cats.

Three regions of the pharynx are recognized (figure 7.7a). However, do not expect to see pronounced anatomical markers delineating the boundaries of each region of the pharynx. Instead, think of the pharynx as a passageway with three general regions. Roughly, the oropharynx lies ventral to the soft palate, the nasopharynx above the soft palate. These two regions join caudally to form the laryngopharynx, between the posterior edge of the soft palate and the end of the larynx. Just anterior to the free border of the soft palate, on the dorsolateral wall, is a pair of palatine tonsils, each slightly bulging from a pit, the tonsillar fossa. Paired folds in the lateral wall of the pharynx

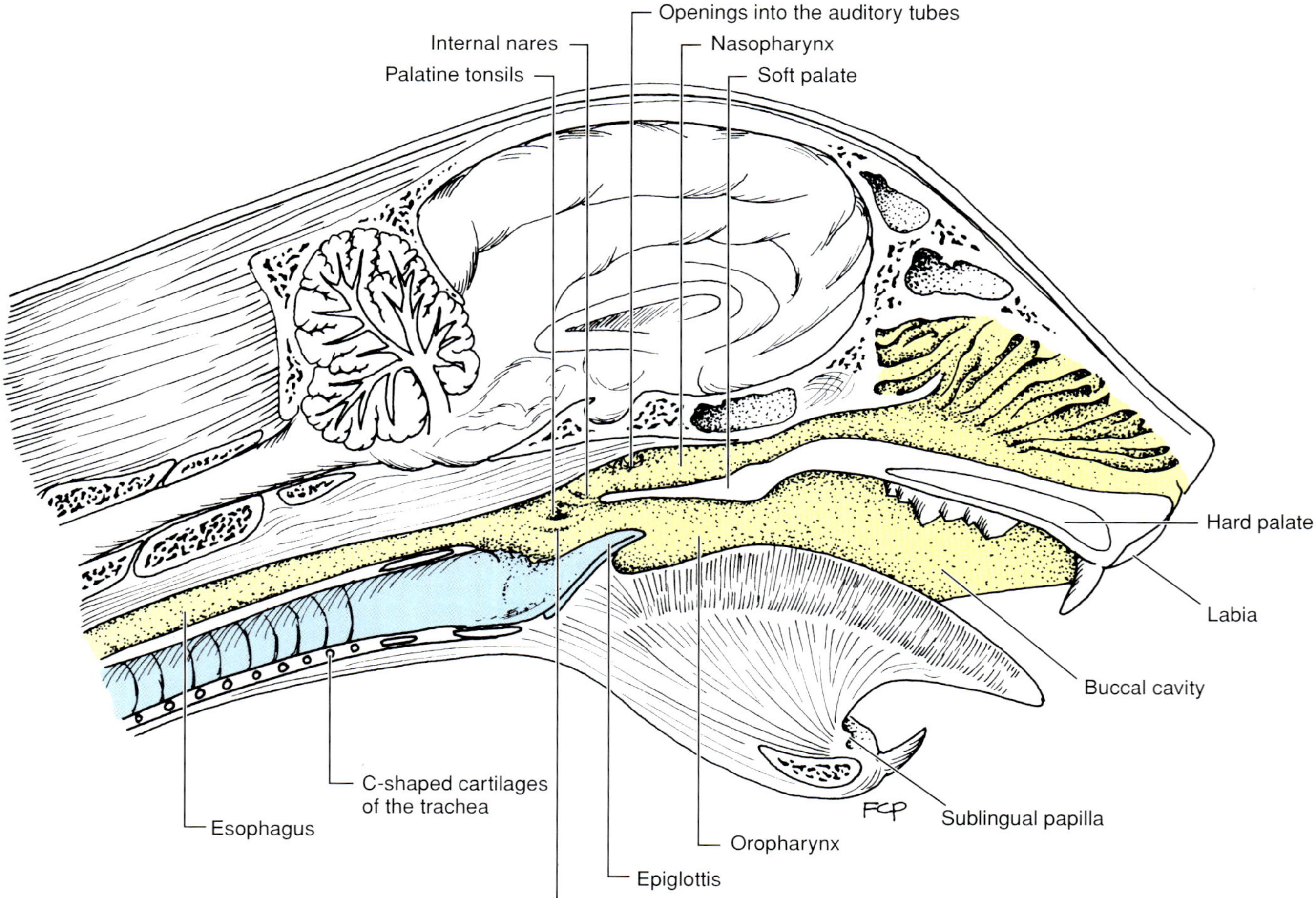

(a)

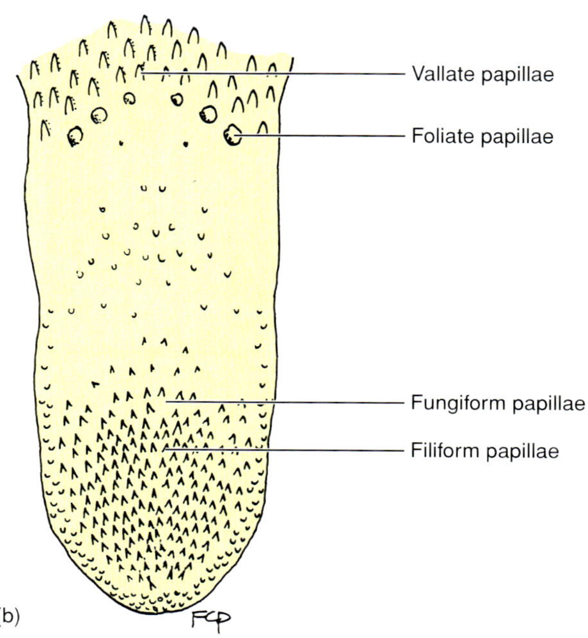

(b)

Figure 7.7
Oral and pharyngeal structures of a cat. (a) Midsagittal section. (b) Dorsal view of the tongue.

immediately bound the tonsillar fossa: the palatoglossal arch in front, the palatopharyngeal arch behind. These folds are important landmarks. The palatoglossal arch marks the official end of the oral cavity. The passage through this arch is the fauces.

Make a midsagittal cut through the soft palate to expose the nasopharynx, which opens anteriorly through two internal nares into the nasal cavity (figure 7.7a). On the dorsolateral walls, about halfway along the nasopharynx, are a pair of slitlike openings into the auditory (= Eustachian) tubes. Carefully probe this region until your blunt probe can be gently inserted into one of these slits.

Just past the rear of the tongue is the cranial end of the larynx bearing a prow-shaped projection the epiglottis (figures 7.7a and 7.8). Continuing beyond the larynx is the trachea, supported by regular C-shaped cartilages. Specifically, the laryngopharynx lies above the larynx and gives way caudally to the esophagus. The opening into the larynx is termed the glottis. Free the larynx from surrounding tissue to better expose the large cartilaginous elements composing it. The epiglottis rests on the large, incomplete thyroid cartilage (figure 7.8) that reaches around the larynx to touch a pair of small, dorsal arytenoid cartilages. At the base of the larynx lies the more caudal cricoid cartilage forming a complete but uneven ring. Cut the larynx along its dorsal midline to find the vocal cords inside, lateral folds running between arytenoid and thyroid cartilages. In some carnivores, such as the cat, false vocal cords are present as well, accessory folds between arytenoids and the base of the epiglottis.

The ventral glandular tissue just posterior to the cricoid cartilage and applied externally to the trachea is the thyroid gland. The paired lateral lobes are joined by a narrow band of tissue, the isthmus. Parathyroid glands are present, buried in the thyroid tissue, but are impractical to see.

Structures of the Thoracic Region

→ Palpate the sternum to discover its most posterior extent beneath the muscles of the chest. With tweezers, lift the end of the xiphisternum and, keeping to the cat's right of the midline, cut forward through the ribs, taking care not to catch the scissors on organs deep within the cavity. At the caudal ends of the thoracic cavity, cut laterally, just anterior to the diaphragm, which forms a thin muscular transverse wall. Muscle fibers contributing to the diaphragm take origin from ribs, sternum, and vertebrae and run inward toward the center of the muscle to insert on a round central tendon, about the size of a dime. The rib cage can thus be opened to expose three coelomic cavities without risking damage to the diaphragm. The pericardial cavity includes the heart and its enclosing membranes (figure 7.4c). The left and right pleural cavities contain the lungs on either side of the pericardial cavity (figure 7.4c). Note the thin partition, the mediastinal septum, between the heart and ventral medial wall. It is formed from the medial walls of left and right pleural cavities.

The thoracic cavity can now be opened further to examine the internal organs (figure 7.8). The inside walls of the pleural cavities are lined by smooth parietal pleura. The surface of the soft, spongy lungs are covered by the thin visceral pleura. The left lung consists of the anterior, middle and posterior lobes. The right lung similarly consists of these lobes (anterior, middle, posterior) plus a fourth, accessory lobe. To find it, lift the apex (blunt posterior end) of the heart to expose a small membranous compartment beneath. Cut through these membranes to discover the accessory lobe within. Finally, the mediastinum is a single and central cavity that encloses the heart, blood vessels, esophagus, and trachea. It is continuous ventrally with the mediastinal septum, which connects to the ventral body wall.

Externally the heart is wrapped in parietal pericardium (= pericardial sac). Cut through this sac. The surface of the heart is covered by the visceral pericardium. The small space between parietal and visceral pericardia is, formally, the pericardial cavity (figure 7.4c).

Cranial to the heart, between anterior lobes of the lungs, lies a mass of glandular tissue, the thymus (figure 7.8), enmeshed in blood vessels. Large in young individuals, it undergoes involution (reduction in size) as an animal grows older. Clearly expose and separate the thymus from the large blood vessels, taking special care not to damage these vessels, which will be identified later.

To expose structures in the dorsal pleural cavity, pull the right lung and heart to the left side of the thoracic cavity. The pulmonary ligament (possibly indistinct), part of the pleura, attaches the lung along most of its length to the dorsal body wall. Medially, a tangle of arteries, veins, and bronchi, the root of the lung or radix, enter the middle of each lung. Spread the heart and right anterior lobe of the lung to reach the large blood vessel, injected red or pink, running directly forward beneath the thymus. Parallel, but dorsal to it is the tubular trachea ribbed with cartilage rings, which feel rough when stroked with the blunt probe (figure 7.8). Dorsal to it is the collapsible and smooth-walled esophagus (figure 7.8).

Turn to the left side of the cat and lift the left lobes of the lung to expose structures along the dorsal midline. A large blood vessel (dorsal aorta) passes along the dorsal wall. Next to the dorsal aorta you can again find the esophagus, which passes dorsal to the trachea and should be followed caudally as it passes through the diaphragm and into the abdominal cavity.

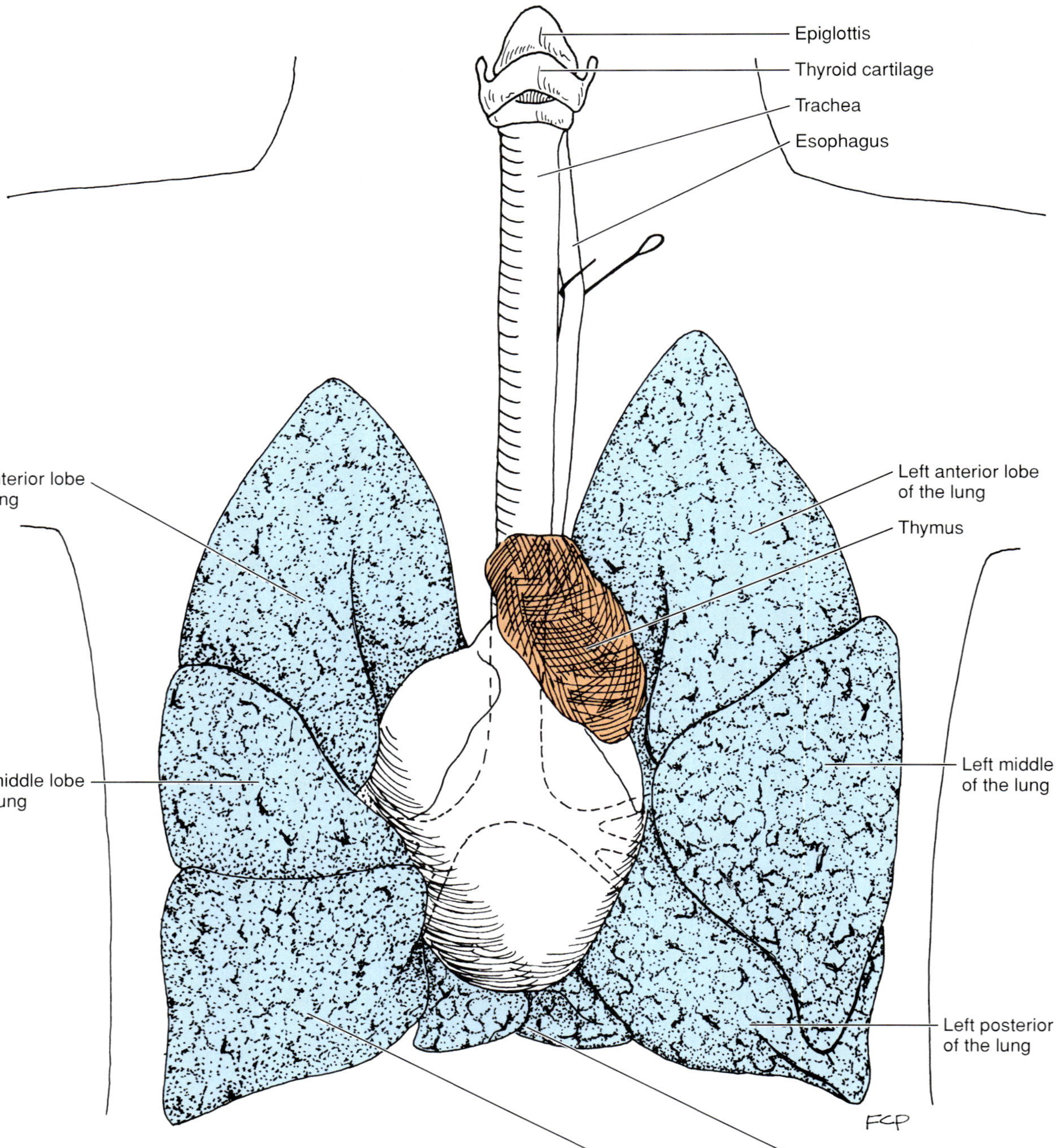

Figure 7.8
Ventral view of the thoracic cavity of a cat.

Abdomen

→ Extend the midventral incision caudally along the abdomen back to the inguinal (groin) region. Be sure to keep to the right of the midline. The cavity so opened is the peritoneal cavity (= abdominal cavity; figures 7.4c and 7.9). Make a pair of lateral cuts along the caudal margin of the diaphragm, such that the body walls may each be folded back. An additional pair of lateral cuts may be made along the posterior margin of the abdomen.

As in the shark and *Necturus*, the parietal peritoneum lines the inner walls of the abdominal cavity, and the visceral peritoneum covers the surfaces of organs. The large spleen lies unpaired, along the left abdominal wall (figure 7.9). The liver fits into the dome-shaped

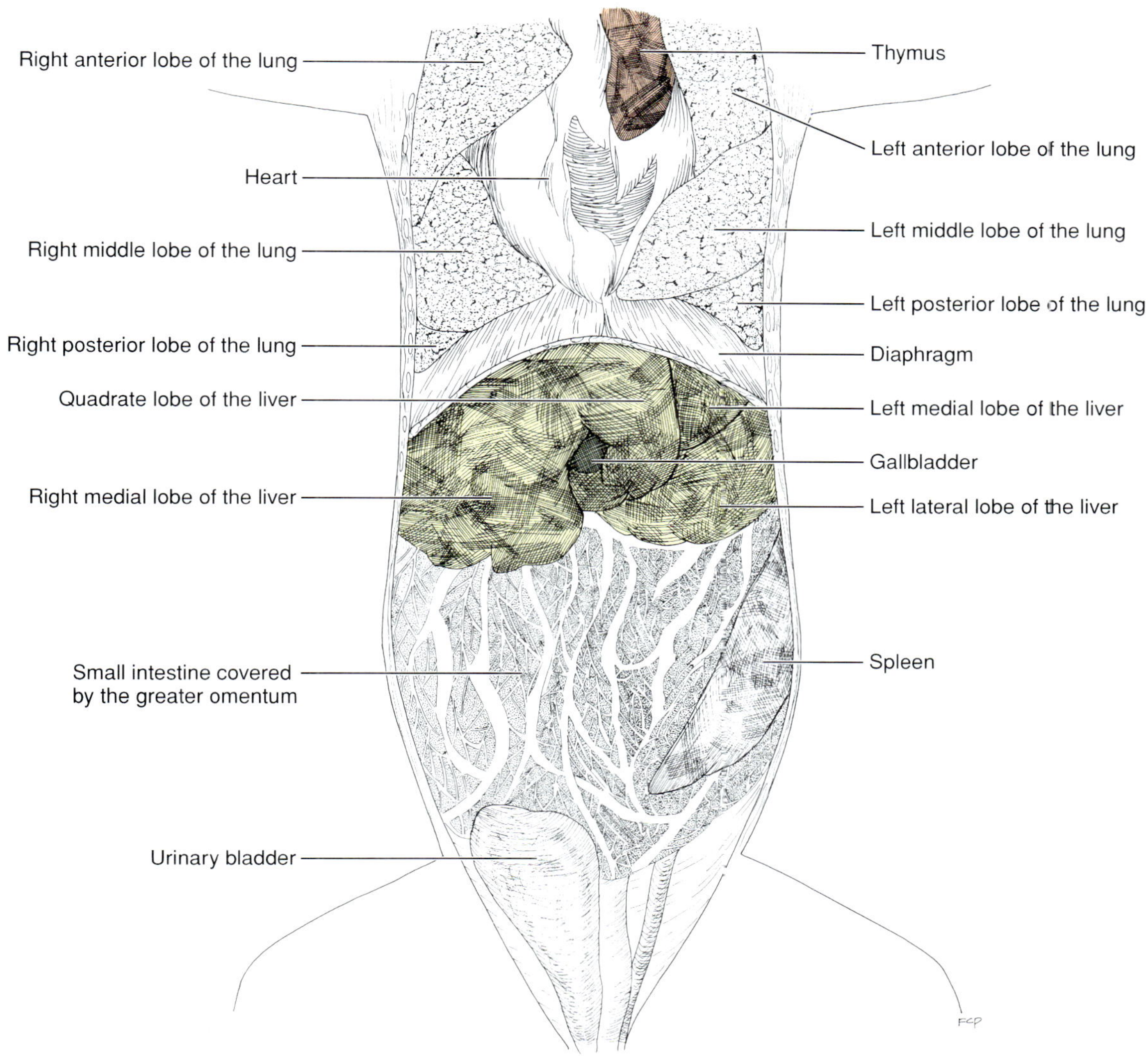

Figure 7.9
Abdominal structures in a cat, viewed ventrally upon initial inspection.

diaphragm. Spread the liver and diaphragm to find the falciform ligament connected between them. In some specimens the ventral free edge of this ligament is thickened. This is the round ligament, an adult remnant of the embryonic umbilical veins that in the fetus carried blood from the placenta. The coronary ligament is connected to the dorsal margin of the falciform ligament. The coronary ligament connects the liver to the central tendon, identified previously as part of the diaphragm.

The falciform ligament joins the liver in a cleft that divides the liver into left and right halves. In turn, each half is divided into medial and lateral lobes (figure 7.10). Of the two, the left lateral lobe is larger than the left medial lobe. On the right the opposite is true; the right medial lobe is larger and covers the smaller right lateral lobe. Additionally, the right lateral lobe splits into two attached parts: an anterior part and a posterior part, which rests on the cranial end of the right kidney. The saclike gallbladder lies within a pocket of the right half of the liver between its medial lobe and the smaller quadrate lobe that occupies a midventral position. Tilt the left and right halves of the liver forward, spreading them from the stomach and coiled intestine. Finally, from the inner bend of the stomach will emerge the small, pointed caudate lobe of the liver, covered by a thin membrane, part of the lesser omentum.

This is also a good time to rediscover the esophagus emerging from the diaphragm. Follow it to where it ends in the expanded stomach, with several regions identifiable in gross structure (figure 7.10). The inner, right side of the stomach is its lesser curvature, the left side the greater curvature. The large, domed cranial end is the fundus, the large part continuing behind this the body or corpus, and the narrowed end the pyloric region that terminates in a

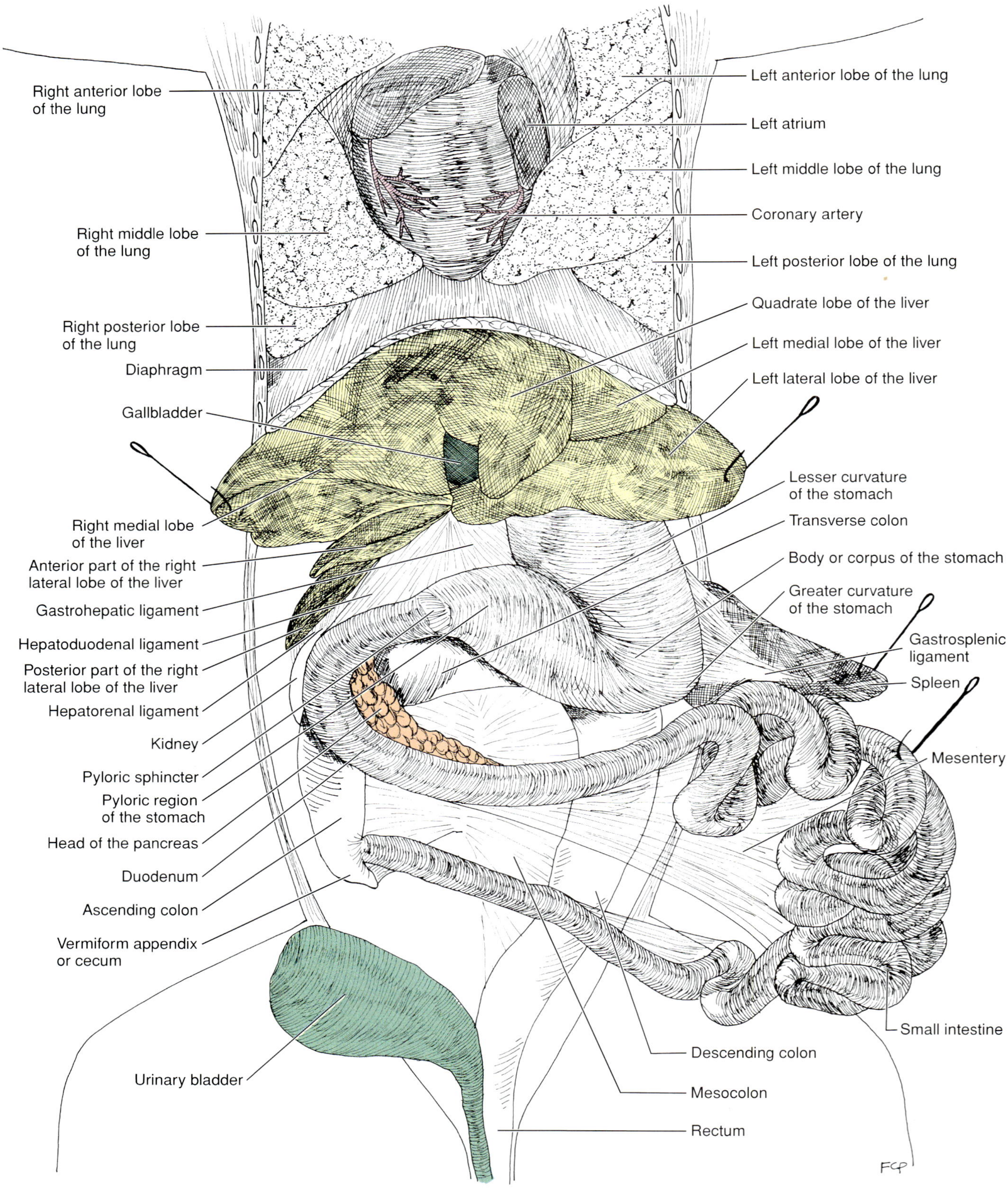

Figure 7.10
Ventral view of the deep abdominal structures of a cat.

constriction, the pyloric sphincter (= pylorus). The point in the stomach where contents from the esophagus first enter is a very short (few millimeters) cardia, but no gross features exist to approximate boundaries. Instead, the histological character of digestive glands in its walls serve to define the cardia. By cutting into the stomach, longitudinal folds, or rugae, can be identified.

Beyond the pyloric sphincter continues the small intestine folded upon itself (figure 7.10). It, like much of the rest of the gut, is draped in mesenteries streaked with fat, often considerable amounts in overfed cats. Sections of this mesentery will be identified shortly, but first delineate the portions of the gut and associated structures. With your fingers, gently lift this mesentery and examine the first segment of the intestine, the duodenum, extending from the pyloric sphincter to about the second caudal bend of the small intestine. In the first bend of the small intestine, along the side of the duodenum, lies the head of the pancreas. The tail of the pancreas extends away from the intestinal wall toward the spleen. The small intestine beyond the duodenum is the jejunum and then ileum, but no distinct gross features mark their boundaries.

Patterns & Connections

The duodenum is positioned very closely to the gallbladder and pancreas in the shark, *Necturus*, and cat. In each of these animals, bile produced in the liver and secretions from the pancreas drain into the duodenum. Thus, the close proximity of these organs reflects their interrelated functions.

The ileum ends at its juncture with an expanded region of the gut, the colon or large intestine (figure 7.10). At this juncture, the end of the colon forms a blunt, small, fingerlike projection, the vermiform appendix or cecum. Three sections of colon are recognized: the ascending colon, the short first part along the right side and arching cranially; the transverse colon crossing to the left; and the descending colon, the final and longest section continuing caudally. The descending colon ends in a short, straight portion, the rectum, bounded by the pelvic girdle. Parts of the urogenital system also converge upon or reside at this point in the abdomen. Thus do not cut or attempt to expose any more of the gut at this point in the dissection. The urinary bladder appears as a muscular bag on the ventral side of the most caudal end of the descending colon.

Make small, short slits in the colon and intestine, wash out debris, and then compare the textures of the internal walls. The wall of the intestine is velvety due to the presence of a mat of tiny projections, called villi. These are absent from the colon and stomach.

Mesenteries

→ The mesentery from liver to the duodenum and lesser curvature of the stomach is the lesser omentum, divisible into two parts: the hepatoduodenal ligament from liver to duodenum, and the gastrohepatic ligament from liver to lesser curvature of the stomach (figure 7.10). The mesentery attached to the opposite side of the duodenum and greater curvature of the stomach is the greater omentum, a derivative of part of the mesogaster (figure 7.9). It drapes over the small intestine. Because the greater omentum is modified in most mammals into a saclike pouch, it is often called an omental bursa. The stretch of greater omentum between the spleen and stomach is the gastrosplenic ligament (= gastrolienic ligament).

Find again the posterior part of the right lateral lobe of the liver under the small intestine (figure 7.10). From this lobe the hepatorenal ligament extends to parietal peritoneum near the right kidney. Note that this large liver lobe extends forward into a small pocket. Ventral to the lobe at this cranial point is the free edge of part of the hepatoduodenal ligament supporting pancreas and blood vessels. This free edge defines the epiploic foramen, the entrance into the omental bursa. Confirm this by returning to the omental bursa and penetrating it with a blunt probe, projecting its point dorsally to exit from the epiploic foramen.

The small gastrocolic ligament passes from the greater omentum dorsal to the spleen to the mesentery suspending the intestines. The dorsal mesentery supporting the gut is subdivided into three regions: the mesoduodenum supports the duodenum, the mesocolon the large intestine, and the mesentery proper the jejunum and ileum (figure 7.10). Note the numerous, smooth glandular lymph nodes on the mesentery. These are found elsewhere throughout the body as well.

The urinary bladder is supported by the single median ligament extending to the midventral abdominal wall, and by paired lateral ligaments at its base.

Return to the region of the duodenum to trace the ducts entering this region of the gut. Care and extra patience will now be helpful. The gallbladder narrows into a duct, the cystic duct, which joins ducts coming from various lobes of the liver, hepatic ducts. Cystic and hepatic ducts merge into a single common bile duct, which passes to the intestine in the hepatoduodenal ligament. When this duct reaches the duodenum, it slightly enlarges, forming the hepatopancreatic ampulla (= ampulla of Vater). Joining the common bile duct in this ampulla is the main pancreatic duct. Pick away tissue to follow this duct a short distance into the pancreas. A second, accessory pancreatic duct enters the duodenum independently about 1–2 cm further along (caudally), but is not easy to locate.

8 Circulatory and Respiratory Systems

Introduction

Respiratory System

The primary function of a respiratory system is to serve as the site of gaseous exchange between an animal and its environment. But because gases diffuse slowly through vertebrate body tissues, vertebrate circulatory and respiratory systems have coevolved to help distribute gases throughout the body. In addition to their primary roles in aiding the diffusion of gases, respiratory systems serve as sites of excretion, promote osmoregulation (gills), and help detect scents (chemoreception). Here we focus on the functions related to gas exchange within the circulatory system. These two systems are thus addressed together in this chapter in respect of their interrelated functions.

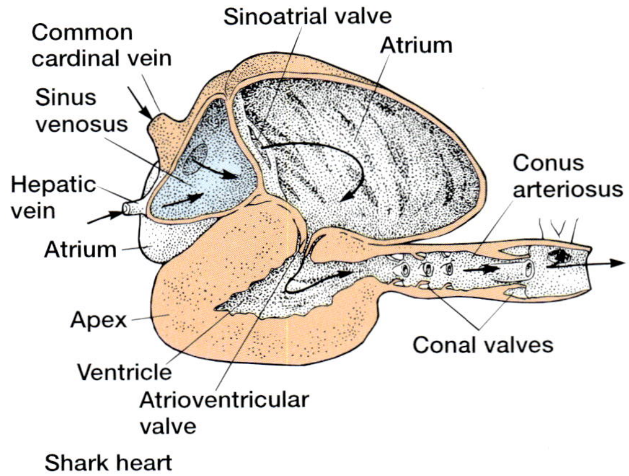

Figure 8.1
Midsagittal section of a shark heart. Cranial is to the right.
After Lawson.

Circulatory System

Blood vessels, especially veins, can show considerable variation in branching pattern within a species. The descriptions that follow are based upon the most common arrangements. The best criterion to identify a blood vessel is the specific area it supplies (arteries) or drains (veins). Thus, when working through this laboratory manual, be sure to note the region served as well as the general shape of each blood vessel.

Peripheral arteries and veins tend to pass together in the walls of an organ. Thus you will find that veins often bear the same name as their companion arteries (e.g. ventral gastric artery/ventral gastric vein). Besides helping with naming, this tip should also help you locate veins that may be poorly injected.

Shark

Pericardial Cavity

→ When examining the peritoneum, the pericardial cavity was reached through hypobranchial muscles. Return to the pericardial cavity seen through this previous entry and identify the four chambers of the heart: in order, sinus venosus, atrium, ventricle, and conus arteriosus (figure 8.1). Cut the transverse septum and clear ventral musculature as needed to reveal the region of the common cardinal and heart (figure 8.2). The triangular ventricle lies ventrally and is continuous with a large tubular chamber, the conus arteriosus, that passes forward through the wall of the pericardial cavity. Dorsal to the ventricle lies the atrium. Pull the apex (blunt posterior corner) of the ventricle forward to observe the sinus venosus where it empties into the atrium—the sinus venosus lies in the transverse septum.

Note the small artery on the wall of the conus arteriosus. This is the coronary artery supplying the heart and should be disrupted no more than necessary to expose the following vessels.

Respiratory System and Branchial Arteries

Afferent Branchial Arteries

→ Where the conus arteriosus emerges from the pericardial cavity, it becomes the ventral aorta and immediately gives off lateral branches, the paired afferent branchial arteries (figure 8.3). Note that the afferent branchial arteries course between distinct bands of muscles, the coracobranchials, not identified previously. Part these muscles as you trace the afferent branchial arteries. The first (most caudal) afferent branchial immediately divides into the 4th and 5th afferent branchial arteries. Make a single cut into the overlying musculature along the ventral midline to follow the ventral aorta forward. Keep in mind that the ventral aorta is not injected and so is more fragile than similarly sized vessels so far dissected.

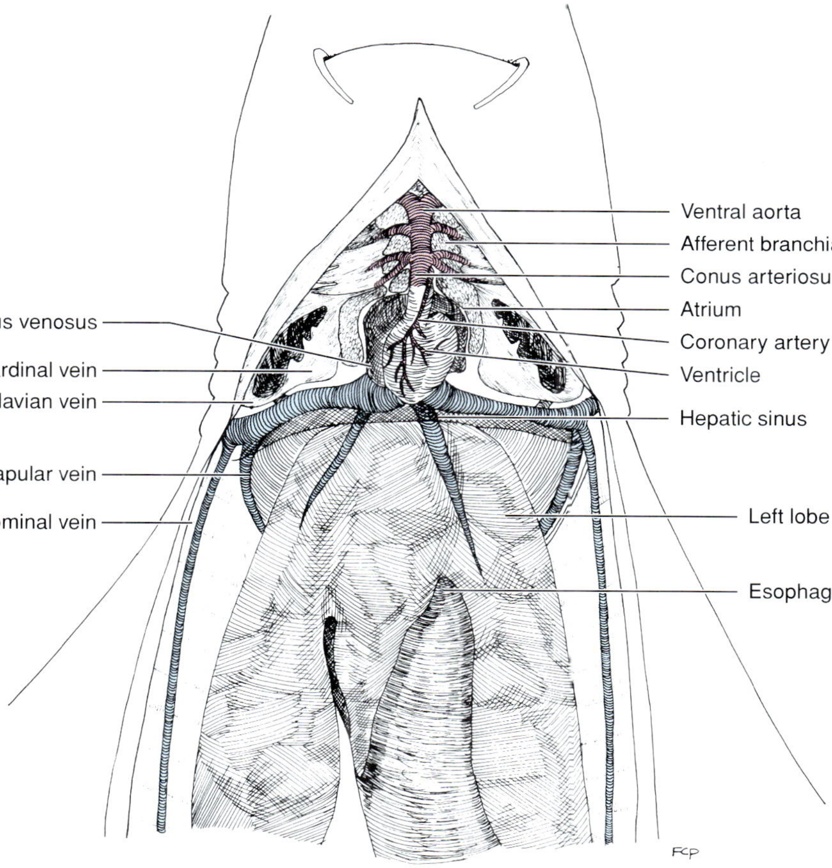

Figure 8.2
Ventral view of the shark heart and associated blood vessels.

Shortly, the 3rd afferent branchial artery exits, and finally the ventral aorta terminates in a bifurcation, which should be followed laterally on the left side. Eventually this fork itself divides, forming the 1st and 2nd afferent branchial arteries to supply the first two gills.

Gill Structure

→ To expose the respiratory system and arteries associated with the gills, you will need to make a major cut along the right side of the head and across the throat, to swing open the lower jaw like a door (in this analogy, the hinge of the door will be the gill region along the left side of the head). Begin at the right corner of the mouth and extend the cut longitudinally and caudally through the middle of the right gills. Then turn the scissors medially and ventrally to cut across the throat just anterior to the coracoid bar. Deviate slightly at the heart to pass through the conus arteriosus, then continue to the base of the opposite gills on the left side of the head. Force open the jaws and observe the transected gills on the right side.

Each gill is supported at its base by an epibranchial cartilage and an adductor muscle just medial to it (figure 8.3). Just lateral to the gill arch cartilage are several small arteries. On the anterior side of the gill arch cartilage reside fingerlike processes, gill rakers, guarding the internal gill slit, the opening between adjacent gill bars. Spread two gills to reveal the gill pouch, a pocket that opens laterally through the external gill slit. Note the soft, fan-shaped gill lamellae extending laterally from the epibranchial cartilage. Trace the path of water as it is drawn into the mouth, forced through the internal gill slit, past the lamellae, and out the external gill slit.

Each gill is considered a complete gill, holobranch, if gill lamellae cover both sides. The first gill is the exception, for lamellae lie only on its caudal side. Thus the first gill is a hemibranch or half gill. On the holobranch gills, note that gill lamellae are separated by, and supported on, an interbranchial septum that continues laterally to form the division between external gill slits.

> **Form & Function**
>
> The many fingerlike processes of gill lamellae increase the surface area of this respiratory surface. More surface area favors gaseous exchange between the capillaries in the gill lamellae and the water ventilating the gills.

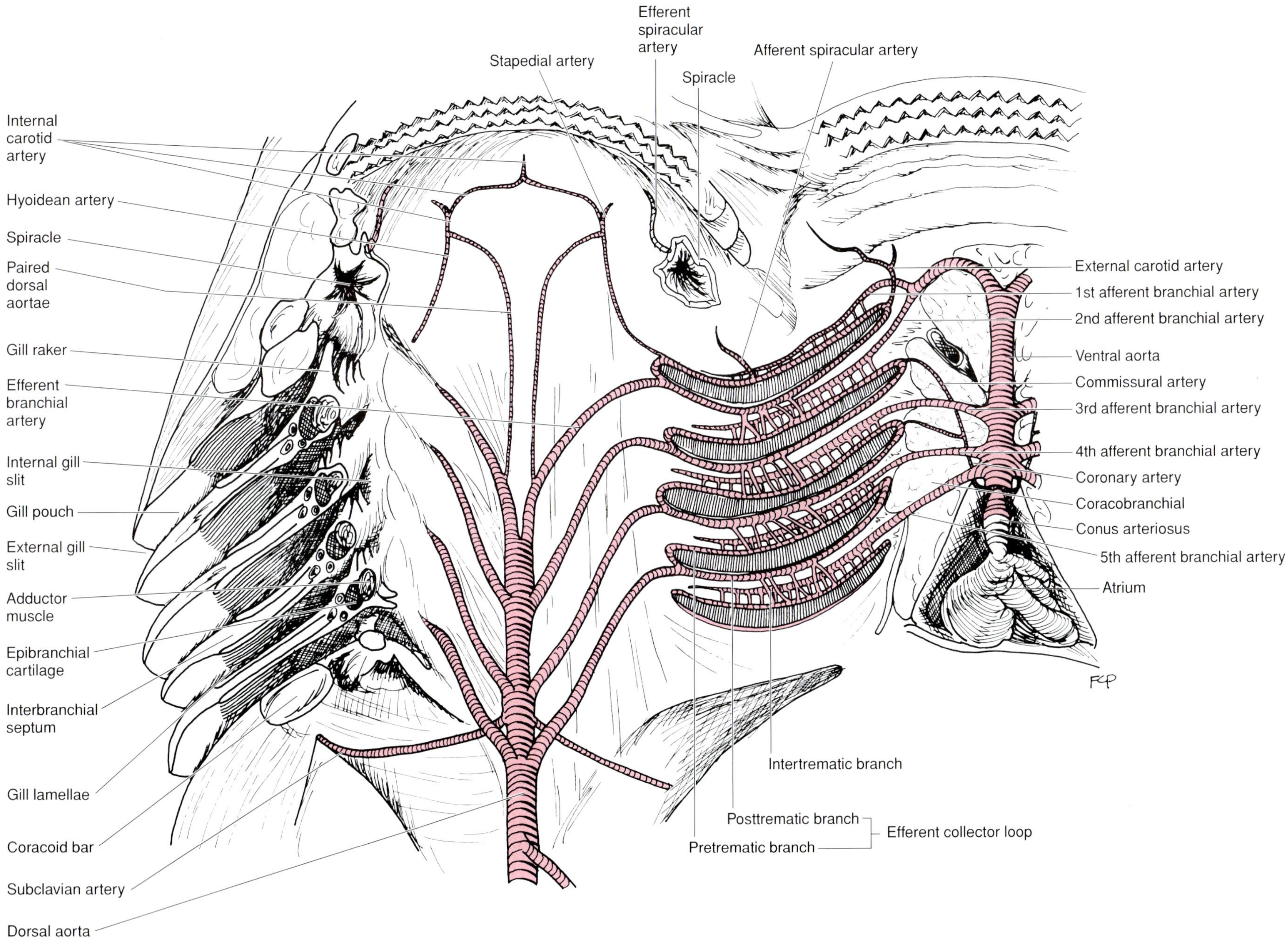

Figure 8.3
Ventral view of the branchial structure and vascular supply in a shark. The floor of the pharynx has been cut free caudal to the heart and along the right pharyngeal side.

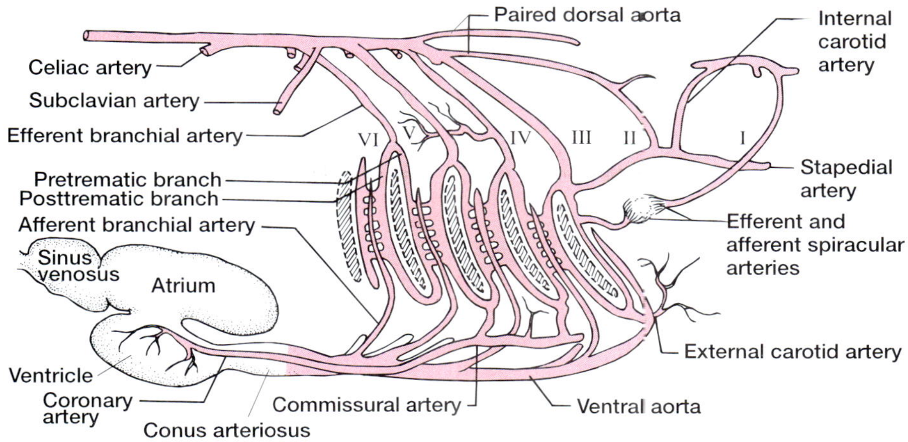

Figure 8.4
Aortic arches and related blood vessels in a shark. Roman numerals indicate aortic arches.
After George C. Kent and Larry Miller, Comparative Anatomy of the Vertebrates, 8th edition, 1997. McGraw-Hill Company, Inc., Dubuque, Iowa.

Efferent Branchial Arteries

→ Now expose the efferent arteries. With tweezers, pull away the mucous membrane on the roof of the mouth taking care not to also pick up underlying blood vessels. If you grasp only the loose mucous membrane, you can be fairly certain to avoid damaging the arteries. Four pairs of large blood vessels, the efferent branchial arteries, are so exposed (figure 8.3). These carry blood, oxygenated in the gills, posteriorly. Eventually they merge to form the single dorsal aorta. Note that the thin paired dorsal aortae arise from the anterior pair of efferent branchial arteries and pass rostrally on each side of the dorsal midline. Clear away connective tissue over all efferent branchial arteries, then follow them all to the cut right side.

Tracing the Blood Flow Past the Gills

→ The efferent branchial arteries disappear dorsal to gill cartilages. On the right side of the head, cut away these protruding medial ends of the gill cartilages, bit by bit, until the efferent arteries can be visually traced back to their respective gill lamellae. Each efferent branchial artery begins in the dorsomedial corner of the gill by fusion of two vessels: the thin pretrematic branch from gill lamellae on the rostral side of a gill pouch, and the slightly thicker posttrematic branch from the lamellae on the caudal wall of the same gill pouch (figure 8.3). Spend time carefully exposing these branches.

Now turn your attention to the intact left side of the head. Make a neat and informative dissection of afferent and efferent vessels to these gills. Pick one of the middle efferent branchial arteries (on the left, intact side) and follow it to the gill, clearing away connective tissue and cartilage as you did for the right side. As you reach and clear the pretrematic and posttrematic branches, notice that, in fact, they form a complete arterial loop around the gill cleft, the efferent collector loop. Pretrematic and posttrematic branches riding on the same interbranchial septum occasionally interconnect directly by a few short intertrematic branches. As you reach the ventromedial corner of the gill, you should find the corresponding uninjected afferent branchial artery from the ventral aorta. Complete your dissection of the branchial region so that the relationship of one corresponding afferent and efferent branchial artery and collecting loop is clearly revealed.

Associated Branchial Arteries

→ **First Collector Loop** The first collector loop is the source of several arterial branches (figures 8.3 and 8.4). From its dorsomedial corner comes the hyoidean artery that passes along the roof of the mouth medial to the spiracle. Each hyoidean artery is joined by the rostral end of a branch of the paired dorsal aorta identified previously. Anterior to this juncture with the paired dorsal aorta, the hyoidean artery becomes the internal carotid artery. The internal carotid continues rostrally and turns medially where it gives off a small lateral branch, the stapedial artery. (Only the base of the stapedial is typically visible.) The paired internal carotids next join at the dorsal midline to form a single internal carotid vessel that enters the chondrocranium through the carotid foramen.

Patterns & Connections

Spelling counts! You may have heard this before, but in some of the circulatory and respiratory terms above, a single letter can make a big difference. Compare the terms "branchial" and "brachial." The word "branchial" addresses structure associated with gills. But remove the "n," and the reference is instead to the chest or arms. Within the circulatory system, you have seen the terms "afferent" and "efferent." The "af" in afferent indicates an association with or toward something, while the substitution of an "ef" for the "af" suggests just the opposite, a direction away from something.

The afferent spiracular artery originates from the middle of the pretrematic branch of the first collector loop and passes to capillary beds in the spiracle (figures 8.3 and 8.4). The efferent spiracular artery continues rostrally from the spiracle (eventually uniting with the internal carotid within the cranium), but is impractical to see here as it is rarely sufficiently injected.

Finally, from the ventromedial corner of the first collector loop exits the short external carotid artery supplying the lower jaw region (figure 8.4).

Second Collector Loop The commissural (= hypobranchial) artery arises from the ventral corner of the second collector loop. Often it is joined by additional commissural arteries that arise similarly from the third and fourth collector loops. The commissural arteries enter the pericardial cavity to become the coronary arteries, identified earlier, to supply the heart (figure 8.4).

Subclavian Arteries Return to the efferent branchial arteries. Follow them posteriorly as they arch inward to join the unpaired dorsal aorta passing to the pleuroperitoneal cavity. Just before the last pair of efferent branchial arteries join the dorsal aorta, the paired subclavian arteries emerge laterally from the dorsal aorta (figures 8.3 and 8.4). You will have to probe deeply under the transverse septum to find them. The subclavian artery supplies the lateral body wall and pectoral fin, but these branches will not be followed.

Pleuroperitoneal Cavity

Four Branches of the Dorsal Aorta

→ Pull viscera to the right to find the dorsal aorta, injected with red or pink latex, running along the dorsal midline. Many tiny, paired branches, intersegmental arteries, leave the dorsal aorta dorsally to supply axial muscles. But only four major, unpaired branches leave the dorsal aorta ventrally to supply specific regions of viscera. In order, they are the celiac, anterior mesenteric, gastrosplenic, and posterior mesenteric arteries.

Celiac Artery The first of four major branches off of the dorsal aorta is the celiac artery, departing well forward in the pleuroperitoneal cavity, cranially, above the liver (figures 7.3 and 8.5). At this point, it is impractical to clear tissue and viscera to get an unobstructed view of the origin of the celiac artery, but you should try to find the approximate location of this branching point. The celiac artery runs caudally without branching until it emerges dorsal to the stomach along the lesser curvature. Here it divides into the gastric artery to the body of the stomach and the pancreaticomesenteric artery primarily to the intestine. After tracing the branches of the pancreaticomesenteric artery, the gastric artery will be followed further.

Near the pyloric sphincter, the pancreaticomesenteric artery sends a small branch to the pylorus and ventral lobe of the pancreas (unnamed), and at about the same point a larger branch, the duodenal artery, to the duodenum. Beyond this point, the artery continues as the anterior intestinal artery distributing along the right, anterior side of the valvular intestine.

Return to the base of the gastric artery to find a thin branch, the hepatic artery, passing to the liver within the gastrohepatoduodenal ligament and parallel to the common bile duct. Beyond the point of the hepatic artery, the gastric artery divides into dorsal gastric and ventral gastric arteries, which branch and distribute across dorsal and ventral walls of the stomach, respectively.

Anterior Mesenteric Artery Along the caudal third of the pleuroperitoneal cavity, two adjacent arteries, the anterior mesenteric and gastrosplenic arteries, depart (figures 7.3 and 8.5). After leaving the dorsal aorta, they cross and pass in the posterior free edge of the mesentery proper. The first of these, the anterior mesenteric artery, divides into branches that course in the spiral creases of the left, middle, and posterior sides of the valvular intestine.

Gastrosplenic Artery Just caudal to the anterior mesenteric artery, the dorsal aorta give off the lienogastric (= gastrosplenic) artery that passes to the spleen, stomach, and dorsal lobe of the pancreas (figures 7.3 and 8.5).

Posterior Mesenteric Artery The fourth median branch of the dorsal aorta is the posterior mesenteric artery that runs along the cranial free edge of the mesorectum to supply the rectal gland (figures 7.3 and 8.5).

Caudal Branches of the Dorsal Aorta

→ Usually, poor injection of latex and obstructing viscera of the urogenital system prevent clear exposure of more posterior vessels. In suitable specimens, or on demonstration, the dorsal aorta caudal to the posterior mesenteric artery can be seen to give off paired iliac arteries that pass to the body wall and then into pelvic fin musculature. Beyond the exit of the iliac arteries, the dorsal aorta continues into the tail as the caudal artery.

Hepatic Portal System

The hepatic portal system consists of vessels that collect blood from capillaries in the gut, pancreas, and spleen and carry it to capillaries in the liver. Because of its unique position between these capillary beds, the hepatic portal system must be injected separately from the arterial and venous systems. This is also true for the *Necturus* and cat circulatory systems. If the hepatic portal system of your animal has been injected with latex, the color will likely be distinct from the color used to inject the veins. Typically, blue and yellow are used for these two venous systems, but variation exists in which of these colors is used for the hepatic portal system. Your instructor will be able to confirm if the hepatic portal system was injected, and what color was used.

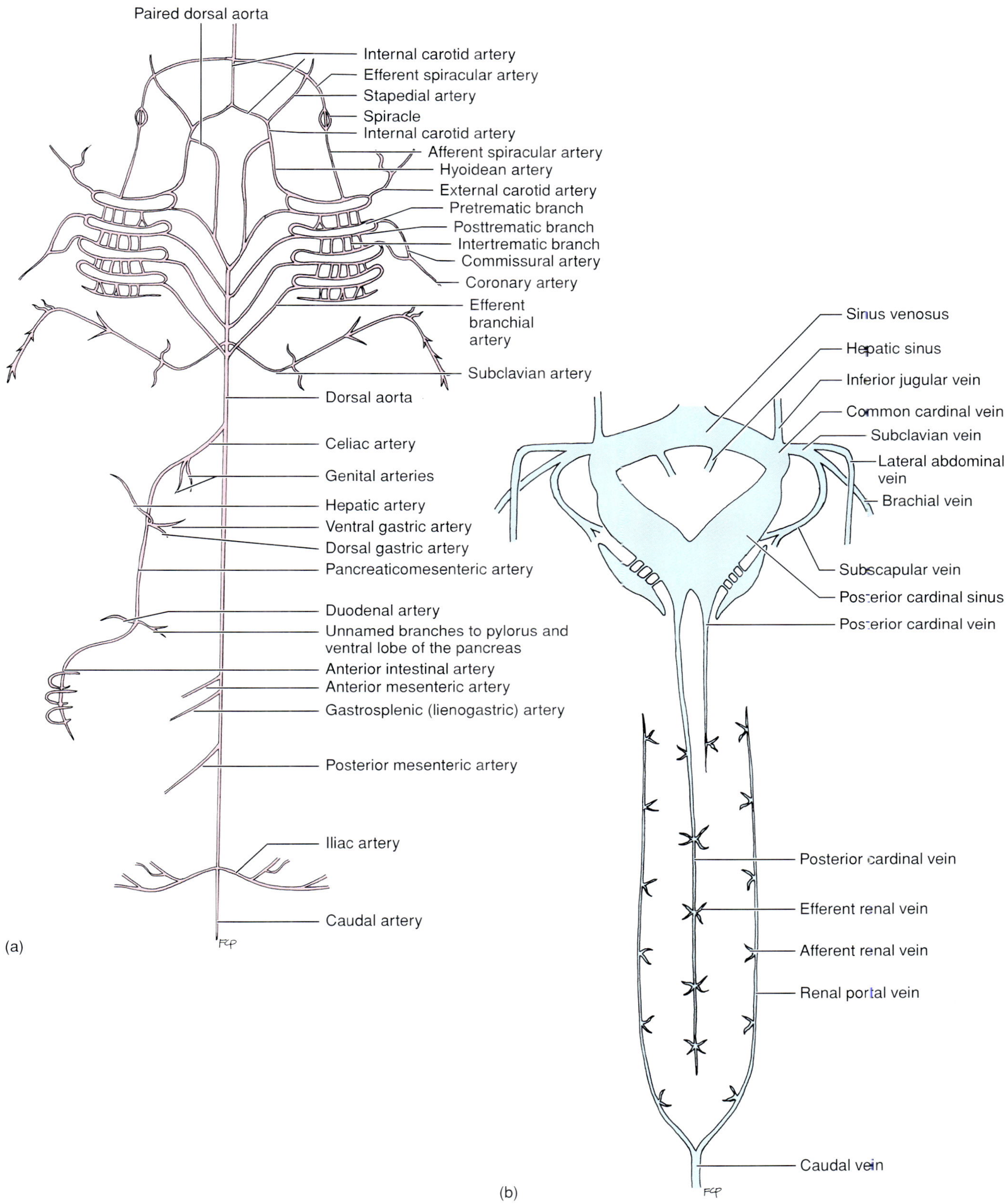

Figure 8.5
Artery (a), vein (b), and hepatic portal vein (c, next page) maps in a shark, ventral view. The floor of the pharynx was rotated to the viewer's right.

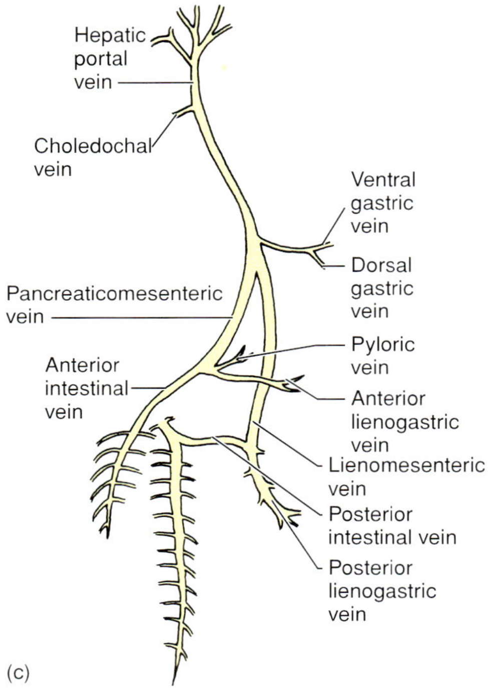

Figure 8.5
Continued.

→ The large vessel running in the hepatoduodenal ligament next to the tiny hepatic artery is the hepatic portal vein (figures 7.3 and 8.5c). Note that anteriorly it divides and distributes to the liver. Following it posteriorly from the liver, it is joined by a thin vein that runs along the common bile duct, the choledochal vein. (Keep in mind that some veins may be poorly injected with color, requiring you to search for the plain, tan-colored vessel.) The caudal end of the hepatic portal vein forms by the confluence of three major branches: gastric, pancreaticomesenteric, and lienomesenteric veins.

→ **Gastric Vein** The gastric vein comes from the stomach and is itself formed by union of the dorsal gastric and ventral gastric veins (figures 7.3 and 8.5).

→ **Pancreaticomesenteric Vein** The second branch joining the hepatic portal, the pancreaticomesenteric vein, begins posteriorly as the anterior intestinal vein extending from the right, anterior side of the valvular intestine (figures 7.3 and 8.5). Trace the anterior intestinal vein forward to where it is joined by a longer thin vein, the anterior lienogastric, from the spleen and portions of the pylorus. A small tributary from the ventral side of the pylorus, the pyloric vein, also joins at about this same point. Cranially from the juncture of these three veins, the vessel is properly termed the pancreaticomesenteric vein.

→ **Lienomesenteric Vein** The third branch joining the hepatic portal, the lienomesenteric vein, begins posteriorly as the posterior intestinal vein draining the left, posterior valvular intestine (figures 7.3 and 8.5). It passes away from the intestine and runs freely toward the pancreas, where it is joined by a short vessel, the posterior lienogastric vein from the spleen and bend of the stomach. Cranially from this juncture with the lienogastric vein, the vessel is properly the lienomesenteric vein. It passes in the surface of the dorsal lobe of the pancreas, from which it receives several small vessels, pancreatic veins. Pick away pancreatic tissue to expose several of these pancreatic veins.

(Notice the quirk in naming arterial and venous vessels to the intestine. Anterior and posterior intestinal veins exist, but only an anterior intestinal artery is identified. The expected posterior arterial "counterpart" is termed the anterior mesenteric artery whose counterpart in turn is the posterior mesenteric artery to the rectal gland.)

Veins

→ Along each lateral body wall runs a lateral abdominal vein returning blood from the fins and adjacent areas (figures 8.2 and 8.5b). Trace it forward to the region just caudal to the transverse septum. Here, the lateral abdominal vein is joined by the subscapular vein running ventrally from the dorsolateral side of the body wall. Clear away the parietal peritoneum covering the subscapular vein and note the deep brachial vein, from the fin, joining the subscapular vein. Together, the lateral abdominal and subscapular veins empty into the short subclavian vein, which immediately supplies the more medial and expanded common cardinal vein. Left and right common cardinals are each joined by a hepatic sinus, from the ventral end of the liver, before emptying into the single sinus venosus, the first heart chamber.

Turn now to the pleuroperitoneal cavity dorsal to the stomach. Take care not to damage the kidney. Next to the dorsal aorta runs the paired posterior cardinal veins, which receive many efferent renal veins draining the kidneys. Follow the posterior cardinal veins forward to where they empty into the very large posterior cardinal sinuses.

Necturus

The Heart, Branches of the Ventral Aorta, and Gills

→ The pericardial cavity has already been entered through the throat musculature. The left atrium and right atrium lie on the cranial base of the single large muscular ventricle (figures 8.6a and 8.7). Lift the apex of the ventricle to find the thin-walled sinus venosus. Poor injection usually makes it difficult to identify veins that contribute to the sinus venosus. Generally, blood returning from the forelimbs and anterior body enters the sinus venosus laterally through the paired common cardinal vein; blood from the posterior body returns from the long, single postcaval vein that passes through the transverse septum and upon

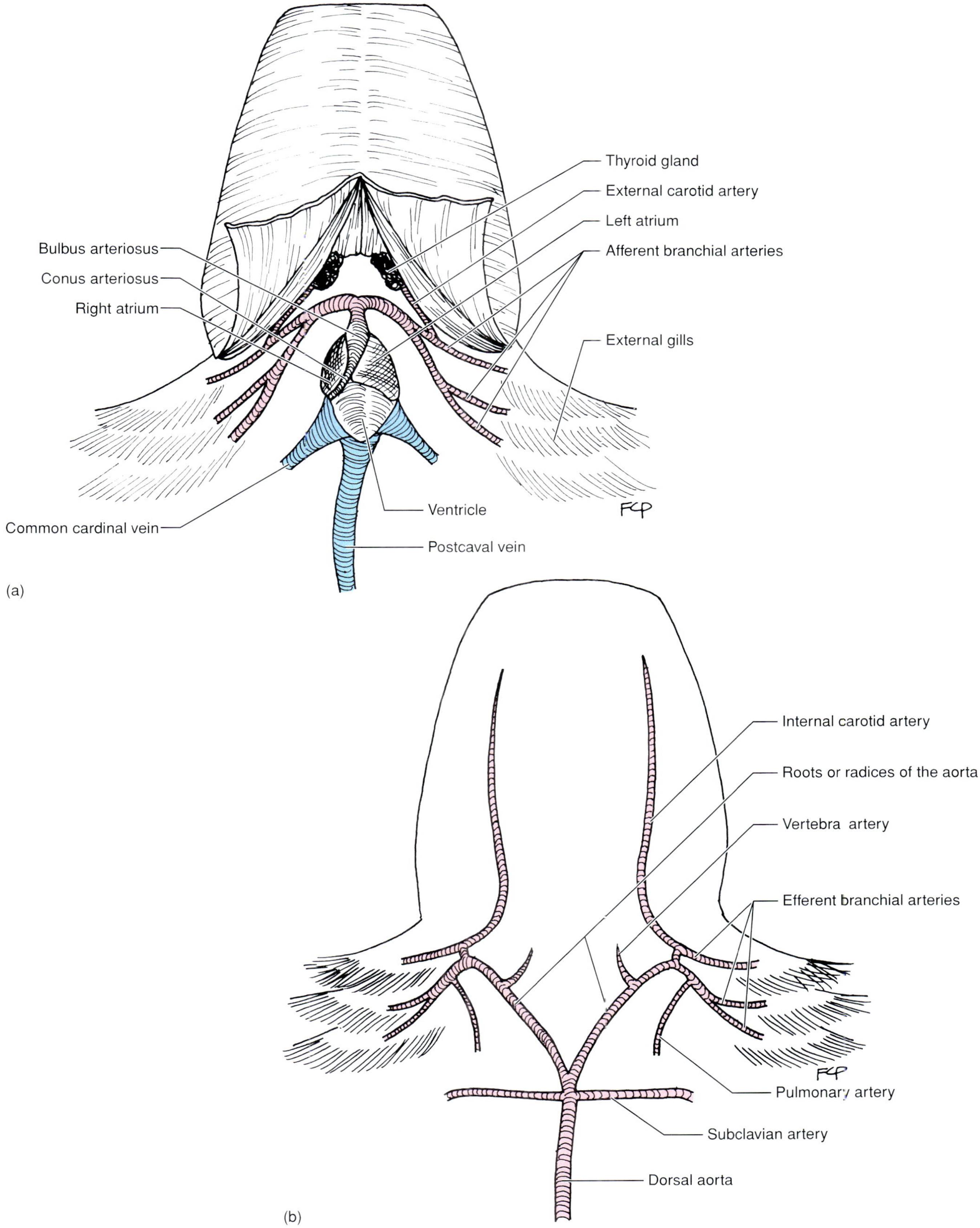

Figure 8.6
Heart chambers and associated blood vessels in *Necturus.* (a) Ventral view of heart and associated blood vessels. (b) Ventral view of blood vessels in the roof of the mouth.

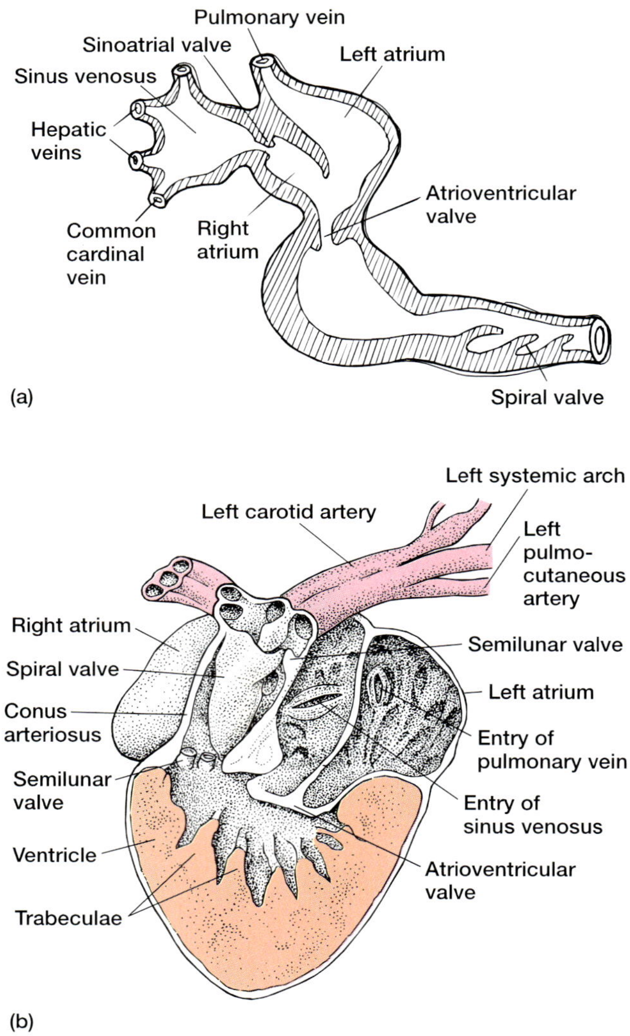

Figure 8.7
Amphibian hearts. (a) Diagram of a typical amphibian heart. Cranial is to the right. Notice that the atrium is divided into left and right chambers but the ventricle lacks an internal septum. (b) Ventral view of a front section of a bullfrog (*Rana catesbeiana*) heart.
(b) After M.H. Wake, ed., Hyman's Comparative Vertebrate Anatomy, 3d ed., ©1979 by The University of Chicago.

Form & Function

Can a vertebrate live on land without lungs? Many species in the salamander family Plethodontidae do, by using their moist skin as the only respiratory surface (cutaneous respiration). As these salamanders have no lungs, the interatrial septum doesn't develop, sending all of the blood through just the single atrium and single ventricle. Increased vascularization of the skin and a smaller body size also help make this system effective. Many other amphibians and some reptiles and fish obtain a significant amount of oxygen through their skin (10% or more). But unlike plethodontid salamanders, these animals rely primarily on lungs.

entering the pericardial cavity divides into hepatic sinuses that join the common cardinal veins in entering the sinus venosus (figure 8.8). From the right, anterior corner of the ventricle exits the narrow conus arteriosus, which between the atria expands into the swollen ventral aorta here termed the bulbus arteriosus (figures 8.6–8.8).

Immediately upon exiting from the pericardial cavity, the bulbus arteriosus divides, sending a single branch to each side to distribute to the feathery, external gills (three on each side; figures 8.6a and 8.8). Follow each branch, left and right, by carefully removing all the overlying throat musculature. Each branch soon divides, the anterior fork going to the first gill, the posterior fork branching again to supply the middle and posterior external gills. These three separate supplies to the external gills are the afferent branchial arteries. Before reaching the gill, the first afferent branchial artery gives off a thin branch, the external carotid artery, that runs forward between gill arch cartilages and branchiohyoideus muscle. At the fork between external carotid and first afferent branchial artery lies the small, lobed thyroid gland, supplied by the external carotid artery.

Clear muscle and connective tissue so that the distribution of afferent branchial arteries on both sides can be followed into the external gills.

Dorsal Aorta and Lungs

→ To expose the afferent branchial arteries and their convergence into the dorsal aorta, it will be necessary to reach the roof of the mouth. Cut through the angle of the jaws on both sides up to the gill arches. On the left side, continue this cut laterally through the bases of the gill arch cartilages by keeping one blade of the scissors in the esophagus, the other external. Remove the mucous membrane from the roof of the mouth.

At the back of the mouth, two large vessels, the roots or radices of the aorta, arch inward, pass posteriorly, and unite to form the dorsal aorta (figures 8.6b and 8.8). On the right side, locate the source of these roots, the three efferent branchial arteries in the external gills. Note that the middle and posterior efferent branchial arteries join immediately upon leaving the external gills. Near their union, the pulmonary artery exits, passing posteriorly to supply the right lung. (The lungs were previously identified in the digestive system chapter. They are slender and saclike running along the dorsolateral sides of the pleuroperitoneal cavities.) The first efferent branchial artery leaves the gill, runs parallel to the root of the aorta joining it by a short unnamed shunt, then swings forward in the roof of the mouth as the internal carotid artery, eventually disappearing into the cranium. The thin external carotid artery, identified earlier, should be located again and followed dorsally around the side of the gills beneath the skin to find where it leaves the first afferent branchial artery.

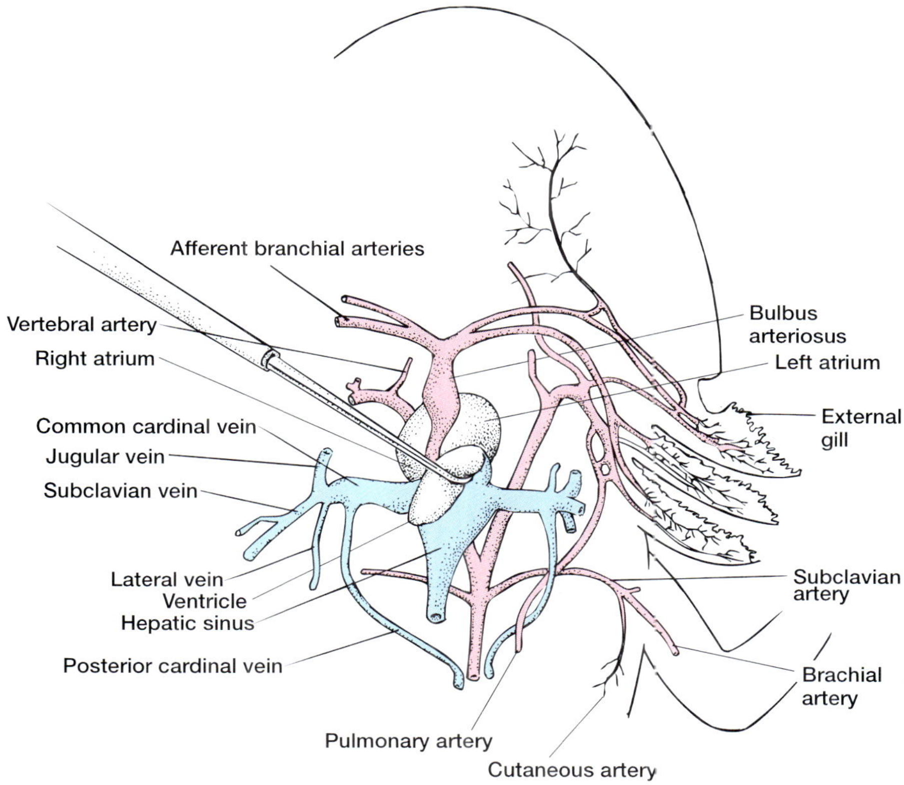

Figure 8.8
Aortic arches in *Necturus*. This figure integrates the separate views indicated in figure 8.6a,b.

Patterns & Connections

Like most larval salamanders, *Necturus* simultaneously possess gills and lungs! In larval salamanders about to undergo metamorphosis, the gills will soon be lost as the lungs become the primary respiratory structures (with the exception of plethodontid salamanders). But why is this adaptive for *Necturus*, an animal that never transforms to a life on land? For *Necturus* living in occasionally oxygen-poor water, lungs can supplement gill respiration. By gulping air at the surface and forcing it into the lungs, *Necturus* obtains supplemental oxygen for this relatively large and occasionally active body. In addition, lungs filled with air also give *Necturus* a means to control of its underwater buoyancy.

→ Just as the root of the aorta turns caudally, it gives off a branch running deep into neck muscles, the vertebral artery (figures 8.6b and 8.8). Immediately following the union of the roots of the aorta, the dorsal aorta gives off a paired lateral branch running into the arm, the subclavian artery. The subclavian sends one branch along the body wall and several branches into the shoulder and arm.

Pull viscera to the right and locate the dorsal aorta along the dorsal body wall (figures 8.9 and 7.5). Dorsally, it gives off parietal (= intercostal) arteries that pass segmentally to the axial musculature. After the subclavian, the next major branch to the viscera is the unpaired gastric artery, which soon forks into dorsal and ventral gastric arteries to respective sides of the stomach. The ventral gastric artery also supplies the spleen.

The next, midventral branch of the dorsal aorta, further posteriorly, is the celiacomesenteric artery, followed immediately by numerous mesenteric arteries to the intestine (figures 8.9 and 7.5). The celiacomesenteric artery divides into four branches: the splenic artery passes cranially to supply the posterior end of the spleen; the pancreaticoduodenal artery supplies and is embedded in the pancreas and eventually reaches the pylorus The remaining two, best seen on the right side, are the short mesenteric branch of the celiacomesenteric artery to the intestine following the duodenum; and the hepatic artery passing to the posterior end of the liver.

The numerous mesenteric arteries supply the remainder of the small and all of the large intestine (figures 8.9 and 7.5). Near the bases of the first few mesenteric arteries are the genital arteries (spermatic arteries or ovarian arteries) passing laterally to the gonad. Posterior to this the dorsal aorta gives off numerous, but very short, renal arteries to the kidney that are usually difficult to find because the kidney is so tightly applied to the aorta.

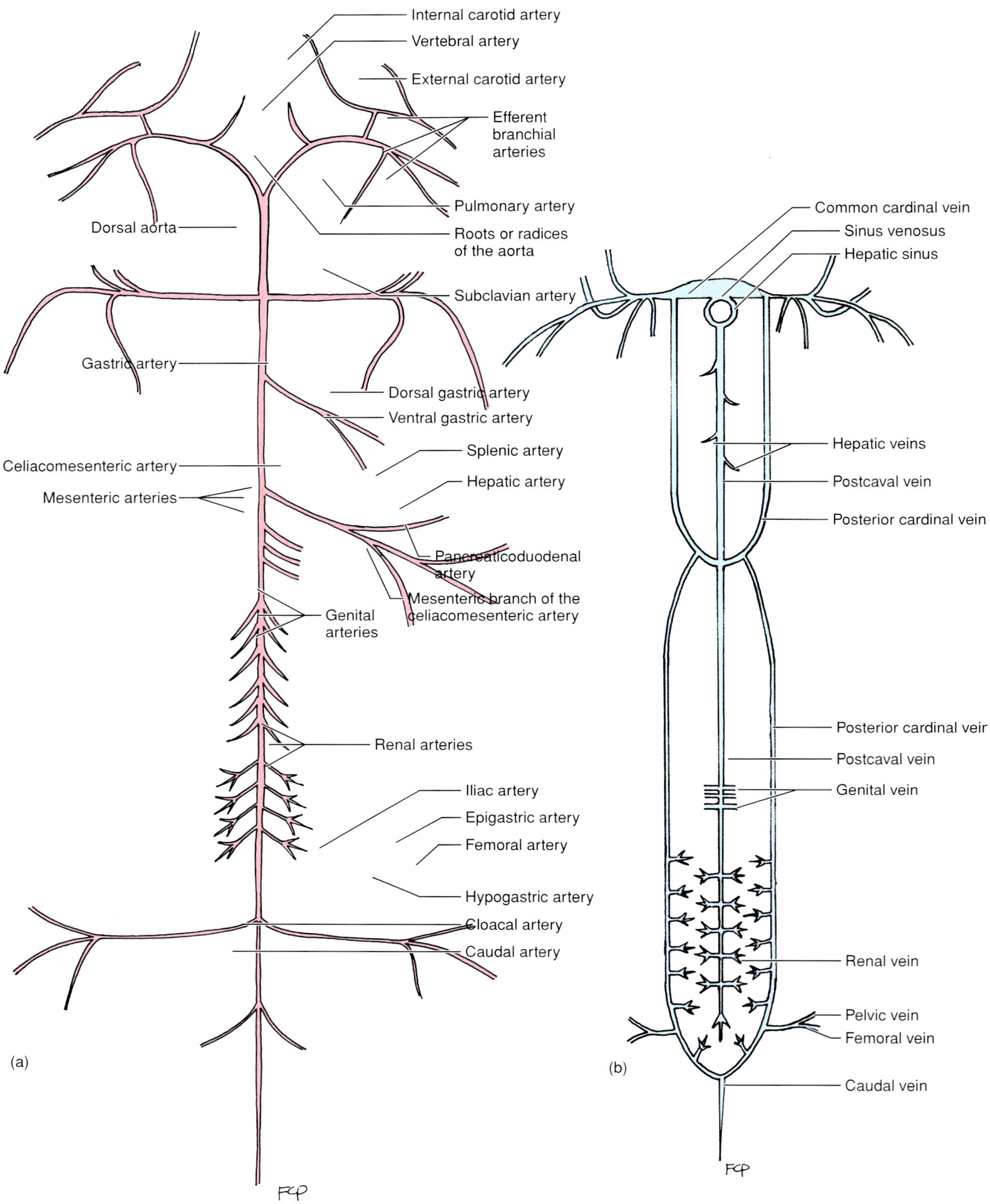

Figure 8.9
Artery (a), vein (b), and hepatic portal vein (c) maps in *Necturus*, ventral view.

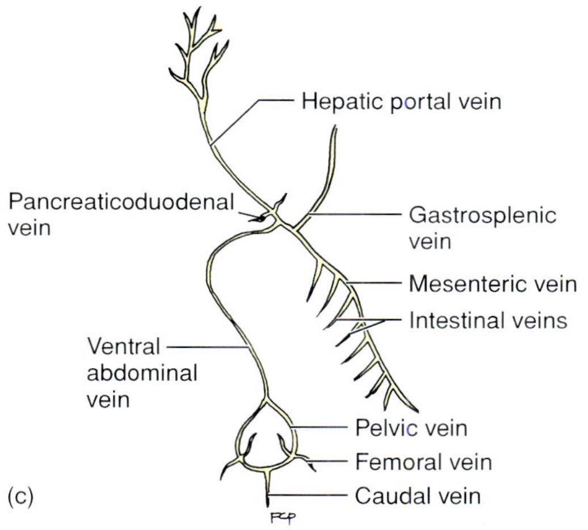

Figure 8.9
Continued.

Near the hindlimb, the dorsal aorta gives off a paired lateral artery, the iliac artery (figures 8.9 and 7.5). It has three major branches. The first, the epigastric artery, passes cranially in the body wall. The second, the femoral artery, penetrates the body wall to enter the limb. The third, the hypogastric artery, swings medially to the urinary bladder and cloaca.

Beyond the iliacs, the dorsal aorta gives off cloacal arteries and becomes the caudal artery to supply the tail. These two arteries need not be followed.

Hepatic Portal System

→ Next to the epigastric artery runs the pelvic vein, which joins its partner of the opposite side to form the unpaired ventral abdominal vein that passes in the ventral body wall forward to the liver (figures 8.9c and 7.5). In most specimens, neither pelvic nor ventral abdominal veins are sufficiently injected to permit clear identification, although uninjected sections can certainly be discovered. Often the cranial part of the ventral abdominal vein can be found injected running past the gallbladder on the most posterior tip of the liver.

The hepatic portal system begins in the mesenteric vein from the large intestine (figures 8.9 and 7.5). Short, numerous intestinal veins join the mesenteric vein on its course forward. As the mesenteric vein approaches the liver, it is joined by three veins all at about the same point. Part of the pancreas may need to be removed to expose the junction of these veins. The first is the pancreaticoduodenal vein from duodenum and pancreas. Next, the gastrosplenic vein, from spleen and stomach, and the ventral abdominal vein join at about the same point. Anterior to this juncture, the collective vein becomes the hepatic portal vein. It soon ramifies to distribute across the dorsal surface of the liver, delivering blood to liver sinuses.

Systemic Veins

→ The postcaval vein begins posteriorly between the kidneys that empty into it through many small renal veins (figure 8.9b). As the postcaval vein journeys forward, genital veins (spermatic veins or ovarian veins, specifically) contribute. The postcava leaves the dorsal body wall and passes to the dorsal surface of the liver, coursing as a vessel through the organ and receiving more veins that drain the liver, hepatic veins (figure 7.5). Pick away liver tissue to follow the postcava on its course. Note, for instance, the prominent hepatic veins on the midventral surface of the liver that join the postcava. The postcava is large as it exits from the cranial corner of the liver (now, by the way, in the coronary ligament) to pierce the transverse septum and, as it enters the pericardial cavity, to divide into hepatic sinuses identified previously. Again remember that these hepatic sinuses and the sinus venosus into which they enter are often poorly injected.

The thin, paired posterior cardinal vein receives blood from the tail and hindlimb and can be identified where it runs along the lateral edge of the kidney delivering some of its blood through small tributaries to this organ (figure 8.9). The posterior cardinal vein, through a short shunt, joins the postcava where this large vessel leaves the dorsal body wall and slants toward the liver. The still paired posterior cardinal vein does not end here, but continues forward parallel with the dorsal aorta. Segmental parietal veins join the posterior cardinal vein dorsally. Nearing the transverse septum, it diverges laterally, piercing the septum, and enters the common cardinal vein.

Pulmonary Vein

→ Both pulmonary artery and vein are likely injected with pink latex. You have already traced the pulmonary artery from the efferent branchial artery to the surface of the lung opposite its mesenteric attachment. The pulmonary vein runs along the lung next to this mesenteric attachment. It joins its partner from the opposite side and enters the left atrium.

Cat

We begin with the dissection of the major arteries, emanating from the heart. The veins, usually more poorly injected, will be traced later. Most veins accompany respective arteries of the same name; thus, knowing arteries will help with finding veins. However, be careful not to destroy the veins in the process of finding the arteries!

Recall that the *deep* dissection of the muscular system was confined to the left side of the cat to limit damage to the blood vessels on the right side. Therefore, the right member of paired blood vessels will be identified first.

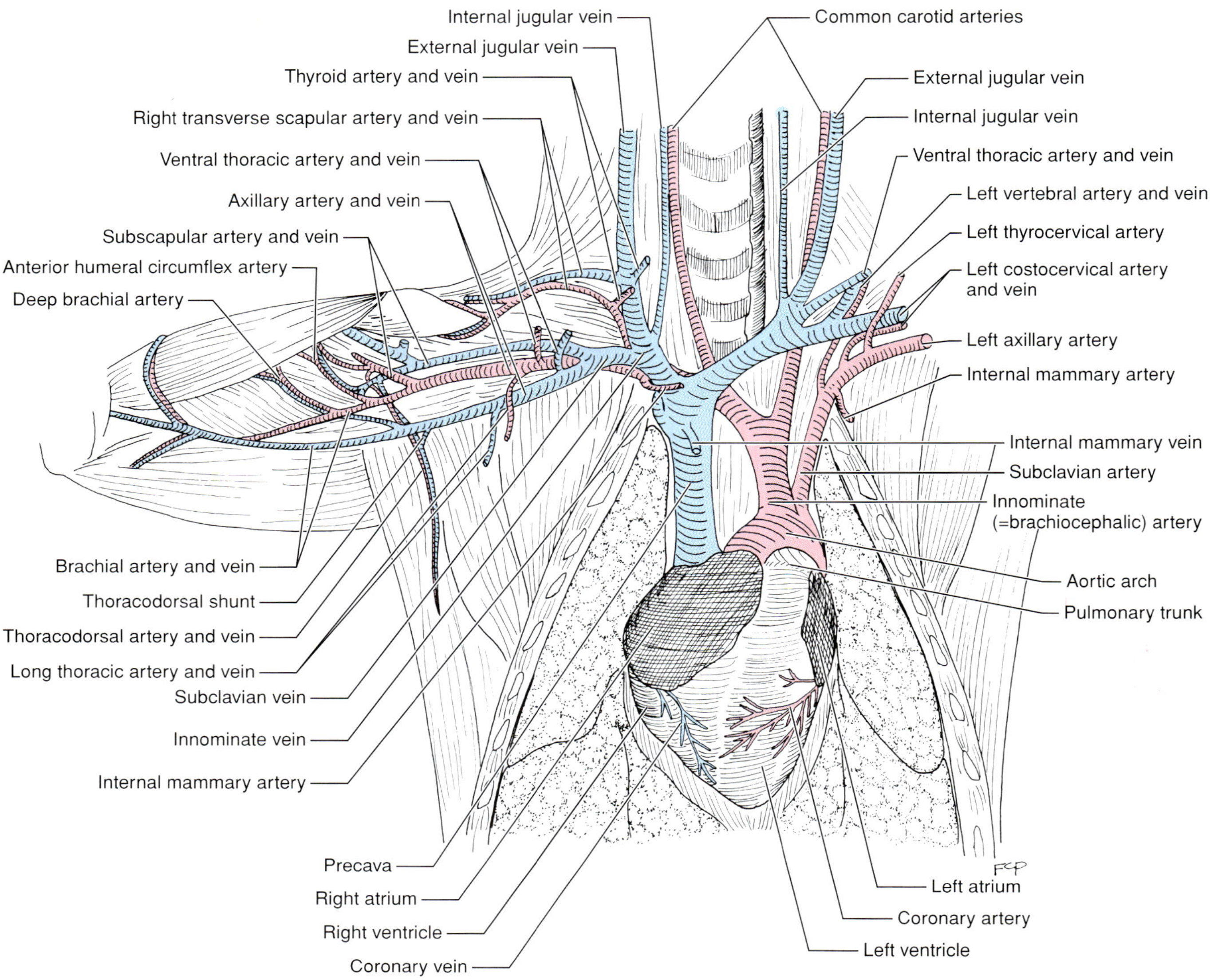

Figure 8.10
Ventral view of thoracic components of the circulatory system in a cat.

Heart and Its Major Vessels

→ If the heart is not already exposed, continue a midventral incision through the sternum anteriorly, until the entire rib cage can be opened. Be careful of large vessels at the anterior corner of the rib cage. A glandular mass, the thymus, lies just cranial to the heart and may hide some of the major arteries and veins. This thymus and associated connective tissue should be cleared away. Cut away completely the parietal pericardium (= pericardial sac) covering the heart to near the base of the anterior vessels entering and leaving the heart. The large muscular ventricles, left and right, are covered by thin coronary arteries and veins that supply the tissues. Across the anterior end of the ventricles lie the left and right atria (figures 8.10 and 8.11). Lift the apex of the heart to find the single large blue vessel, the postcava,[1] passing cranially through the diaphragm to the right atrium. The large, blue precava also enters here and can be seen coming from the anterior region of the body. The aorta exits from between the atria on the cranial side of the heart (figures 8.10 and 8.11). It then gives off two large branches that pass cranially, and then itself arches posteriorly along the dorsal wall of the thoracic cavity. To the left of the aorta, near its base, lies the pulmonary trunk. This large blood vessel passes dorsally from the right ventricle to divide into left and right pulmonary arteries supplying the respective lungs. Note that this artery, named so because it transports blood away from the heart, carries blood with some of

[1]Also termed *posterior* and *anterior*, or *inferior* and *superior venae cavae*, respectively.

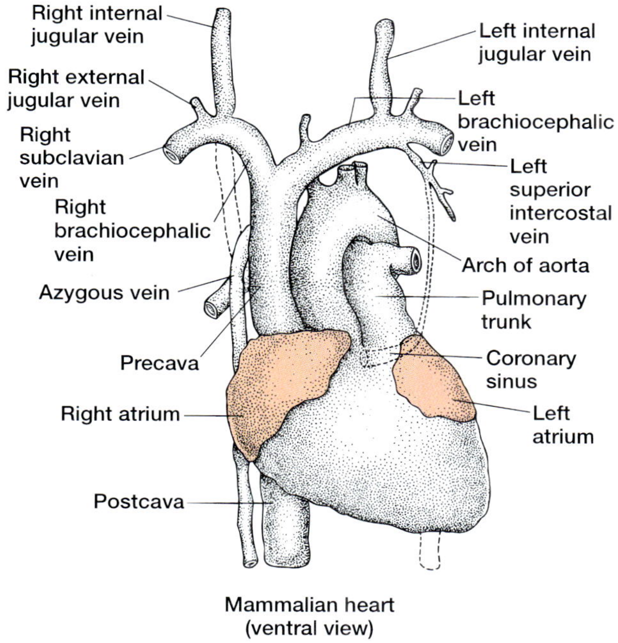

Figure 8.11
Mammalian heart, ventral view, to show the complex venous and arterial components surrounding the heart.
After Lawson.

the lowest levels of oxygen anywhere in the body. Clear any obscuring fat to follow these vessels. A short piece of connective tissue, the ligamentum arteriosum, is often seen extending between the bases of the pulmonary trunk and aorta. This ligament is a remnant of the fetal shunt, the ductus arteriosus.

Aorta

→ To aid location of the aorta, remove and discard the entire left lobes of the lung by cutting cleanly through their bases. Three regions of the aorta are often recognized: the aortic trunk, the base of the vessel where it leaves the heart; the curved aortic arch; and the long dorsal aorta running posteriorly along the dorsal midline.[2]

In the cat, two vessels project from the aortic arch. The first and slightly larger on the cat's right is the single innominate (= brachiocephalic) artery. Next, adjacent to it, is the left subclavian artery to the left arm (figure 8.10).

Innominate Artery Follow the innominate first. At about the same point, it divides into three branches. The first two branches to come off are the left then right common carotid arteries, which lie next to and parallel the trachea. The third branch, the right subclavian artery, passes laterally into the right arm (figures 8.10 and 8.12).

Right Subclavian Artery Four branches exit at about the point where the subclavian passes the first rib (figures 8.10 and 8.12a). First, the internal mammary artery arises from the ventral surface of the subclavian, passes to the chest wall in the company of the corresponding vein, and runs caudally along the ventral side of the thoracic cavity. Second, the vertebral artery arises opposite the internal mammary artery, off of the dorsal side of the subclavian. Lift the subclavian to find its point of departure and passage deep into the vertebrarterial canal. Third, near the base of the vertebral artery, is the costocervical artery, which passes dorsally and forks immediately. The cranial fork runs forward and deep along the outer side of the rib cage, reaching epaxial muscles of the neck; the caudal fork runs posteriorly along the inner wall of the rib cage. Fourth, the thyrocervical artery springs from the anterior side of the subclavian, passing forward for a short time in the company of the external jugular vein. It gives off a thin thyroid artery to the thyroid gland and continues as the transverse scapular artery to the shoulder.

Beyond the point of exit of these four arteries, the subclavian is termed the axillary artery. Continue tracing the branches of the axillary artery on the right side (figures 8.10 and 8.12a). The ventral thoracic artery, the first to spring from the axillary, passes with its corresponding vein to the ventral pectoralis muscles. Next arises the long thoracic artery passing posteriorly with its venous counterpart into pectoralis muscles. The axillary next gives off the large subscapular artery extending anteriorly to the upper arm. The thoracodorsal artery exits from the posterior side of the subscapular and passes caudally along the medial side of the teres major and latissimus dorsi muscles in the company of its venous counterpart. Just lateral to the base of the subscapular is the smaller anterior humeral circumflex artery running dorsally over the biceps muscle near its origin. Beyond the point of departure of the anterior humeral circumflex, the axillary becomes the brachial artery passing toward the elbow. About midway along the upper arm, the brachial gives off a lateral branch, the deep brachial artery, crossing the dorsal belly of the biceps muscle.

Common Cartoid Artery

→ Trace the two common carotid arteries cranially from their departure from the innominate artery. Each gives off tiny branches to the trachea and to neck muscles. Near the cranial end, the superior thyroid artery extends medially to the thyroid gland, and opposite it, the muscular artery extends laterally to the neck musculature (figures 8.10 and 8.12a). At the level of the larynx, a ventral branch, the laryngeal artery (possibly destroyed), supplies the larynx, and at the same point a dorsal branch, the occipital artery, passes around the tympanic bulla and into the neck muscles. Just anterior to the occipital artery, the ascending pharyngeal artery[3] exits also from the dorsal side of the common carotid. The ascending pharyngeal artery dips

[2] These regions are roughly comparable to the ascending aorta, aortic arch, and descending aorta of human anatomy.

[3] A true internal carotid artery is absent, being replaced in the cat by the ascending pharyngeal artery, its functional equivalent.

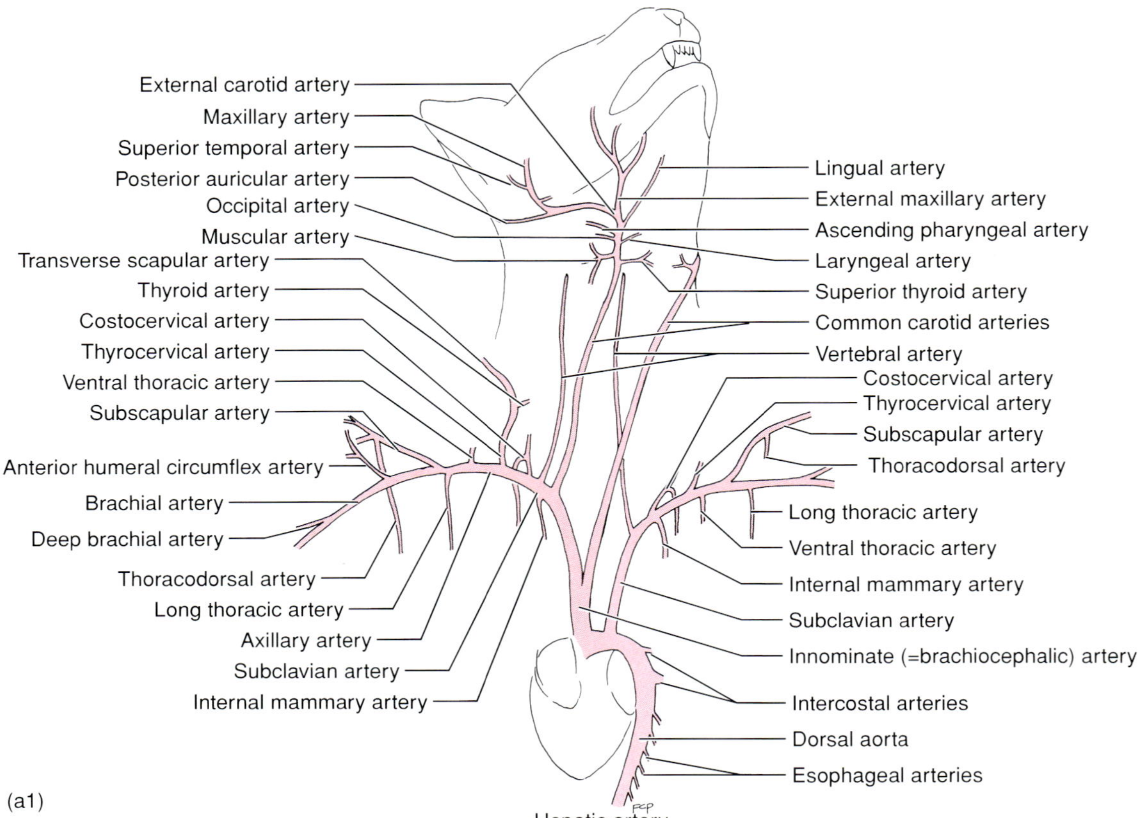

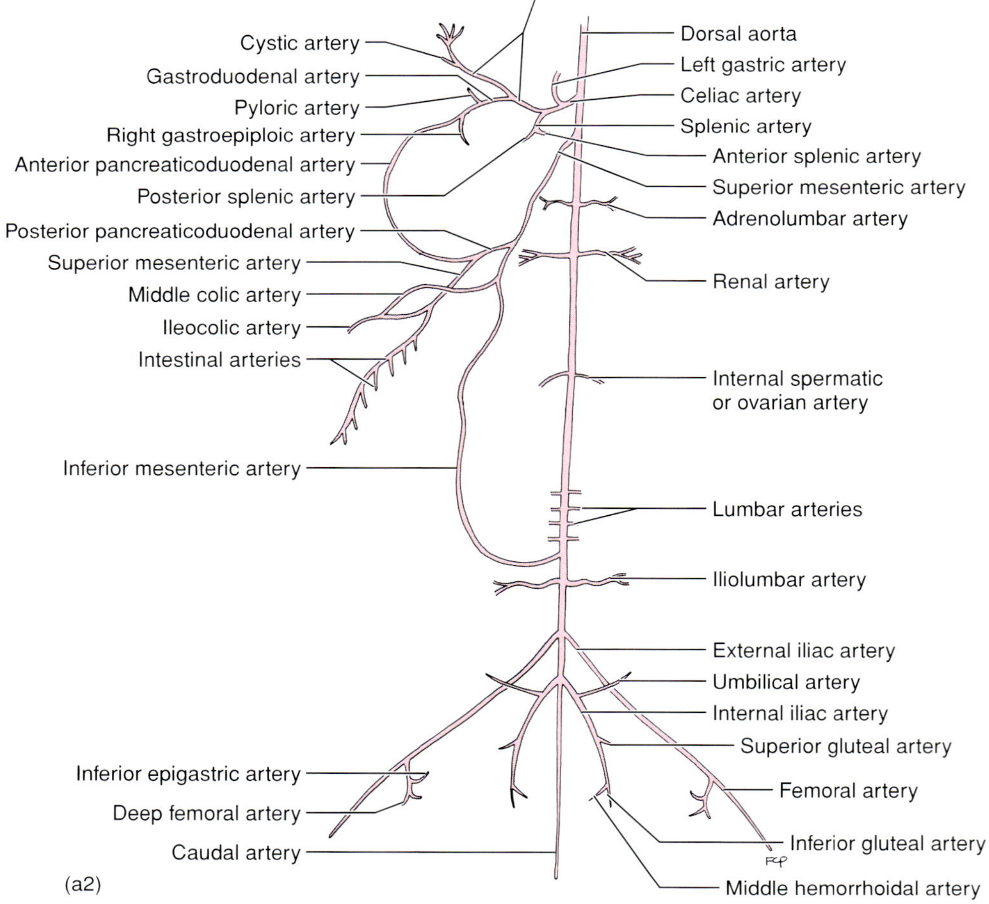

Figure 8.12
Artery (a), vein (b), and hepatic portal vein (c) maps in a cat, ventral view.

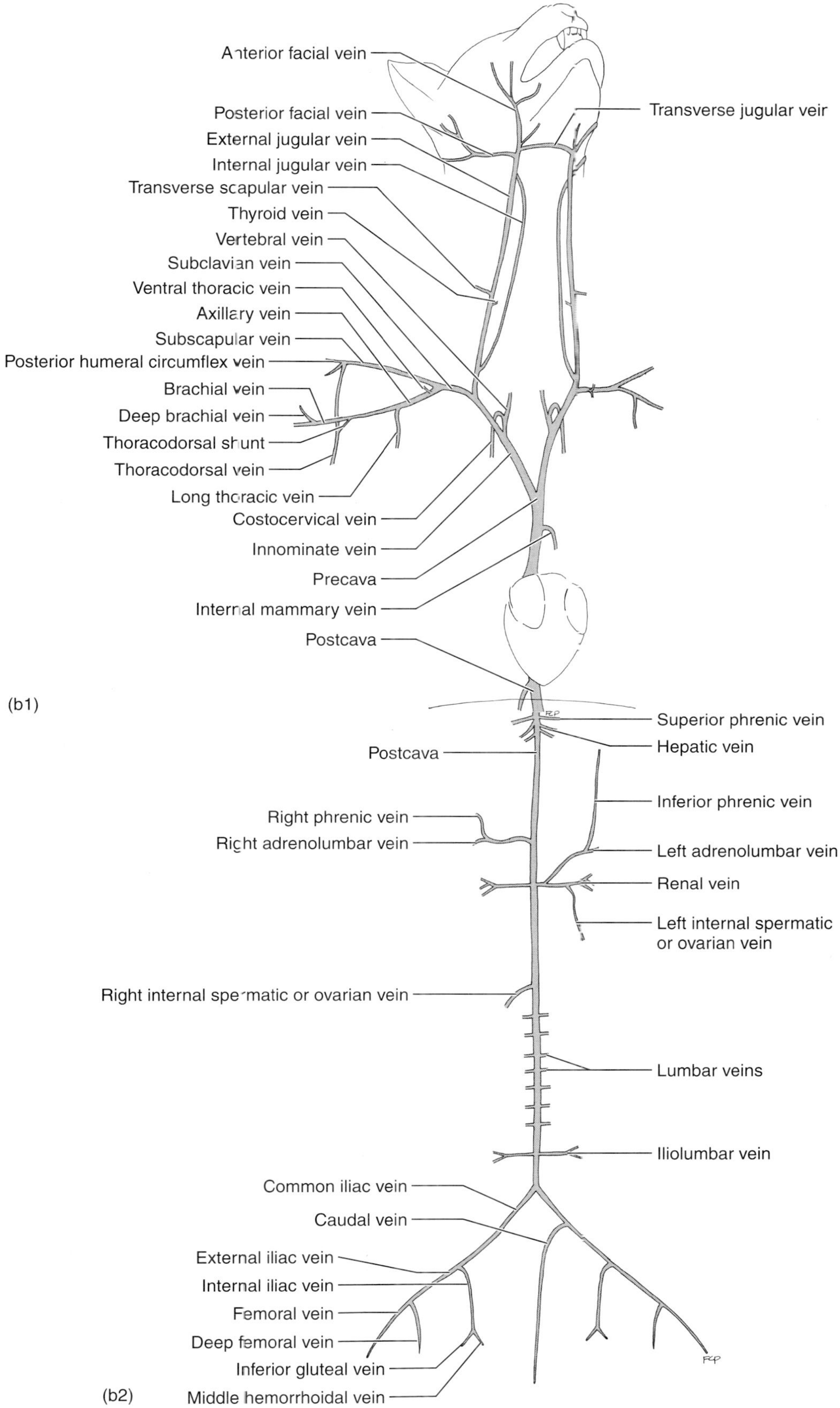

Anterior facial vein
Posterior facial vein
External jugular vein
Internal jugular vein
Transverse scapular vein
Thyroid vein
Vertebral vein
Subclavian vein
Ventral thoracic vein
Axillary vein
Subscapular vein
Posterior humeral circumflex vein
Brachial vein
Deep brachial vein
Thoracodorsal shunt
Thoracodorsal vein
Long thoracic vein
Costocervical vein
Innominate vein
Precava
Internal mammary vein
Postcava
Transverse jugular vein
(b1)
Superior phrenic vein
Hepatic vein
Postcava
Inferior phrenic vein
Right phrenic vein
Right adrenolumbar vein
Left adrenolumbar vein
Renal vein
Left internal spermatic or ovarian vein
Right internal spermatic or ovarian vein
Lumbar veins
Iliolumbar vein
Common iliac vein
Caudal vein
External iliac vein
Internal iliac vein
Femoral vein
Deep femoral vein
Inferior gluteal vein
Middle hemorrhoidal vein
(b2)

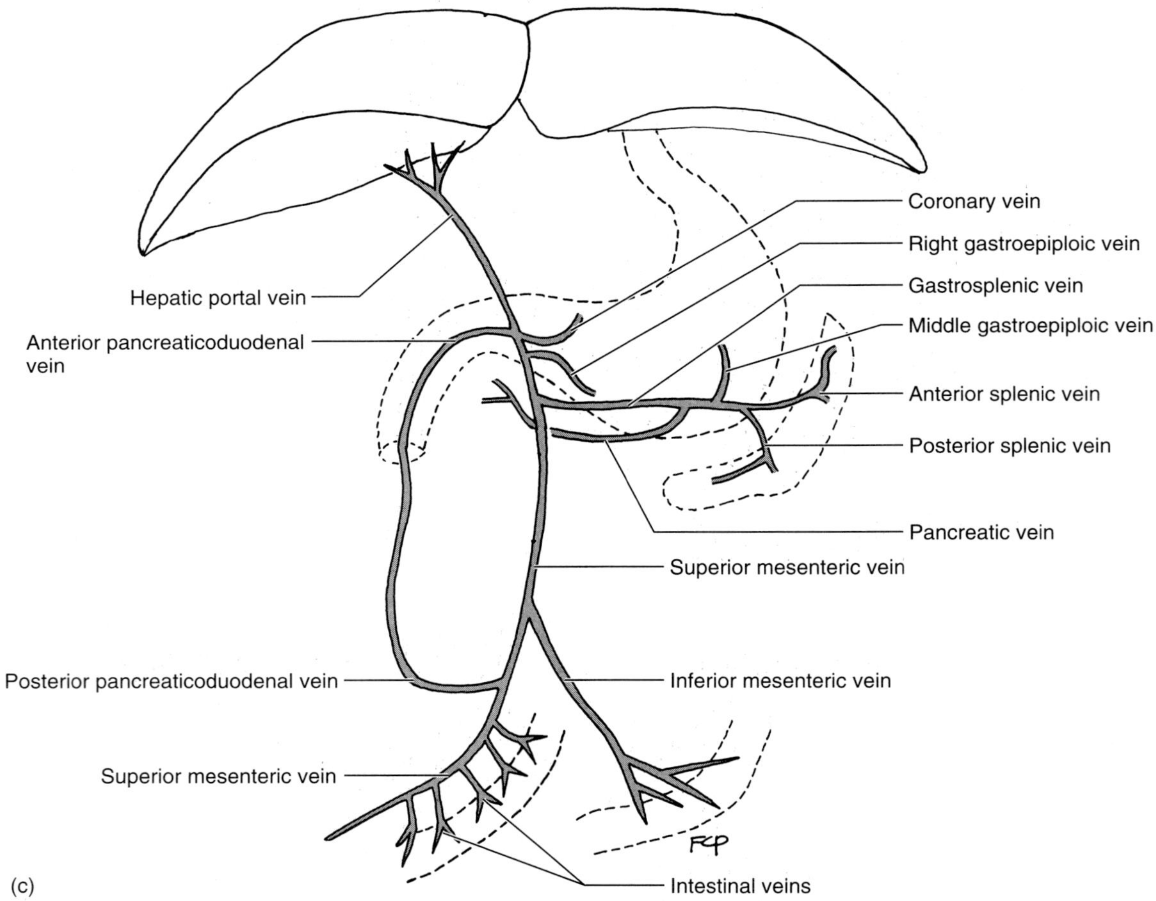

Figure 8.12
Continued.

deep and eventually enters the skull. Next anteriorly, the common carotid artery forks. The ventral fork, the lingual artery, supplies the tongue. The dorsal fork, the external carotid artery, passes toward the pinna. The external carotid, near its base, gives off the external maxillary artery that courses along the ventral border of the masseter. The external carotid itself passes in the opposite direction along the posterior border of the masseter, gives off first the posterior auricular artery extending dorsally behind the ear, then the superior temporal artery extending dorsally in front of the ear, and thereafter the external carotid becomes the maxillary artery passing from view into the masseter muscle.

Branches of the Left Subclavian Artery

→ Names and distribution of arteries from the left subclavian are the same as already discussed for the right subclavian, although the first few branches exit at slightly different points. Near the first rib, the left subclavian gives off four arterial branches: internal mammary (ventral, runs to the interior thoracic cavity), vertebral (dorsal, runs deep to the vertebrarterial canal), costocervical (to inner and outer side of the rib cage), and thyrocervical (anteriorly, into shoulder as the transverse scapular artery).

Passing into the axilla, the subclavian becomes the axillary artery, whose subsequent distribution is similar to that already followed along the right side.

Respiratory System

Anatomy

→ Most of the structure of the respiratory system of the cat has already been identified at various points in the digestive and circulatory system dissections. Now return to these regions for an integrated look at this system. Trace the flow of air as it passes into the body to the lungs.

Air is inhaled through the external nares into the nasopharynx, the general region dorsal to the secondary palate. It then passes through the laryngopharynx into the larynx through the glottis. After passing through the larynx, air travels down through the trachea into primary, then secondary, and tertiary bronchioles, which continue to branch until reaching the blind-ended respiratory surfaces called alveoli. Gently probe the structure of the trachea to detect the many

Form & Function

Why does the trachea but not the esophagus have cartilaginous rings? Food is propelled down the esophagus by rhythmic muscular contractions that force the esophagus open as necessary. But air is drawn through the trachea by negative pressure, resulting from a lowering of the diaphragm and an expansion of the chest. This negative pressure in the trachea is similar to the air flow through a vacuum cleaner hose. In both the trachea and the hose, a series of stiff rings (C-shaped cartilages in the trachea) give it flexibility to bend with changes in body posture and to keep the tube from collapsing during breathing.

C-shaped cartilaginous rings that help keep the trachea open. Return to the cut base of the left lobes of the lung. Probe the stumps of the remaining structure to distinguish the hollow bronchioles from the surrounding blood vessels.

Pulmonary Arteries and Veins

→ The pulmonary trunk exits from the right ventricle, next to the aortic trunk, and divides into left and right pulmonary arteries supplying respective lungs (figure 8.11).

The left and right pulmonary veins receive blood returning from the lungs and empty into the left atrium. Find this point of entry on the back (dorsal side) of the heart by carefully picking away extraneous tissue and fat.

Precava

→ Follow the precava cranially to identify tributaries. First, at the base of the precava, enters the azygous vein (figures 8.11 and 8.12b). Lift the precava to find the base of the azygous vein. Next, dig deep to follow the azygous dorsally to the roof of the thoracic cavity where it turns posteriorly and runs parallel along the right side of the dorsal aorta. Along this length, the azygous receives regular intercostal veins from between ribs of both sides. Next along the midventral side of the precava is an unpaired vessel, the internal mammary vein, draining the ventral wall of the thoracic cavity.

Anterior to the internal mammary, the left and right innominate (= brachiocephalic) veins join to begin the precava (figures 8.10, 8.11, and 8.12b). Each innominate has similar, but not symmetrical, tributaries. Thus, dissect and follow each side simultaneously. About halfway along the innominate, it is joined dorsally by the vertebral vein coming from the vertebrarterial canal anteriorly in the neck collecting from the spinal cord. (Frequently, the vertebral vein first joins the precava.) Trace the vertebral vein dorsally from its base, locating the costocervical vein, from more posterior muscles of the back, that joins it. The innominate is first formed near the first rib, by union of the lateral subclavian vein from the limb and the more cranial external jugular vein.

Right Subclavian Vein

→ Follow the subclavian first, spreading pectoralis muscles and ribs as needed (figures 8.10 and 8.12b). The short subclavian itself forms from two major branches: the subscapular vein from the upper arm and dorsal side of the humerus, and the more posterior axillary vein.

Axillary Vein The base of the axillary vein is joined by the ventral thoracic vein from medial regions of the pectoralis muscles (figures 8.10 and 8.12b). Further laterally, the axillary is joined by a long thoracic vein from other portions of the pectoralis and then by a short shunt from the thoracodorsal vein draining the latissimus dorsi muscle. The thoracodorsal vein continues cranially beyond this shunt to connect to the subscapular vein. These three tributaries to the axillary vein (ventral thoracic, long thoracic, and thoracodorsal) roughly parallel their arterial counterparts. Beyond juncture with the thoracodorsal vein, the axillary becomes the brachial vein passing along the upper arm. It receives a vessel, the deep brachial vein, that courses closely along with the deep brachial artery, from above the biceps muscle.

Subscapular Vein The subscapular vein first forms laterally at the junction of the thoracodorsal and lateral posterior humeral circumflex vein (figures 8.10 and 8.12b). The posterior humeral circumflex vein drains the lateral side of the upper arm.

Form & Function

Why does the thoracodorsal vein drain into both the axillary and subscapular veins? Throughout the circulatory systems of vertebrates, we find many "alternate" blood pathways. Different body postures during sleeping may exert pressure to one area of the body and require rerouting of blood past a compressed region. Junctions like the thoracodorsal shunt provide such useful alternatives.

External Jugular Vein

→ Return to the base of the external jugular vein and from here follow it anteriorly (figures 8.10 and 8.12b). It is joined medially by the slender internal jugular vein (often difficult to identify). Farther forward, a thin thyroid vein from the thyroid gland enters ventromedially; and at about the same point a larger branch from the shoulder, the transverse scapular vein, enters laterally. The transverse scapular vein can be followed over the lateral side of the shoulder, anastomosing with other veins, and through one circuit, connects with the posterior humeral circumflex vein. (Often, on the left side, the brown, chainlike thoracic lymphatic duct can be seen to join the venous system near the base of the subscapular, external jugular, or subclavian veins.)

Near the head, the external jugular forms by the confluence of three veins previously identified with the salivary glands (see figures 7.6 and 8.12b). Ventrally, the transverse jugular vein passes across the throat to connect with the opposite side. The middle tributary, anterior facial vein, runs along the ventral margin of the masseter muscle. The dorsal tributary, the posterior facial vein, comes from beneath the parotid gland and passes superficially across the round submaxillary gland.

Dorsal Aorta

→ Three categories of vessels depart from the dorsal aorta: somatic branches to the body wall, median visceral branches to the digestive tract and its associated glands, and lateral visceral branches to the urogenital system.

Somatic Branches

→ Pull viscera to the right to permit a left side view of the dorsal aorta. Shortly beyond the departure of the left subclavian artery, tiny paired intercostal arteries emerge segmentally from the top of the dorsal aorta to supply the wall of the thoracic cavity (figure 8.12a). Further caudally, after the dorsal aorta enters the abdominal cavity and passes the level of the kidneys, it gives off several more dorsal pairs of arteries to the adjacent abdominal wall; these are the lumbar arteries.

Return now to the area of the aorta in the thoracic cavity so the course of the dorsal aorta can be followed. Note the several tiny esophageal arteries to the esophagus. Also note the two arteries that supply the diaphragm, the phrenic arteries, one to the anterior and one to the posterior surface of the diaphragm.

Median Visceral Branches

→ Median visceral branches are unpaired. The first, the celiac artery, arises as the dorsal aorta enters the abdominal cavity; the second, the superior mesenteric artery, arises immediately next to it; and the third, the inferior mesenteric artery, arises further posteriorly above the colon, which it also supplies (figures 8.12a and 8.13).

Celiac Artery Follow the celiac artery first. It soon divides into three branches: the left gastric, hepatic, and splenic arteries (figure 8.12a).

The left gastric artery passes to the lesser curvature of the stomach. The lesser omentum should be selectively removed for a better view.

The hepatic artery swings cranially into the liver via the hepatoduodenal ligament. The first branch of the hepatic artery is the gastroduodenal artery, that itself almost immediately sends off a branch, the pyloric artery, to supply the pyloric region. The gastroduodenal artery next divides. One branch, the right gastroepiploic artery, veers to the left along the pylorus and the greater curvature of the stomach. The other branch of the gastroduodenal artery, the anterior pancreaticoduodenal artery, continues on to supply the duodenum and adjacent pancreas. Finally, just prior to joining the liver, the hepatic artery sends a short cystic artery to the gallbladder.

The splenic artery, the third and largest branch of the celiac artery, courses in the greater omentum. Near the spleen, it forks. The first fork, the anterior splenic artery, distributes to the anterior end of the spleen. The other fork, the posterior splenic artery, passes through and supplies the pancreas, to eventually reach the posterior end of the spleen.

Superior Mesenteric Artery Clear away fat to expose this vessel and its branches that in large part lie in mesenteries of the gut (figures 8.12a and 8.13). Note the prevalent lymph nodes. First to leave is the middle colic artery passing to the transverse and descending parts of the colon. Next to it, the superior mesenteric gives rise to the posterior pancreaticoduodenal artery that loops forward through the pancreas, supplying it and the adjacent duodenum. Follow it forward to confirm that it joins the previously identified anterior pancreaticoduodenal artery, a branch of the hepatic artery. The superior mesenteric next gives rise to the ileocolic artery to the caecum and terminal portion of the ileum; some branches loop to the left along the colon, anastomosing with the middle colic artery. Finally, numerous small looping branches, intestinal arteries, leave the superior mesenteric to distribute to the small intestine.

Inferior Mesenteric Artery The inferior mesenteric artery supplies the descending colon and rectum, and joins anteriorly with the middle colic artery (figures 8.12a and 8.13).

Lateral Branches of the Dorsal Aorta

→ **Adrenolumbar and Inferior Phrenic Arteries** Just behind the departure of the superior mesenteric artery, the dorsal aorta sends off laterally, at slightly different levels, the adrenolumbar arteries each supplying a small, hard, oval-shaped adrenal gland (figures 8.12a and 8.13). Before the adrenolumbar reaches the adrenal glands, the thin inferior phrenic artery departs cranially to supply the diaphragm.

Renal Arteries The next caudal branches off of the dorsal aorta are the large renal arteries extending to the kidneys (figures 8.12a and 8.13). A small branch to the adrenal gland may be located leaving the anterior side of each renal artery.

Internal Spermatic or Ovarian Arteries Past the kidneys, the dorsal aorta gives rise to arteries supplying the gonads (figures 8.12a and 8.13). In males, the slender internal spermatic arteries make the long journey posteriorly out of the abdominal cavity and into the scrotum containing the testes. In females, the larger, convoluted ovarian arteries pass laterally to the ovaries.

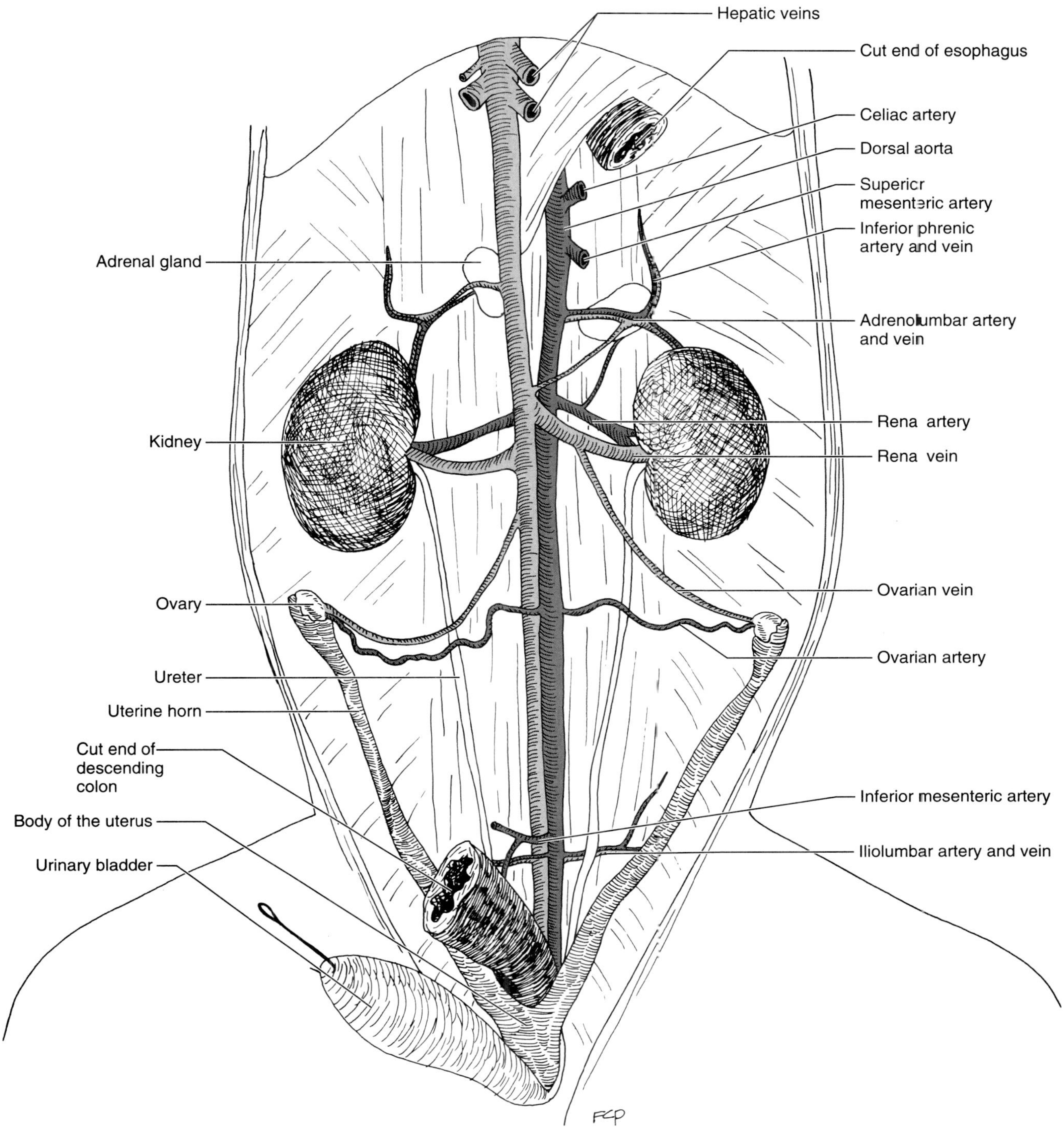

Figure 8.13
Ventral view of branches off the dorsal aorta and tributaries to the postcava in the trunk of a cat.

Iliolumbar Arteries Near the inferior mesenteric artery, the dorsal aorta gives off laterally the iliolumbar arteries to the adjacent abdominal wall, crossing first the psoas minor muscle, then iliopsoas muscle (figures 8.12a and 8.13).

External Iliac, Internal Iliac, and Caudal Arteries Shortly after giving rise to the iliolumbar arteries, the dorsal aorta forks into two external iliac arteries that pass to the hindlimbs (figures 8.12a and 8.14). Past this major fork, the dorsal aorta is considerably reduced in size and almost immediately sends off another lateral pair of branches, the internal iliac arteries (= hypogastric arteries). Beyond this point, the dorsal aorta is termed the caudal artery (= median sacral artery), eventually running into the tail.

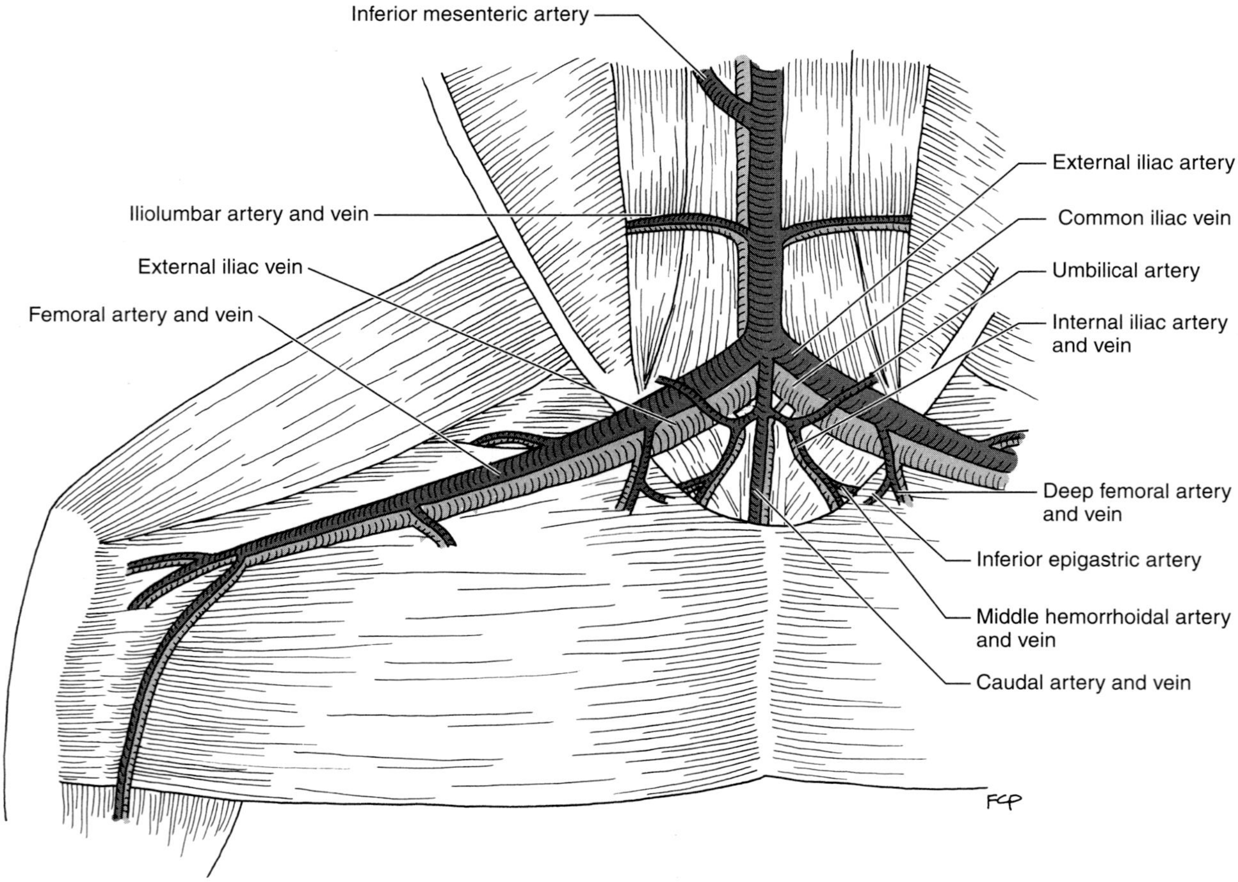

Figure 8.14
Blood vessels of the medial hindlimb of a cat., ventral view.

Branches of the Internal Iliac Artery Near its base, the internal iliac gives off the umbilical artery to the urinary bladder, the superior gluteal artery passing deep to gluteal and other thigh muscles, and finally, the internal iliac ends in a bifurcation (figure 8.12a). One branch, the middle hemorrhoidal artery, swings ventrally to supply the rectum and itself sends, in females, a uterine artery to the uterus. The other branch, the inferior gluteal artery, dips dorsally passing next to the sciatic nerve.

Branches of the External Iliac Artery The external iliac passes out of the abdominal cavity and into the thigh (figures 8.12a and 8.14). Its first branch is medial—the deep femoral artery that passes the adductor longus on its way to supply deep muscles of the thigh. At the base of the deep femoral arises, among other branches, the inferior epigastric artery that swings forward to run along the rectus abdominis, eventually joining the superior epigastric artery, a continuation of the internal mammary. Beyond this point of departure of the deep femoral, the external iliac becomes the femoral artery coursing along the medial side of the thigh, giving off branches that will not be followed.

Postcava

→ Small veins joining the postcava and its large tributaries parallel their arterial counterparts and bear the same names.

The postcava first forms, at the rear of the abdominal cavity, by the union of the two common iliac veins, one from each hindlimb (figures 8.12b and 8.14). The caudal vein (= median sacral vein) joins one common iliac, usually the left. In turn, the common iliac forms by union of the large external iliac vein and smaller, medial internal iliac vein (= hypogastric vein). (The external iliac receives the femoral and deep femoral veins. The internal iliac receives middle hemorrhoidal and inferior gluteal veins.)

As the postcava passes forward, it receives first the lateral iliolumbar veins, then the right internal spermatic or the right ovarian vein, and finally the large renal veins (figures 8.12b and 8.13). However, the left internal spermatic or left ovarian vein usually contributes to the left renal vein before reaching the postcava. Small left and right adrenolumbar veins join the postcava at slightly different points just anterior to the entry of the renal veins. Before passing through the diaphragm, the postcava receives several hepatic veins from the liver and the paired superior phrenic vein from the adjacent di-

Form & Function

Within the circulatory system of the cat, many arterial loops have been identified. For example, the anterior joins the posterior pancreaticoduodenal artery, the pyloric artery (gastroduodenal) meets the left gastric artery (celiac), and the anterior connects to the posterior humeral circumflex vein. What could be the function of these loops? Like the thoracodorsal shunt to the axillary vein, these loops provide alternate routes for blood to flow to a particular region. When blood flow through a particular route is decreased, perhaps by pressure, alternate pathways can be used.

aphragm. The inferior phrenic vein, as you might guess from identification of its arterial counterpart earlier, joins the adrenolumbar vein.

The postcava, while in the abdominal cavity, regularly receives dorsally, unpaired veins from a median groove between muscle masses, the lumbar veins. Pull the postcava to one side of its course along the roof of the abdominal cavity to see the entry of these vessels.

Hepatic Portal System

Veins of the hepatic portal system drain the gut and its associated glands. However, the hepatic portal veins empty not into the postcava, but into the liver sinuses. From the liver, blood is collected into the hepatic veins flowing eventually to the inferior vena cava. The hepatic portal system must be specifically injected, since latex added to the arterial or venous systems will not reach these blood vessels through the capillaries of the gut. Latex injected into the hepatic portal system is usually a different color than the latex in the arterial or venous systems. Your instructor will tell you how your cats were injected.

Hepatic Portal Vein

→ The hepatic portal vein is primarily formed along the midline near the lesser curvature of the stomach by the merger of the superior mesenteric and gastrosplenic veins (figure 8.12c). The tributaries to these two veins are followed below. Just cranial to this junction, the hepatic portal is joined by three additional veins: the coronary vein, draining the lesser curvature of the stomach, the right gastroepiploic vein, from the greater curvature of the stomach, and the anterior pancreaticoduodenal vein, draining the duodenum and pancreas.

Tributaries of the Gastrosplenic Vein

→ Four veins combine to form the gastrosplenic vein (figure 8.12c). As their names imply, the anterior splenic vein drains the anterior end of the spleen and the posterior splenic vein the posterior end. Two smaller veins also join: the pancreatic vein from the pancreas and the middle gastroepiploic vein from the lesser curvature of the stomach.

Tributaries of the Superior Mesenteric Vein

→ The superior mesenteric vein, the largest contributor to the hepatic portal vein, primarily forms by the fusion of the intestinal, inferior mesenteric, and posterior pancreaticoduodenal veins (figure 8.12c). Begin tracing these tributaries by pulling the small intestine away from the body cavity. There, nestled within the mesentery, are many small intestinal veins that contribute to the beginning (the most caudal extent) of the superior mesenteric vein. As the superior mesenteric continues forward, it receives the inferior mesenteric vein draining the large intestine. Finally, the superior mesenteric is joined by the posterior pancreaticoduodenal vein from the posterior end of the pancreas and duodenum.

Form & Function

Think back to the thin esophageal arteries and the prominent visceral arteries to parts of the gut (e.g., celiac artery). Why do the diameters of these blood vessels vary so greatly? The differences in arterial size reflect differences in size and activity of the organs supplied. The esophagus primarily serves as a mere conduit to the stomach. The stomach, however, churns and mixes food with gastric juice and engages in some absorption, which requires additional blood flow to support these activities. But the intestines, through their many large arteries, receive the greatest blood supply. Like the stomach, the intestines engage in extensive muscular activities, churning and moving food along. But in addition, the intestines are the primary site of absorption of digested molecules into the blood supply. Here, as with the earlier muscular and skeletal systems, differences in function are correlated to differences in form.

The volume of blood flow to parts of the respiratory system show similar correlations. The trachea and bronchi, serving primarily as gas conduits, have only a minimal vascular supply. But the lungs, the site of gas exchange, are one of the most extensively vascularized regions in the body.

9 Urogenital System

Survival of an individual depends upon many factors, including adaptations to escape predators, obtain food, and withstand harsh environmental conditions. But to pass adaptations to descendants, the individual must reproduce successfully, the primary biological role of the genital system.

The reproductive systems of vertebrates must produce and store viable **gametes** (eggs or sperm) prior to mating. In males, the secretions of many glands mix with the sperm to form **seminal plasma**, the fluid used to stabilize and transfer the sperm to the female. Male vertebrates typically have **copulatory structures** (e.g., claspers in sharks, a penis in mammals) used to transfer sperm. However, some male fish and most male salamanders (including *Necturus*) transfer sperm using **spermatophores**, gelatinous structures capped with sperm. Cloacal glands in these vertebrates produce components of the spermatophore.

Female reproductive systems vary according to the site of development of the embryos. **Viviparous species** are "live" bearing, giving birth to young not encased in a shell. The dogfish shark and cat are viviparous. Their young develop within a uterus. Their reproductive tracts are therefore adapted to receive sperm and support embryonic development. **Oviparous species**, including *Necturus* and most other amphibians, are egg laying. The embryo emerges from the mother wrapped in a shell or jelly coat secreted by the uterus. Their reproductive tracts include ducts and glands that produce the jelly coats (or shells, in reptiles and birds) around the eggs.

The prominence of reproductive components of the urogenital system varies with maturity and seasonal cycle. Be prepared for differences in size and development of the systems. Also, remember to move around the laboratory to see dissections and appreciate the significant differences in male and female sexual anatomy.

Compared to the reproductive system, the urinary system of vertebrates is devoted to quite different functions, including the elimination of waste products, the regulation of water balance, and the control of electrolyte levels. We treat the urinary and reproductive systems together as the urogenital system because they share so many of the same ducts.

Shark

Male Urogenital System

Pelvic Fin

→ The pelvic fins of males possess enlarged medial bars, the claspers, derived from pterygiophores (figure 9.1).

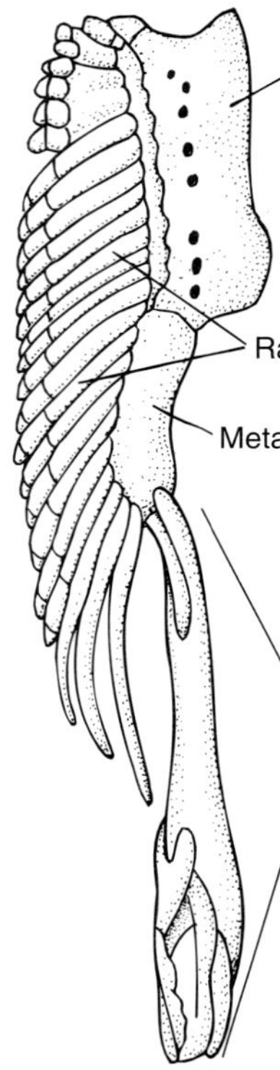

Figure 9.1
Ventral view of the left pelvic girdle, pelvic fin, and intromittent organ of the shark *Chlamydoselachus*.
After vanTienhoven.

The medial side bears a groove that carries seminal fluid during copulation. At the ventral base of each pelvic fin, just under the skin, lies the collapsed, muscular, siphon, usually dark brown in color (figure 9.2). Open the siphon to find a medial pore. Fluid secreted by the siphon passes into the groove, travels along the clasper, and thus contributes to seminal fluid. Slide a blunt probe along the groove and into the pore to discover its course.

In the male, remove and discard the liver by cutting the lobes close to their points of attachment. When necessary, you may also want to cut through the puboischiac bar to expose parts of the urogenital system.

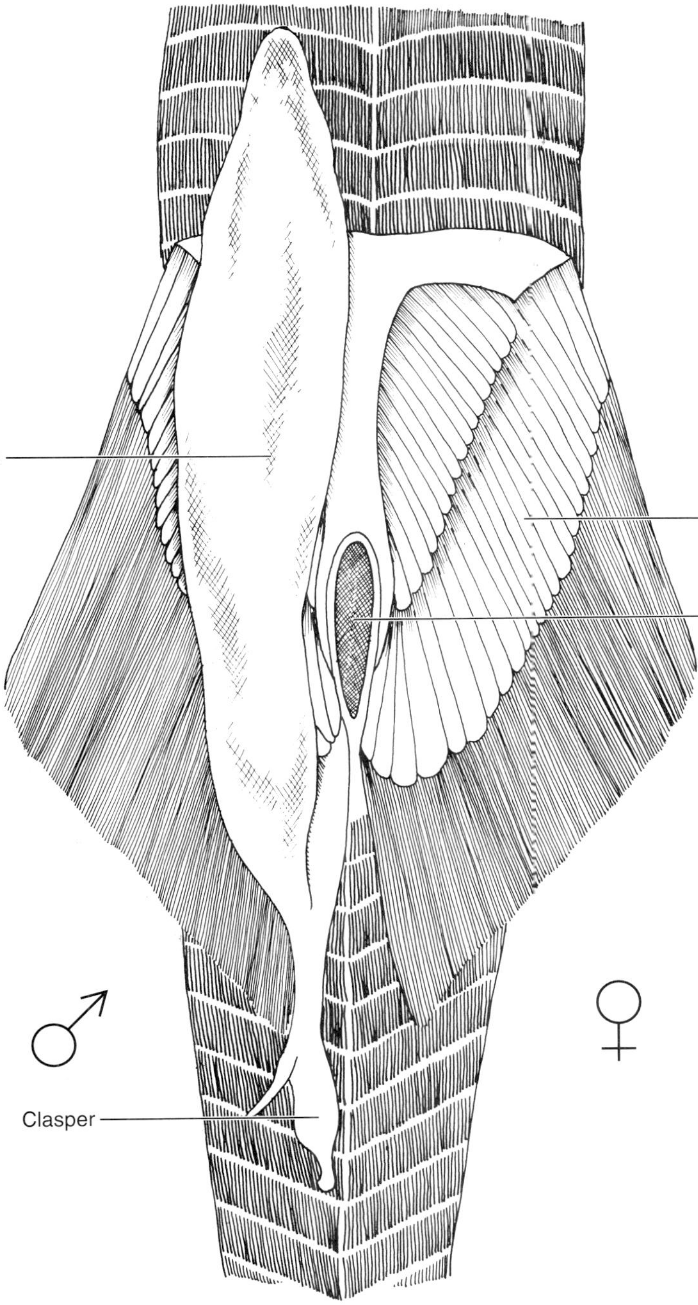

Figure 9.2
Ventral view of the female and male shark pelvis.

Testis and Its Ducts

→ Each oval, soft testis is suspended from the anterior roof of the pleuroperitoneal cavity by the mesorchium (figure 9.3). The tightly convoluted, paired duct along the dorsal body wall is the ductus deferens (= vas deferens, wolffian duct, mesonephric duct, or archinephric duct). Several threadlike, efferent ductules pass in the mesorchium and carry sperm from the testis to the epididymis within the anterior end of the kidney. These small efferent ductules may be seen by shining a light through the mesorchium from behind. The epididymis is composed of a knot of many small tubules concentrated just lateral to the mesorchium. From the epididymis, sperm enter the larger but still coiled ductus deferens (figure 9.3). As the ductus deferens extends caudally, it enlarges and straightens, forming the seminal vesicle. Trace the seminal vesicle caudally. Just where it looks as though the seminal vesicle will enter the cloaca, it instead terminates in a bilobed sperm sac.

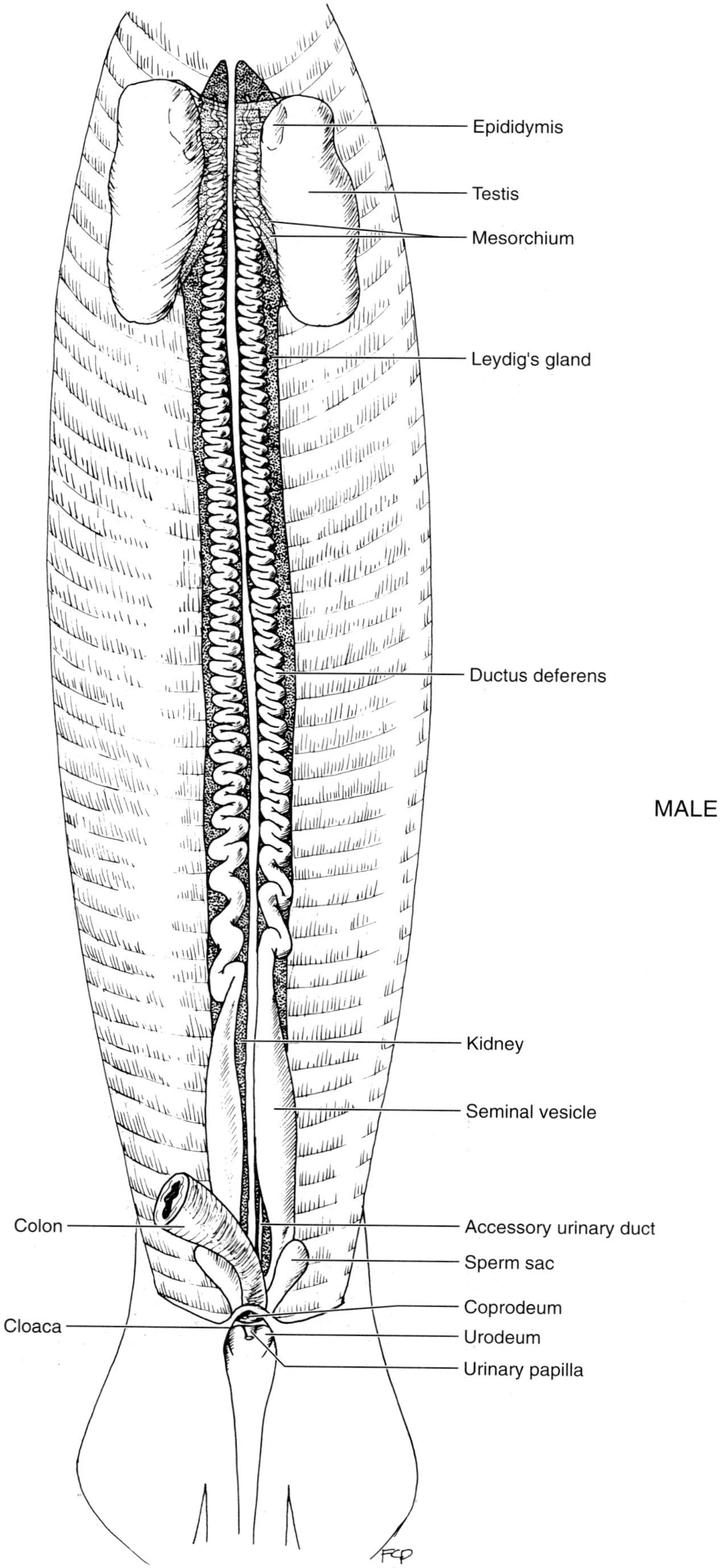

Figure 9.3
Ventral view of the urogenital system of a male shark, with the digestive tract removed.

Patterns & Connections

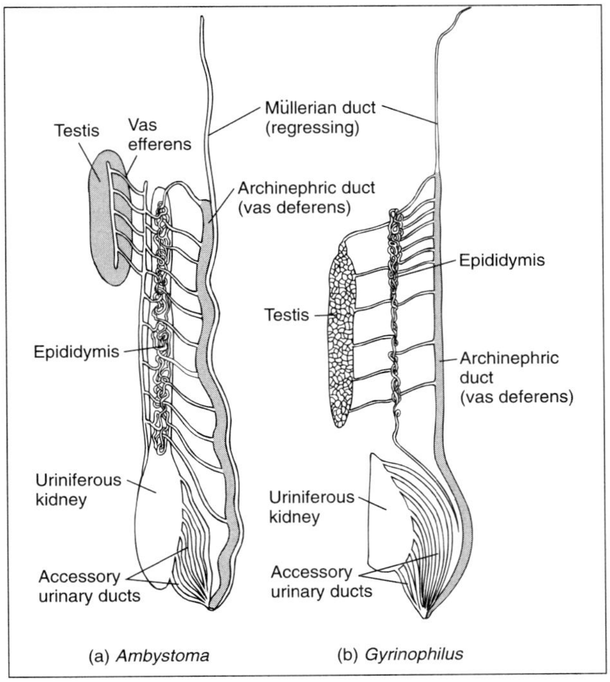

Urogenital systems of male salamanders. (a) Salamander *Ambystoma*. (b) Salamander *Gyrinophilus*. In both animals, the left testis and associated ducts have been reflected to the other side. The right testis and ducts are not shown.
(a) After C.L.Baker and W.W.Taylor, Jr., 1964, "The urogenital system of the male Ambystoma," Jrnl.Tenn. Acad. Sci. 39:2.
(b) After P. Strickland, 1966, "The male urogenital system of Gyrinophilus Danielsi dunni," Jrnl.Tenn.Acad. Sci. 41:1.

Most of the urogenital system of embryonic vertebrates typically begins as gender neutral. That is, parts are present, but whether they will develop, and how they will develop, will be influenced by the sex of the individual. One example is the müllerian and archinephric ducts (= wolffian duct, pronephric duct, mesonephric duct, opisthonephric duct, ductus deferens, vas deferens).

In males, the müllerian ducts rarely develop. Instead, the archinephric ducts develop into parts of the urogenital system variously associated with sperm and urine transport, depending upon the particular vertebrate.

In females, the archinephric ducts tend to function only within the urinary system. Instead, the müllerian ducts variously form the oviduct, uterus, and vagina, again, depending upon the particular vertebrate. Watch for these names throughout the dissections of the urogenital systems.

Kidney

→ The shark kidney, or opisthonephros, is positioned outside the peritoneum within a long, narrow groove along the roof of the pleuroperitoneal cavity. The anterior region of the kidney, called Leydig's gland, serves no urinary function, but instead helps in sperm transport. It is primarily composed of the epididymis and ductus deferens identified previously. Glands around the ductus deferens produce a milky fluid that contributes to the seminal fluid. Cut a small slit in the wall of the ductus deferens to reveal these milky secretions.

The remaining posterior region is termed the kidney or opisthonephros proper. In males, urine produced by the opisthonephros is collected into the accessory urinary duct, running along the medial side of the kidney. Cut through the middle of one seminal vesicle and reflect the ends. This exposes the microscopic accessory urinary duct, which runs caudally to enter the base of the sperm sac.

Cloaca

→ The cloaca receives products of the urogenital and digestive systems. Open the cloaca with a short medial cut. A partial fold transversely divides the cloaca into ventral (coprodeum) and dorsal (urodeum) regions. The intestine drains into the coprodeum and the urogenital system empties into the urodeum. The single, nipplelike process in the urodeum is the urinary papilla. All urinary and genital products passed to the sperm sac empty into the cloaca through this urinary papilla.

Female Urogenital System

Pelvic Fin

The pelvic fin of females lacks claspers, siphon, or any other copulatory specializations found in males.

Ovary and Its Ducts

→ Each ovary hangs from the anterior roof of the pleuroperitoneal cavity by the mesovarium (figure 9.4). In immature females, the ovaries are soft and oval structures. The oviducts in immature females are slender tubes running free without mesenteries along the dorsal pleuroperitoneal cavity next to the dorsal aorta.

In sexually mature females, the ovaries are usually lumpy and the oviducts are now suspended by the mesotubarium (figure 9.4). Follow the oviducts cranially. They pass dorsal to the liver, arch toward the midline, pass into the falciform ligament, and join in a common opening, the ostium. The ostium is slitlike and its rim is usually collapsed. If you have difficulty finding it, again follow the arching oviducts forward to where they unite in the falciform ligament.

As with the male, remove and discard the liver at its base, leaving undisturbed the regions immediately associated with the falciform ligament, the oviducts, and the ostium. When necessary, you may also want to cut through the puboischiac bar to expose parts of the urogenital system.

Beginning at the ostium, trace the oviduct posteriorly. Just above the ovary in mature specimens, the oviduct slightly enlarges, forming the shell gland

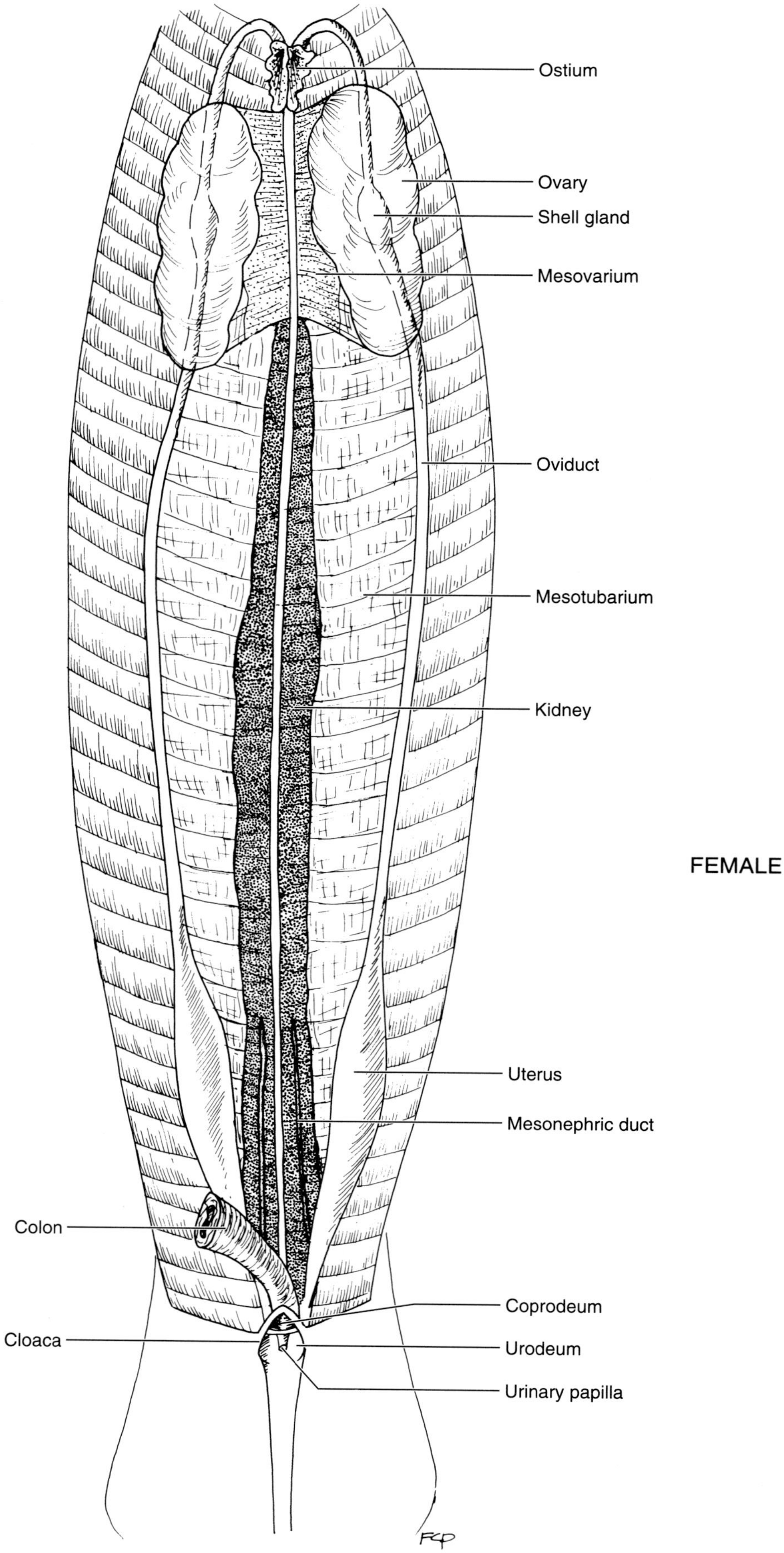

Figure 9.4
Ventral view of the urogenital system of a female shark, with the digestive tract removed.

Form & Function

The caudal region of the ductus deferens of *Necturus* and most larval salamanders can potentially conduct both urine and sperm in the sexually mature adult. However, that dual role is not typically found in adult salamanders. Instead, the connection between the urinary collecting tubules and ductus deferens shifts caudally, so that urine is passed either separately or only at the most caudal end of the ductus deferens. That *Necturus* males retain this juvenile trait is yet another example of their unusual paedomorphic morphology.

But why do the urinary collecting tubules shift caudally in the typical adult salamander? Consider the roles of the adult ductus deferens. Sperm are stored within the ductus deferens and epididymis of male salamanders just prior to the breeding season. But storing sperm in a region of the ductus deferens that also must transport urine may harm the sperm or the sperm may interfere with the passage of urine. Thus, by shifting the connections between the ductus deferens and urinary collecting tubules more caudally, additional space is available in the ductus deferens to store sperm.

(= nidamental gland) that secretes a thin membrane around eggs as they pass (figure 9.4). Caudal to this gland, the oviduct again narrows for a short distance, then enlarges greatly into the uterus, which contains developing embryos. In immature specimens, the uterus is only slightly wider than the oviduct.

Kidney

→ In the female, the anterior part of the kidney has no urinary or reproductive role and fails to develop. What persists is the middle and posterior portions that form the adult kidney (figure 9.4). Transect the oviduct of one side and reflect the ends. Next, cut through the peritoneum to expose the kidney that runs outside the parietal peritoneum next to the dorsal aorta (as in the male). No accessory urinary duct of males exists in females. Instead, the kidney is drained by the mesonephric duct (= wolffian) lying along the ventral surface of the kidney. Before emptying into the cloaca, the mesonephric duct expands, forming the urinary vesicle (often impractical to find in immature females).

Cloaca

→ As in the male, the cloaca is divided horizontally into a ventral coprodeum, receiving contents from the intestine, and a dorsal urodeum, receiving the products of the urogenital system. Make a short midventral cut in the cloaca to see these regions. Within the urodeum is the single, nipplelike urinary papilla formed by union of mesonephric ducts. Urinary products eventually enter the cloaca through this urinary papilla. The oviducts enter the cloaca through pores on either side of the urinary papilla.

Necturus

The kidneys and the arrangement of the associated ducts are similar, but not identical, in the shark and *Necturus*. For example, both the shark and *Necturus* possess **opisthonephric kidneys.** As in the male shark, the *Necturus* **ductus deferens** (= vas deferens) serves in transport of sperm and seminal fluids. But unlike in the shark, **urinary collecting tubules** transport urine from the kidneys into the caudal end of the ductus deferens. Thus, part of the ductus deferens in male *Necturus*, but not male sharks, transports both urine and sperm. In females, the **mesonephric duct** only drains the kidney. The **oviduct** (= müllerian duct) transports eggs between the ovary and cloaca and deposits jelly coats along the way.

Male Urogenital System

Testis, Kidney, and Related Ducts

→ Each elongate testis hangs from the middorsal wall of the pleuroperitoneal cavity by the mesorchium (figure 9.5). Running through the mesorchium are efferent ductules. These ducts transport sperm from the testis to the cranial end of the kidney, or epididymis. (As in the male shark, the kidney is divisible into a narrower anterior region, the epididymis, involved in sperm transport, and a larger posterior region that produces urine.) These tiny efferent ductules may be seen by shining a light through the mesorchium from behind, as was done in the male shark. Numerous spermatic arteries and veins will be prominent within the mesorchium.

From the epididymis, sperm are transported to the ductus deferens, a larger and somewhat coiled duct running parallel to the kidney along its lateral margin. At the cranial end of the ductus deferens, just beyond the epididymis, a thin black cord parallels the ductus deferens and then continues cranially on its own. This cord is what remains of the müllerian duct that fails to develop in males.

Follow the ductus deferens posteriorly (figure 9.5). Near the caudal end of the pleuroperitoneal cavity, it straightens and receives numerous urinary collecting tubules from the urinary kidney. Beyond the kidney it slants medially to empty into the cloaca.

Cloaca

→ The collapsed urinary bladder empties into the terminus of the large intestine just prior to its entry into the cloaca (figure 9.5). If not already done, then make a short midventral incision through the cloaca to find points of entry of the large intestine and of the paired mesonephric ducts.

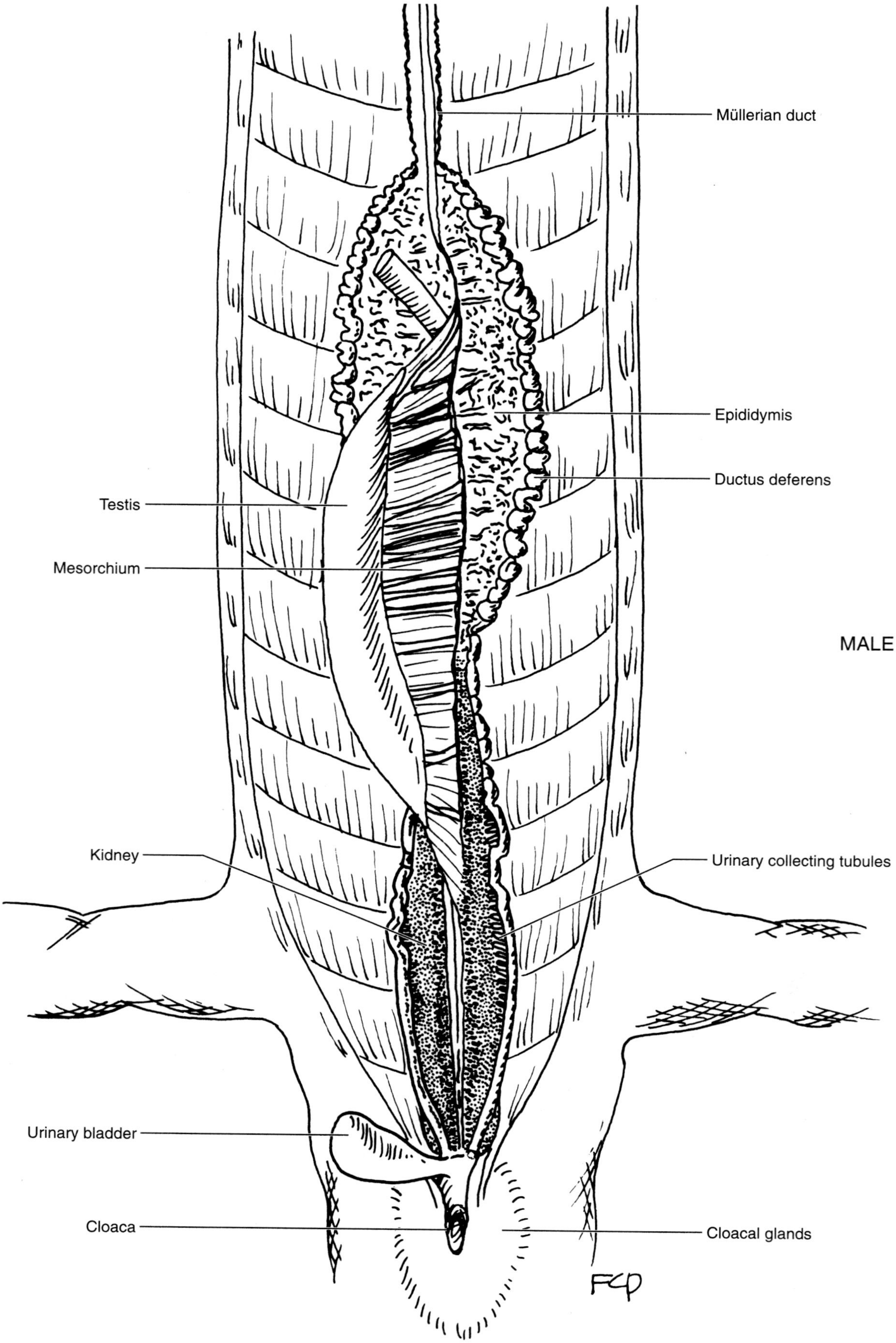

Figure 9.5
Ventral view of the urogenital system of a male *Necturus*, with the digestive tract removed.

Female Urogenital System

Ovary and Its Ducts

→ Each ovary, containing hundreds of beadlike ova or eggs, hangs by the mesovarium from the middorsal wall of the pleuroperitoneal cavity (figure 9.6). Ovarian arteries and veins pass through the mesovarium. The paired, tortuous oviducts run along the roof the pleuroperitoneal cavity. Trace one oviduct forward to find its funnel-shaped beginning and the opening into it, the ostium. Eggs, released from the ovary into the body cavity, are directed into the ostium to begin their transport down the oviduct. At its posterior terminus, the oviduct empties into the cloaca.

Kidney

→ The paired kidney lies in the posterior roof of the pleuroperitoneal cavity. Unlike the male, the female kidneys are composed of only the posterior urinary section. Numerous urinary collecting tubules transport urine from the kidneys to the mesonephric duct (= wolffian), running along the lateral edge of the kidney (figure 9.6). The mesonephric duct drains into the cloaca.

Cloaca

→ As in the male, the unpaired urinary bladder empties into the terminus of the large intestine, which in turn empties into the cloaca (figure 9.6). Make a short midventral incision through the cloaca to find points of entry of the large intestine, the mesonephric ducts, and finally the oviducts through their small nipplelike papillae.

Cat

Male Urogenital System

Kidney, Urinary Bladder, and Associated Structures

→ The general structures of the urinary system will first be explored to provide basic orientation to the pelvic region. Arising from the medial indentation (hilus) on each kidney is a thin white tube, the ureter, that passes posteriorly to the base or fundus of the muscular urinary bladder (figure 9.7). The free cranial end of the urinary bladder is its vertex (= apex). Pull the vertex toward you and thus away from the descending colon. The pouchlike cavity between the bladder and colon is the rectovesical pouch. The fundus of the bladder narrows into the urethra. In this region, the descending colon straightens into the final section of the gut, the rectum, running above the urethra. The rectum will be better exposed and followed shortly.

Inguinal Canal and Associated Structures

→ Return now to the dorsal aorta and postcava. On one side, locate the very thin internal spermatic artery and vein, if proper injection permits. Follow these posteriorly until they pierce the rear wall of the abdominal cavity (figure 9.7). This exit portal is the inguinal canal. Its openings on inner and outer sides are, respectively, internal and external inguinal rings. Note that a white cord arches over the ureter and extends laterally to join the internal spermatic vessels as they enter the inguinal canal. This cord is the ductus deferens, carrying sperm from the testis to the urogenital canal at the base of the penis. Clear out fat and connective tissue to expose the ureter, ductus deferens, and spermatic vessels.

The three structures—ductus deferens, internal spermatic artery, and vein—become bound in a connective tissue sheath as they pass out of the abdominal cavity (figure 9.7). All three tubules so wrapped are collectively known as the spermatic cord. Cut through the inguinal canal to locate the spermatic cord. Continue the cut through the skin following the cord posteriorly into the scrotum, the saclike fold of skin containing the teardrop-shaped testis (figure 9.7). The testis is also wrapped in connective tissue and will be dissected in detail shortly.

Patterns & Connections

The inguinal canals of adult male humans are regions of potential problems. For example, heavy lifting or other activities that generate strong abdominal pressures can cause the intestines to protrude into the inguinal canal, resulting in an inguinal hernia and often great pain.

Reproductive Glands and Associated Structures

→ Now return to the urethra to find the point where the ductus deferens loops medially to join it. Cut through the pubic and ischial symphyses with heavy scissors. Spread the cut by pressing the knees of the cat laterally to the bottom of the dissecting pan. As needed, extend this cut by trimming off more bone adjacent to the symphyses.

The ductus deferens of each side extends cranially from the internal inguinal ring, arches over the corresponding ureter, and then turns posteriorly and medially to join the urethra, thus forming the urogenital canal (figure 9.7). Follow each ductus deferens to its junction with the urethra. At this point of union resides the prostate gland, a small dorsal swelling (figure 9.7). The urogenital canal runs for about 2 cm to where it terminates in the penis. This terminal point of the urogenital canal is marked by two small lateral swellings usually enmeshed within connective tissue, the bulbourethral (= Cowper's) glands (figure 9.7). (Most mammals possess a third gland contributing to the genital ducts, the seminal vesicle, but this is absent in the cat.)

Above the urethra and urogenital canal runs the rectum. Now, with the symphyses cut and spread, the rectum is better exposed. At its terminus is an anal sphincter, a

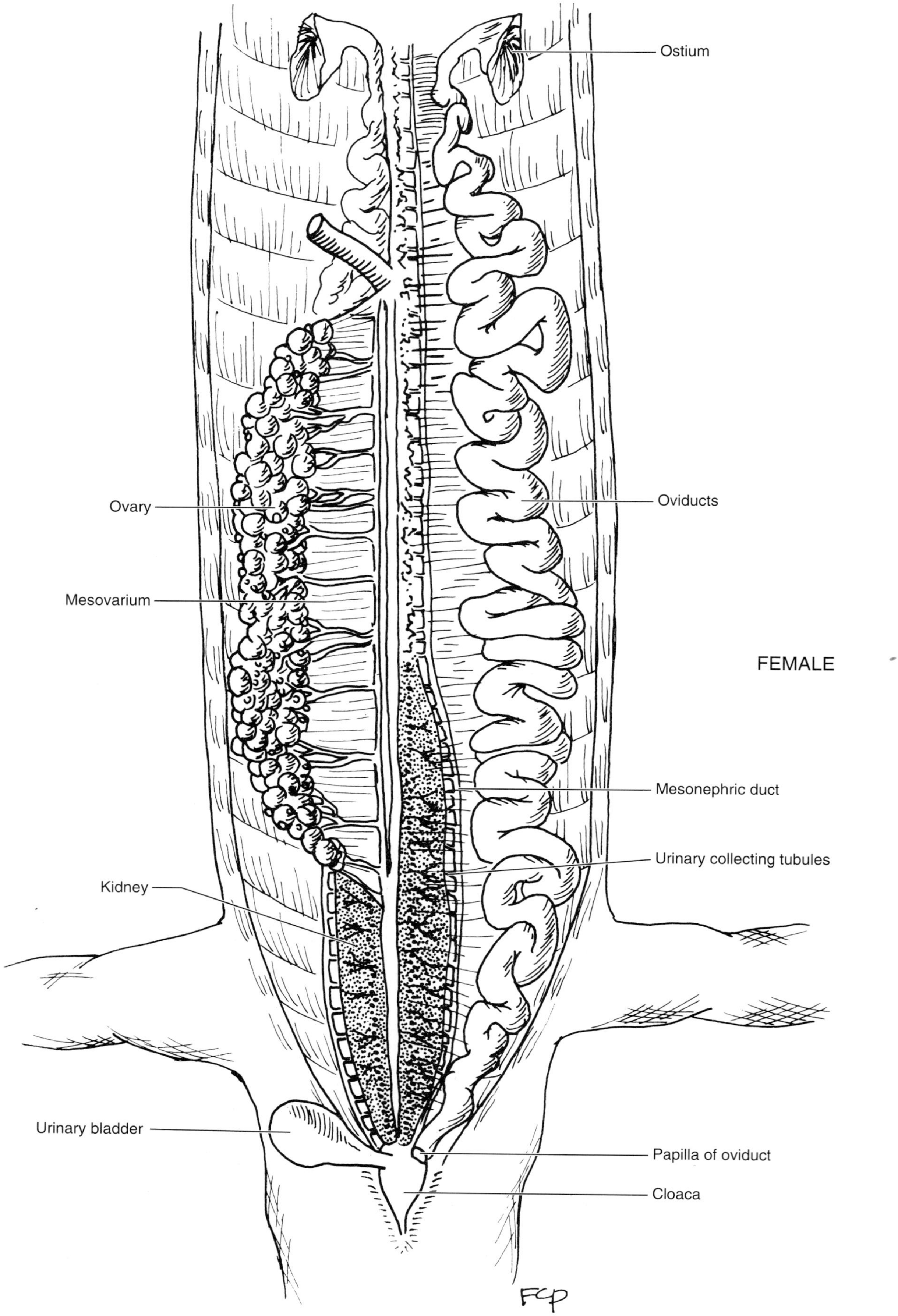

Figure 9.6
Ventral view of the urogenital system of a female *Necturus*, with the digestive tract removed.

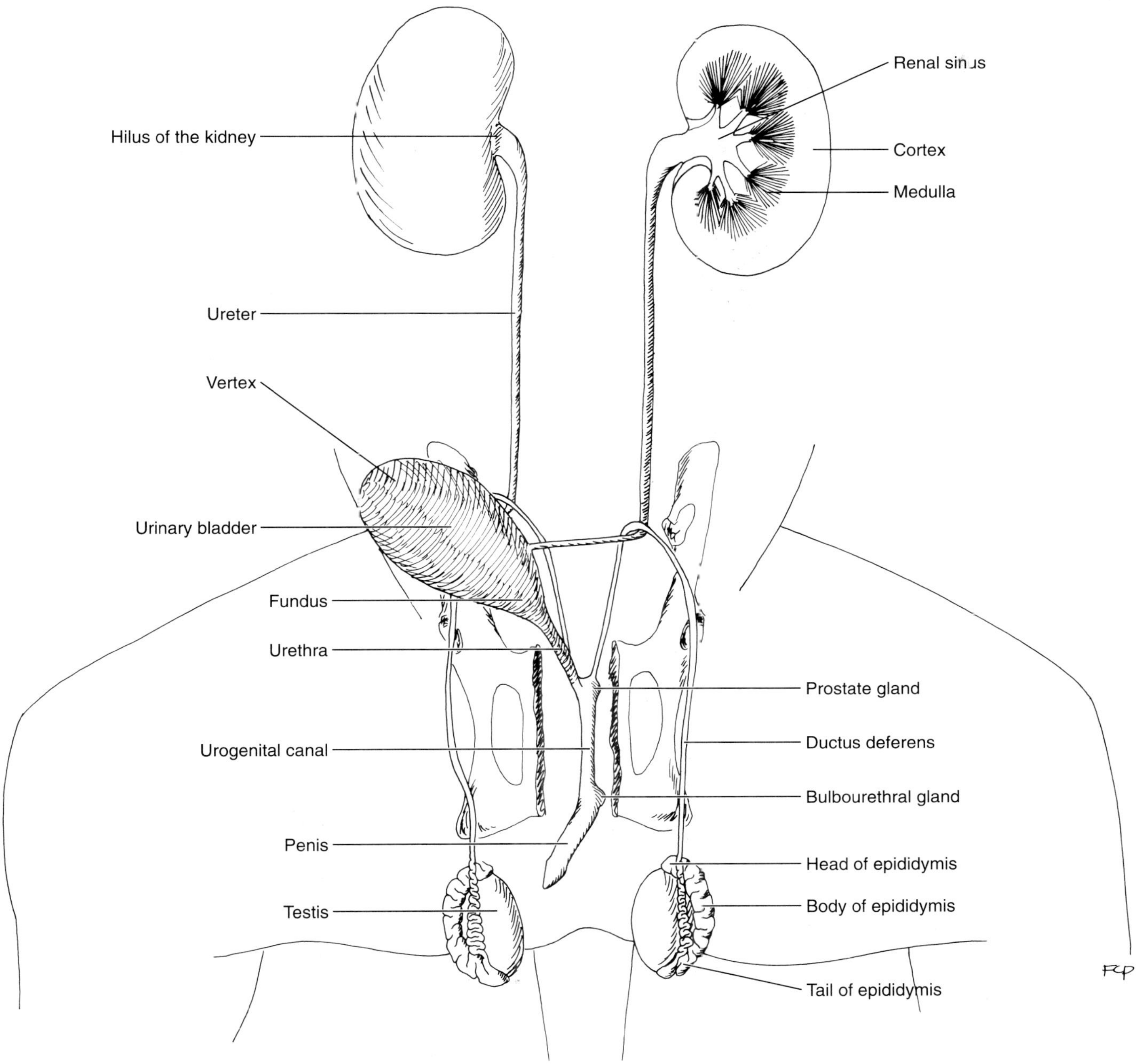

Figure 9.7
Ventral view of the urogenital system of a male cat, with the digestive tract removed.

band of smooth muscle. This sphincter controls the posterior opening of the digestive tract, the anus. A pair of rounded anal glands lie along the sides of the rectum near the anus.

Testis

→ During fetal life, each testis descends from the abdominal cavity, through the then larger inguinal canal, and into a pouch of skin, the scrotum (figure 9.8). The fascial sheath, or cremasteric fascia (= spermatic fascia or cremasteric pouch), encloses components of the spermatic cord and also wraps the testis. Embryologically it arises as an outpocketing of abdominal wall that projects into the scrotum. It is derived specifically from fascia of the external oblique and from the internal oblique muscle. The latter muscle also gives rise to the cremasteric muscle (atrophied in the cat).

The cremasteric fascia forms a sac around the testis. Cut open this sac along the ventral edge. The cavity opened, between the testis and fascia, is the vaginal sac (figure 9.8). The posterior end of the testis attaches to the cremasteric fascia by a short, tough ligament, the

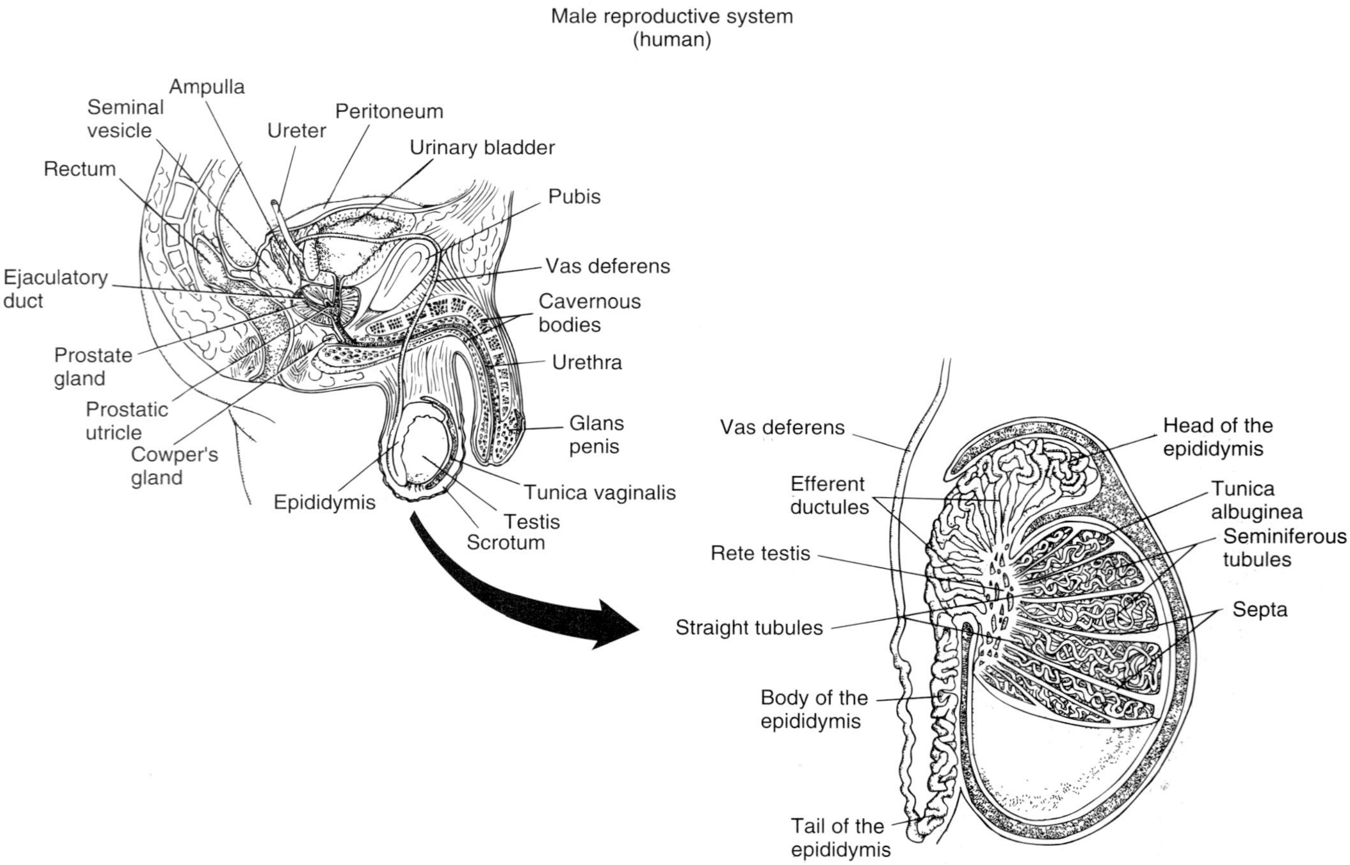

Figure 9.8
Human male reproductive system. (Top) This sagittal section of the male pelvis shows the reproductive organs and their relationships to the urinary and digestive systems. (Bottom) This midsagittal section reveals the testis and its complex duct system. Spermatozoa produced in the seminiferous tubules eventually pass through the straight tubules into the rete testis and enter the epididymis. Fluid is added as spermatozoa are moved through the vas deferens by contractions of sheets of smooth muscle in its walls.

gubernaculum (= caudal ligament of epididymis). The C-shaped epididymis wraps around the ends and back side of the testis. The cranial part of the epididymis is its head, the section along the side of the testis its body, and the region over the caudal end of the testis the tail (figure 9.8). The tail empties into the thin, convoluted ductus deferens. The internal structure of the testis is depicted in figure 9.8. Sperm are produced in the walls of the seminiferous tubules, then pass to the tubuli recti, to the rete testis, then out of the testis into efferent ductules and so enter the head of the epididymis.

Internal Kidney Structure

→ Clear away fat, and note the kidney's legendary bean shape. The medial indentation is the hilus through which renal blood vessels, lymphatic vessels, and ureter pass.

Horizontally cut one kidney to divide it into approximately equal dorsal and ventral halves. The cavity within the kidney entered when passing through the hilus is the renal sinus (figure 9.7). However, because the renal sinus is also occupied by blood vessels, the ureter, and fat, you may not initially find the renal sinus. Note that within the renal sinus, the ureter is expanded, forming the pelvis of the ureter. The projection of kidney tissue into the renal pelvis is the renal papilla. The outer part of the kidney is the cortex; its inner core, including the papilla, is the medulla (figure 9.7).

Female Urogenital System

Urinary Structures

→ Drainage of the kidney is similar to that in the male. The ureter, a white cord, leaves each kidney at its medial indentation, and runs posteriorly to the base or fundus of the urinary bladder (figure 9.9). The free anterior end of the urinary bladder is the vertex (= apex). The fundus narrows into a slender urethra passing above the pubic symphysis and below the uterus (figure 9.9). This will be followed shortly. Now, however, pull the vertex of the bladder toward you to locate the deep pocket between the urethra and uterus, the vesicouterine pouch.

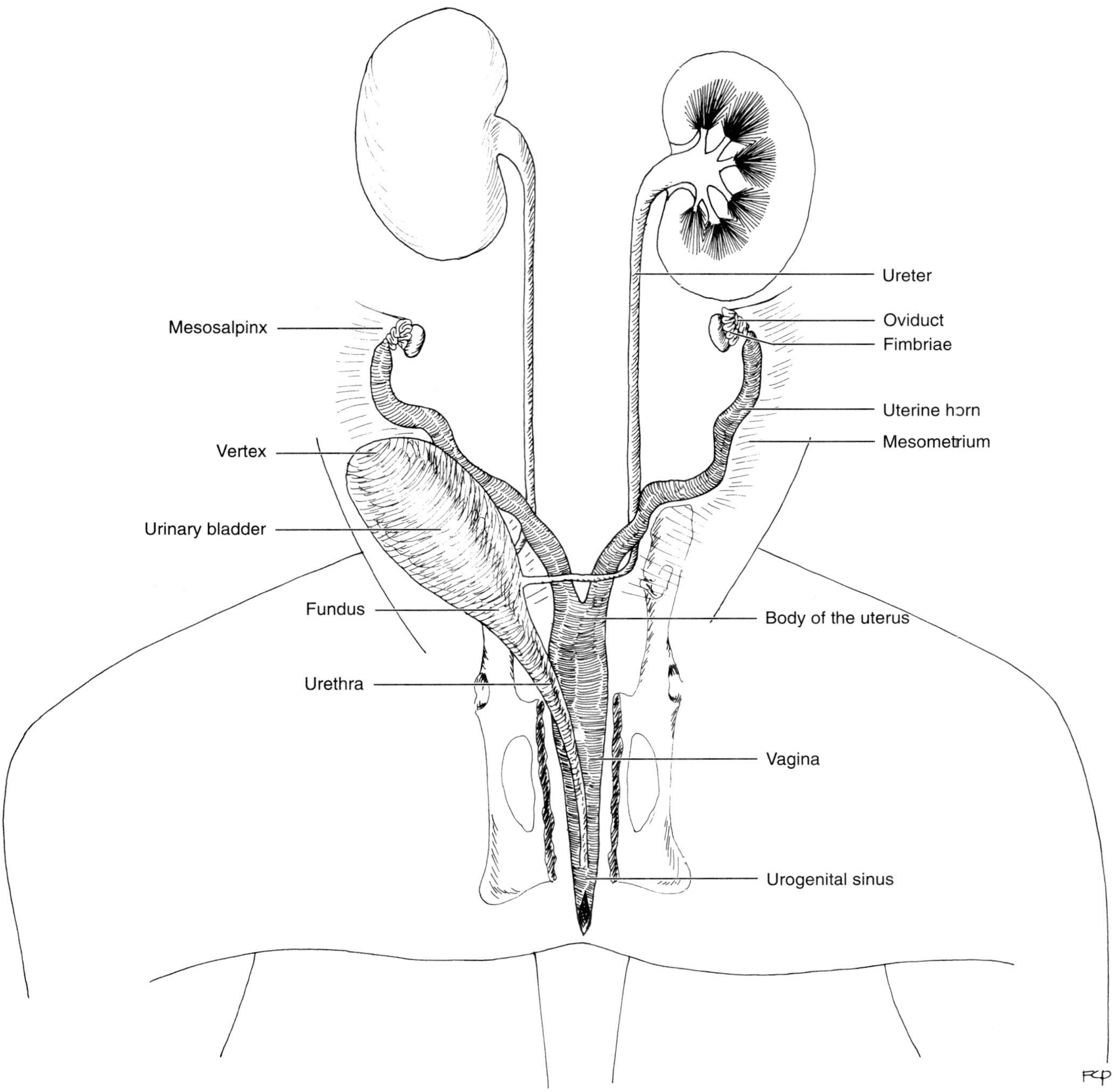

Figure 9.9
Ventral view of the urogenital system of a female cat, with the digestive tract removed.

Refer to the description of the kidney in the male for identification of internal parts and regions.

Reproductive Structures

→ The ovarian artery and vein of one side, if properly injected, can be followed laterally to the small, oval ovary. Ova ovulated from an ovary are collected into the thin, convoluted oviduct (= fallopian or uterine tube), which lies just lateral to the ovary (figure 9.9). The opening into the oviduct is the ostium, whose frilly rim is the fimbriae. The oviduct transports the ova to its associated uterine horn, where development of the embryos occurs. The paired uterine horns, one from each side, join above the urinary bladder to form the single body of the uterus. After identifying the associated mesenteries, the reproductive tract will be followed further.

Mesenteries of the Reproductive Tract The ovary is suspended from the abdominal roof by the mesovarium. The short, cordlike ovarian ligament attaches the posterior end of the ovary to the oviduct. The oviduct is supported by the mesosalpinx, and each uterine horn is supported by a mesometrium. Collectively, the mesovarium, mesosalpinx, and mesometrium form the broad ligament of the mammalian uterus.

Deep Dissection of the Reproductive Tract To follow the reproductive tract further, cut through the pubic and ischial symphyses with heavy scissors. Spread the cut by pressing the knees of the cat laterally to the bottom of the dissecting pan. As needed, extend this cut by trimming off more bone adjacent to the symphyses.

→ The body of the uterus is short and soon gives way to the more posterior vagina (figure 9.9). No external demarcation marks the boundary between these regions. However, internally, a strong muscular sphincter, the cervix, controls the movement of young into the vagina during the birthing process (parturition). Even more caudally, the vagina merges with the urethra to jointly form the urogenital sinus (= vaginal vestibule).

Free the urogenital sinus, vagina, and body of the uterus from the more dorsal rectum. Make a longitudinal cut along the dorsal side to open this part of the reproductive system (figure 9.9). On the ventral side of the urogenital sinus, near its external opening, is a small unpaired process, the clitoris. Anterior to this is the urethral orifice, the point of entry of the urethra. The vagina is the region between the urethral orifice and the more cranial cervix, which should now be located.

The descending colon straightens above the body of the uterus to form the rectum. Smooth muscle, the anal sphincter, controls its terminus and external opening, the anus. The pair of rounded structures lying next to the rectum near the anus are the anal glands.

Form & Function

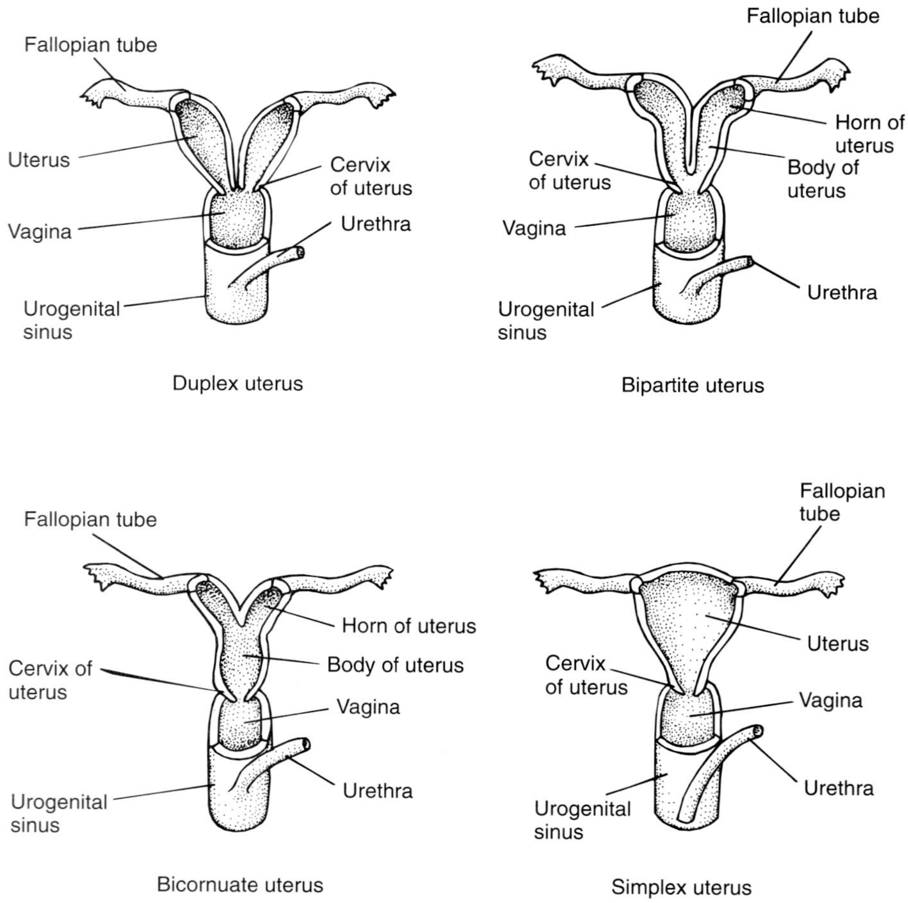

Reproductive tracts of female placental mammals. The four types of uteri vary by the degree of fusion of the paired (lateral) uteri.

The uteri of female placental mammals are quite variable. Cats possess a bipartite uterus and humans a simplex uterus. Why are there such variations? One reason is packing. Recall that development in placental mammals occurs within the uterus. Animals with large litters need more room for the many young to attach and grow. The paired uterine horns in these animals provides for more space than the simplex uterus in humans, who typically have only one or two children during a typical pregnancy.

Nervous System

10

Introduction

The nervous system receives stimuli from one or more **receptors,** such as eyes or ears, that monitor the external and internal environment. It then transmits information to one or more **effectors,** such as muscles or glands, that respond appropriately to the signals. Between receptors and effectors the information undergoes extensive processing within the central nervous system. In general, the more information arriving, the larger is the part of the brain responsible for initial sensory input. For example, where eyes gather extensive visual stimuli, the parts of the brain receiving this information are also large. But in blind cave fish, without eyes, the corresponding part of the brain may be quite small. The nervous system further regulates an animal's performance by integrating immediate incoming sensory information with **stored information,** the results of past experience. Past and present information are then translated into action by effectors.

The nervous system is divided into the **central nervous system** (CNS), which includes the brain and spinal cord, and the **peripheral nervous system** (PNS), which includes all nerves, ganglia, and other nervous tissue outside the CNS. The differentiation and enlargement of the anterior part of the nervous system into a regionalized brain were early events in chordate evolution (figure 10.1). This anterior concentration of distinctive nervous structures and formation of a head, called **cephalization,** seems to be built upon a segmental plan, like the trunk. But the segments contributing to the head fuse, enlarge, and in vertebrates are usually indistinguishable from one another.

Descriptive anatomy takes advantage of this ancestral segmentation as the basis for organizing and relating nervous system structures, especially the cranial nerves, between vertebrate groups. In fact, embryology confirms a general retention of relationships between specific nerves, muscles, and skeletal elements. For example, particular nerves tend to supply homologous muscles from one vertebrate group to the next. The practical benefit for the student is that by learning a general scheme of nerve distribution, any particular vertebrate can be examined with some initial idea of what to expect in terms of nerve supply.

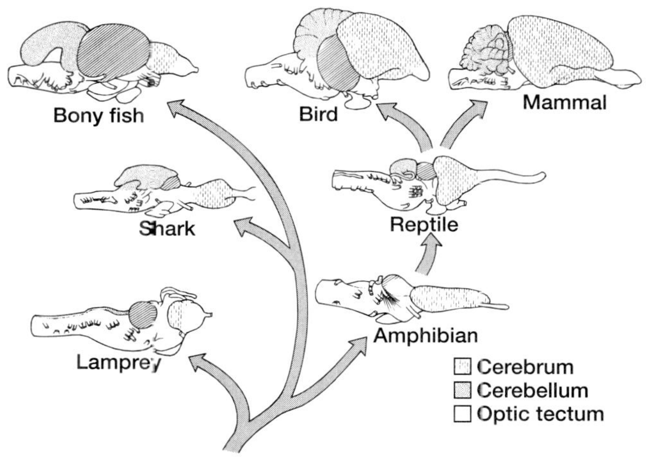

Figure 10.1
Evolution of the vertebrate brain. Note the phylogenetic enlargement of the cerebrum.

Nerve Function

Nerves are classified upon the basis of the information they carry (sensory or motor) and upon the basic type of tissue they serve (visceral or somatic). **Sensory** or **afferent** sensations travel along nerve fibers to the central nervous system (figure 10.2). **Motor** or **efferent** impulses travel away to peripheral effector organs (e.g., muscles, glands). **Visceral tissues** derive from the embryonic gut and adjacent tissue. **Somatic tissues** are all else, namely derivatives of the embryonic body wall (skin, musculature, skeleton). The terms **general** and **special** recognize that some senses or effectors (e.g., taste, sight, olfaction, hearing) became specialized from more general functions. Thus, a nerve carrying information to skeletal muscle would be a **general somatic motor nerve.** A nerve carrying sensations from the gut would be a **general visceral sensory nerve,** and so on.

Spinal Nerves

Spinal nerves are sequentially arranged and numbered (C-1, T-1, L-1, S-1) according to their association with regions of the vertebral column (cervical, thoracic, lumbar, sacral). The particular name of a spinal nerve corresponds to the vertebrae just caudal to the nerve's departure. For example, cervical spinal nerve 3 extends from the spinal cord between cervical vertebrae 2 and 3.

Early anatomists recognized **dorsal** and **ventral roots** of each spinal nerve. Afferent fibers enter the spinal cord via

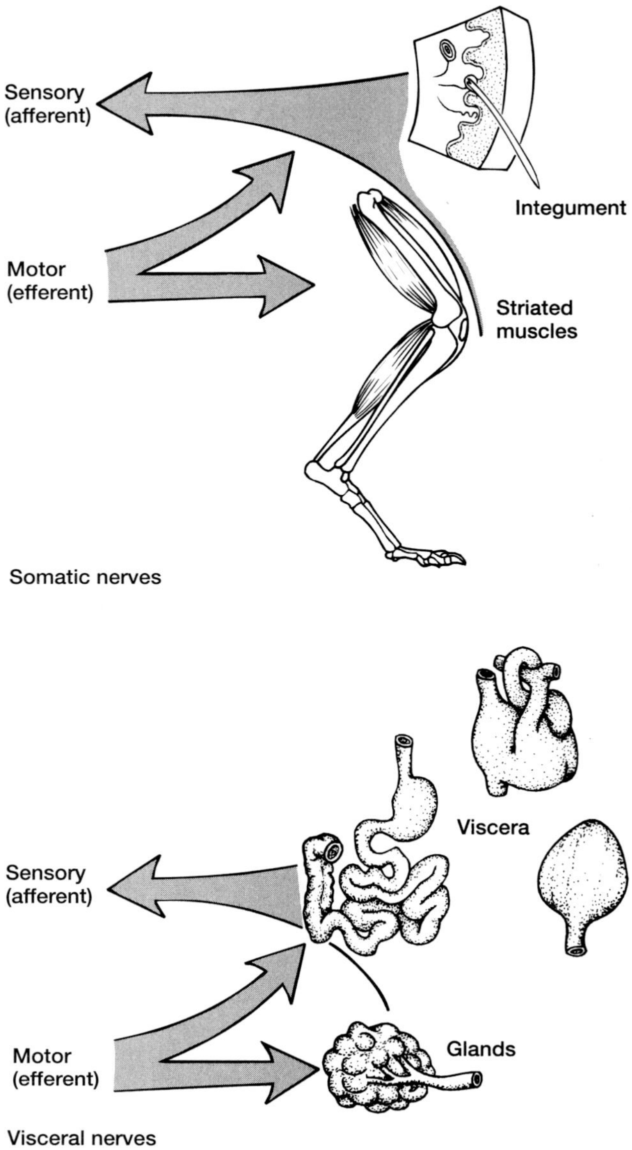

Figure 10.2
Functional categories of neurons of the peripheral nervous system. Some neurons supply somatic tissues (top), others visceral tissues (bottom). They can be sensory and respond to stimuli from these tissues, or they can be motor and deliver stimuli to these tissues.

the dorsal root, and efferent fibers leave by way of the ventral root (figure 10.3). The **dorsal root ganglion,** a swelling on the dorsal root, is a collection of neuron bodies whose axons contribute to the spinal nerve. Parallel to the spinal cord and attached to each spinal nerve through the **ramus communicans** is the **sympathetic chain** of ganglia (paravertebral ganglia), a paired series of linked ganglia adjacent to the vertebral column or notochord. Other peripheral ganglia form the **collateral ganglia** (prevertebral ganglia). The paired **cervical, coeliac,** and **mesenteric ganglia** are examples of the collateral ganglia. The **visceral ganglia** occur within the walls of the visceral effector organs. Thus, there are three types of ganglia: **sympathetic, collateral,** and **visceral.**

Cranial Nerves

Cranial nerves have roots enclosed in the braincase. These are named and numbered by roman numbers from anterior to posterior. The conventional system for numbering these nerves is sometimes inconsistent. For instance, most anamniotes are said to have ten cranial nerves. A few anamniotes and all amniotes are said to have 12. In fact, there is an additional terminal nerve at the beginning of this series. If counted at all, it is numbered 0 to avoid renumbering the conventionally numbered sequence. Further, the second cranial nerve (II) is not a nerve at all, but an extension of the brain. Nevertheless, by convention it is called the optic "nerve."

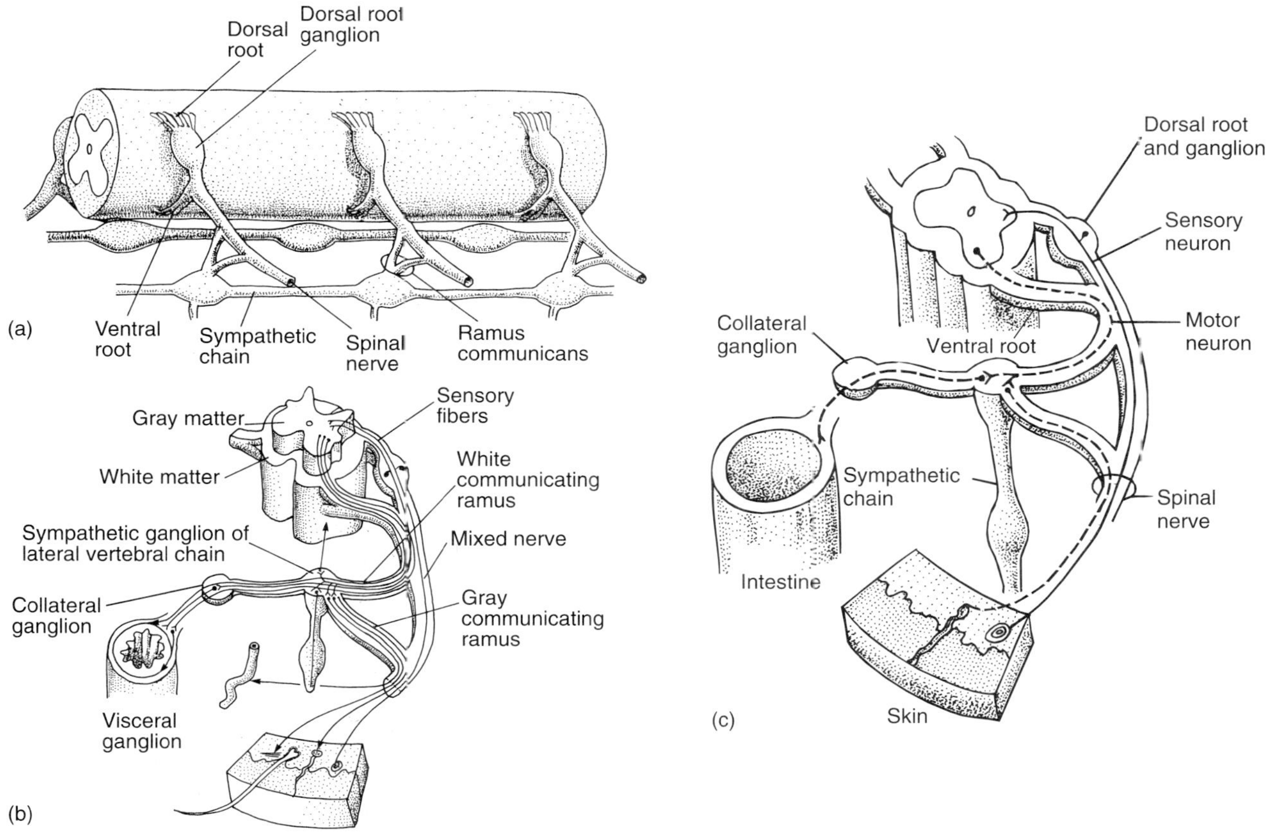

Figure 10.3

Spinal nerve anatomy. (a) Dorsal and ventral roots connect spinal nerves to the spinal cord. A dorsal root is enlarged into a dorsal root ganglion. Spinal nerves join with the sympathetic chain through communicating rami. (b) Configuration of sensory and motor neuronal routes in an adult mammal. (c) Diagrammatic representation of established afferent and efferent neurons within spinal nerves.
After Tuchmann–Duplessis et al.

The eleventh cranial nerve (XI) represents the merger of a branch of the tenth cranial nerve (X) with elements of the first two spinal nerves (C-1 and C-2). Despite its composite structure, it is called the spinal accessory nerve and designated roman numeral XI.

Phylogenetically, the cranial nerves are thought to have evolved from dorsal and ventral roots of a few anterior spinal nerves that became incorporated into the braincase. Like spinal nerves, the cranial nerves supply somatic and visceral tissues and carry general sensory and motor information. Some cranial nerves consist of only **sensory** or only **motor** fibers. Other nerves are **mixed,** containing both types. Cranial nerves concerned with localized senses (e.g., sight, hearing, olfaction, taste) are called **special cranial nerves** to distinguish them from those concerned with the sensory or motor innervation of the more widely distributed viscera, **general cranial nerves.**

Anamniotes

Most anamniotes possess ten cranial nerves (all but cranial nerves XI and XII listed for amniotes below; figure 10.4). The first few spinal nerves behind the braincase become housed in the skull of later derived groups. But in anamniotes, these anterior spinal nerves are still partially outside the skull, although they have lost their dorsal roots and differ from other spinal nerves. In cyclostomes, these anterior spinal nerves outside the skull are called **occipitospinal nerves.** In other fishes and amphibians, the anterior spinal nerves become partially incorporated into the braincase. When this occurs, the nerves exit via foramina in the occipital region of the skull and are termed **occipital nerves.** Occipital nerves unite with the next few cervical spinal nerves to form the composite **hypoglossal nerve** that supplies hypobranchial muscles in the throat.

Amniotes

Amniotes are traditionally said to possess twelve cranial nerves (figure 10.5), although the inaccuracies of this statement have been explained above. The anterior and posterior occipitospinal nerves have become incorporated into the skull, modified, and their nuclei within the spinal cord shifted forward so they now actually issue from the medulla

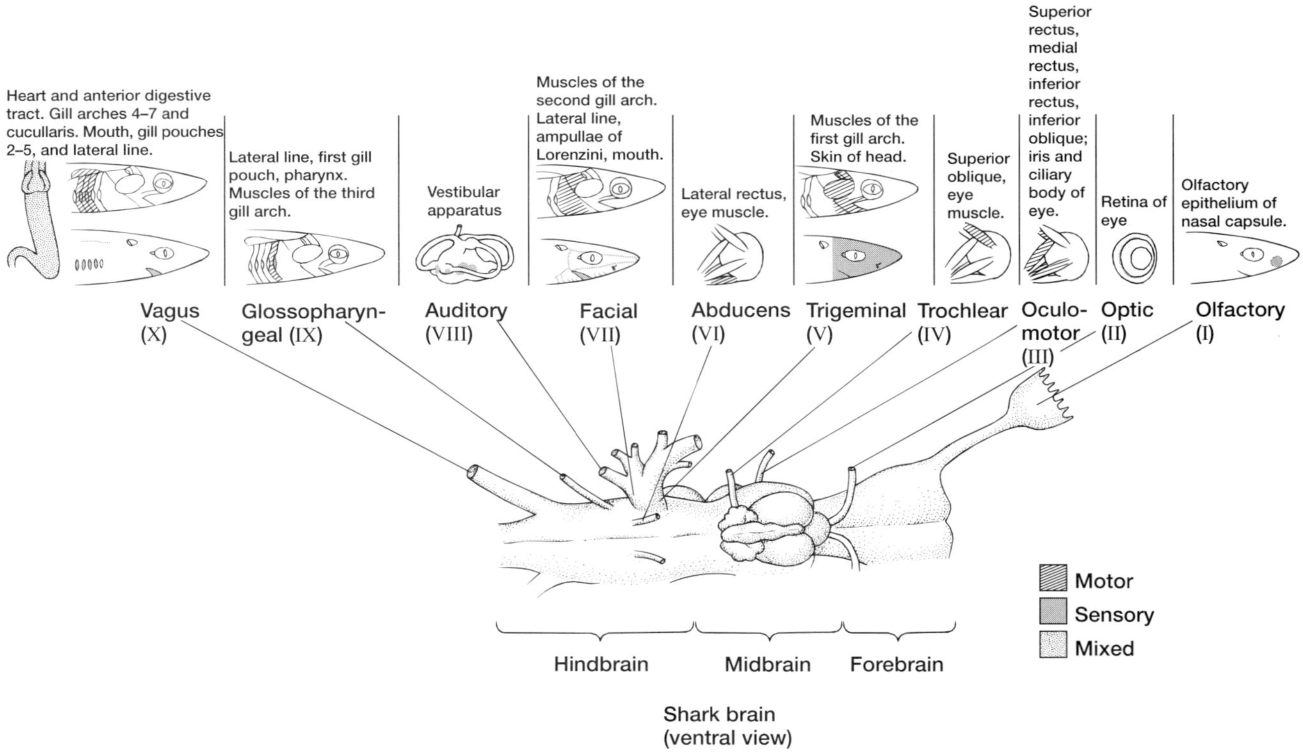

Figure 10.4
Distribution and functions of cranial nerves in the shark *Squalus*.
After Gilbert.

oblongata. Thus, by such a phylogenetic shift, amniotes derive the eleventh and twelfth cranial nerves.

Nervus Terminalis (0)—Mixed The **terminal nerve** is not numbered. It may be testimony to an ancient anterior head segment that has been lost. This nerve is present in all classes of gnathostomes except birds. It runs to blood vessels of the olfactory epithelium in the olfactory sac and most likely carries visceral sensory and some motor fibers.

Olfactory Nerve (I)—Sensory The **olfactory nerve** is a sensory nerve associated with the sense of smell. Olfactory cells lie in the mucous membrane of the olfactory sac. A short axon called an olfactory fiber leads from each cell to the **olfactory bulb.** Collectively, the olfactory fibers form the short olfactory nerve.

Optic Nerve (II)—Sensory As noted previously, the **optic nerve** is not a nerve but a sensory tract. That is, it is not a collection of peripheral axons; it is a collection of fibers in the CNS. Embryologically, it develops as an outpocketing of the brain. However, once it is differentiated, it lies outside the brain. Its fibers synapse in the thalamus and midbrain.

Oculomotor Nerve (III)—Motor The **oculomotor nerve** primarily supplies extrinsic eye muscles (superior rectus, medial rectus, inferior rectus, and inferior oblique muscles) derived from preotic myotomes. It is a motor nerve that also carries a few visceral motor fibers to the iris and ciliary body of the eye. Fibers arise in the oculomotor nucleus in the floor of the midbrain.

Trochlear Nerve (IV)—Motor The **trochlear nerve** is a motor nerve that supplies the extrinsic, superior oblique eye muscle. Fibers arise in the trochlear nucleus of the midbrain.

Trigeminal Nerve (V)—Mixed The **trigeminal nerve,** or **trigeminus,** is so named because in amniotes it is formed of three branches: **ophthalmic** (V1), **maxillary** (V2), and **mandibular** (V3). The **ophthalmic nerve,** sometimes called the **deep ophthalmic nerve** to distinguish it from a more superficial nerve, usually merges with the other two branches. However, in lower vertebrates, the ophthalmic nerve often emerges from the brain separately, possibly representing an ancient condition in which the ophthalmic nerve supplied an anterior arch that has been lost. The other two branches, the maxillary ramus (V2) to the upper jaw and the mandibular ramus (V3) to the lower jaw, presumably represent rami of a typical branchial nerve, the mandibular arch.

The mixed trigeminus includes sensory fibers from the skin of the head and areas of the mouth, and motor fibers to derivatives of the first branchial arch. Sensory fibers of the trigeminus return to the brain from the skin, teeth, and other areas through each of the three branches. The mandibular branch also contains visceral motor fibers to muscles of the mandibular arch.

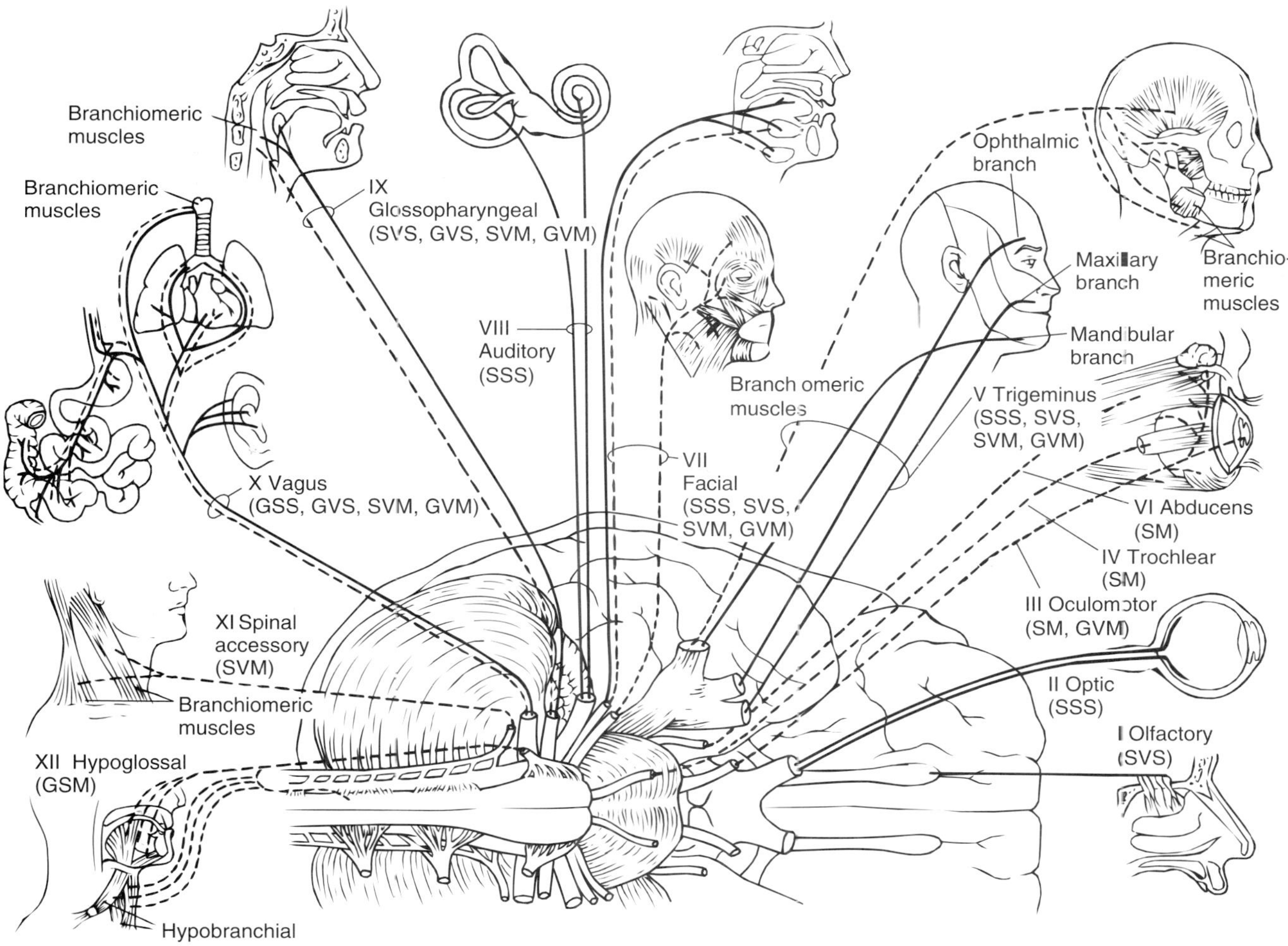

Figure 10.5
Distribution of cranial nerves in a mammal, *Homo sapiens*. Sensory (solid lines) and motor (dashed lines) nerve fibers are indicated. Enlarged views of innervated structures of cranial nerves are shown around the human brain in ventral view. Abbreviations: general somatic sensory (GSS), general visceral sensory (GVS), general somatic motor (GSM), general visceral motor (GVM), special somatic sensory (SSS), special visceral sensory (SVS), special visceral motor (SVM).
After H. Smith.

Abducens Nerve (VI)—Motor The **abducens** is the third of the three cranial nerves that innervates muscles controlling movements of the eyeball. It is a motor nerve that supplies the extrinsic, lateral rectus eye muscle. Fibers arise in the abducens nucleus located in the medulla oblongata.

Facial Nerve (VII)—Mixed The mixed **facial nerve** includes sensory fibers from the lateral line of the head, ampullae of Lorenzini, and taste buds as well as motor fibers to derivatives of the second (hyoid) arch.

Auditory Nerve (VIII)—Sensory The sensory **auditory nerve** (= acoustic, vestibulocochlear, or statoacoustic) carries sensory fibers from the inner ear, which is concerned with balance and hearing. The nerve synapses in several regions of the medulla oblongata.

Glossopharyngeal Nerve (IX)—Mixed The mixed **glossopharyngeal nerve** supplies the third branchial arch. It contains sensory fibers from taste buds, the first gill pouch, and the adjacent pharyngeal lining and lateral line. Motor fibers innervate muscles of the third branchial arch.

Vagus Nerve (X)—Mixed The term **vagus** is Latin for wandering and aptly applies to this mixed nerve. The vagus meanders widely, serving areas of the mouth, pharynx, and most of the viscera. It is formed by the union of several roots across several head segments.

Spinal Accessory Nerve (XI)—Motor In anamniotes, the **spinal accessory nerve** is probably composed of a branch of the vagus nerve and several occipitospinal nerves. In amniotes, especially in birds and mammals, it is a small but distinct motor nerve that supplies derivatives of the cucullaris muscle (cleidomastoid, sternomastoid, and trapezius). A few of its fibers accompany the vagus nerve to supply part of the pharynx and larynx and perhaps the heart. Fibers arise from several nuclei with the medulla oblongata.

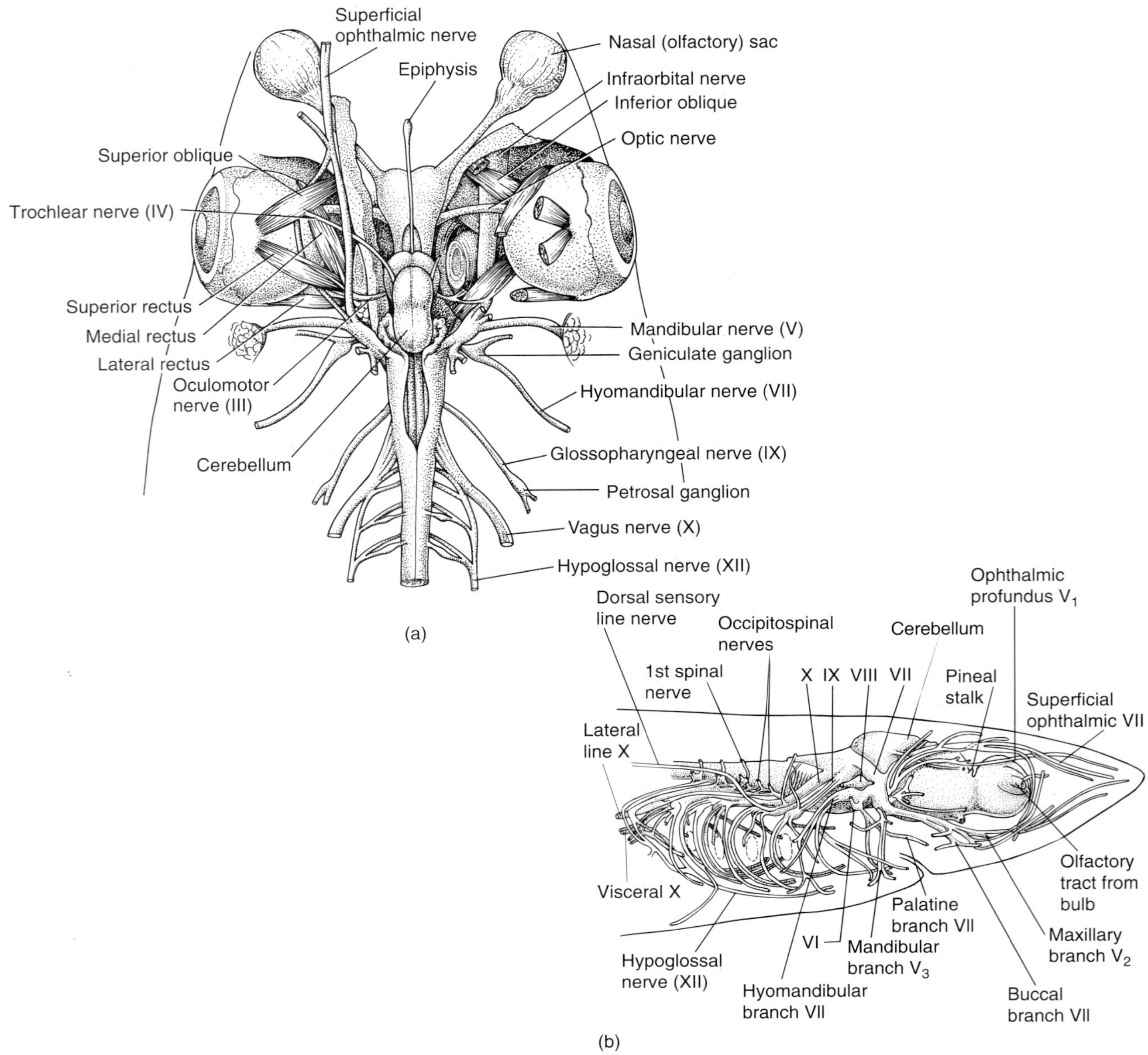

Figure 10.6
Brain and cranial nerves in the shark *Squalus*. (a) Dorsal view. The roof of the chondrocranium has been removed and the four eye muscles have been cut to expose the deeper nerves to the right. (b) Lateral view.
(b) After Norris and Hughes, 1920.

Hypoglossal Nerve (XII)—Motor The **hypoglossal nerve** is a motor nerve that innervates hyoid and tongue muscles. Fibers originate in the hypoglossal nucleus within the medulla oblongata.

Lateral Line Nerves In addition to cranial nerves, fishes and aquatic amphibians possess anterior and posterior **lateral line nerves** that are rooted in the medulla oblongata and supply the lateral line system. They were once thought to be components of the facial and vagal nerves, but they are now recognized as independent cranial nerves. Unfortunately, this late recognition as distinct cranial nerves has left them without an identifying roman numeral!

Shark

Dissection Procedure

→ To expose the brain and cranial nerves, most of the roof of the chondrocranium will have to be removed. This should be done carefully, taking care not to damage major nerves, white or tan strings passing through the chondrocranium. The removal of the roof of the chondrocranium will be done in steps as follows (a–d):

(a) Remove the skin over the head and snout. Clear away the watery tissue and relocate some landmarks of the chondrocranium: rostrum, filled with a gel-like plug; medial epiphyseal foramen; endolymphatic fossa. Remove the upper eyelids. Between the epiphyseal foramen and orbit emerges the superficial ophthalmic trunk (figure 10.6a) from a foramen in the chondrocranium. This nerve trunk runs forward, lateral to the rostrum, branching profusely into many threadlike nerves to the skin.

(b) Pick away the chondrocranium along the dorsal rim starting at the orbit. Note that the superficial ophthalmic trunk runs along the dorsomedial wall of the orbit, over the extrinsic eye muscles. Clear the orbit to expose the extrinsic eye muscles and locate two of the three cranial nerves supplying them (figure 10.6a). The slender trochlear nerve emerges from a foramen in the dorsomedial wall of the orbit and passes under the superficial ophthalmic to innervate the superior oblique muscle. The oculomotor nerve is best seen at the point of convergence of rectus muscles on the medial wall of the orbit. The slender deep ophthalmic nerve can also be seen now passing under the superior rectus muscle, across the surface of the eyeball, under the superior oblique muscle, then out of the orbit.

(c) Remove the skin around the spiracle. The hyomandibular nerve passes along the posterior rim of the spiracle.

(d) Remove enough of the axial muscles along their attachment to the chondrocranium to obtain an unobstructed view of the posterior chondrocranium and anterior first few vertebrae. The roof of the chondrocranium will next be removed by picking away cartilage pieces. The strategy is to remove first the roof, then the sides of the chondrocranium, until the dorsal side of the brain is exposed and the full extent of exiting cranial nerves lies bare. One approach is to begin by dissecting the olfactory sacs dorsally and then working posteriorly. The other is to begin between the orbits. Whichever method is used, be sure not to damage the soft, even mushy, shark brain within. One further note. Between the spiracles, next to the brain, lie the semicircular canals embedded in calcified cartilage of the otic capsules. These semicircular canals can be removed or picked away as needed. Now remove the roof, then sides of the chondrocranium.

Brain (Dorsal Aspect)

→ In the middle of the brain is the large, unpaired cerebellum (figures 10.6a and 10.7a). Adjacent to the posterior end of the cerebellum are the two folded, earlike auricles (figure 10.6a). The region behind and continuous with the cerebellum is the medulla oblongata. In life, the medulla oblongata is covered dorsally by a thin pigmented roof, a tela choroidea, one area where cerebral spinal fluid is elaborated. Remove the tela choroidea, if it is still present. Where the medulla oblongata narrows posteriorly, it gives way to the spinal cord.

Now examine the floor of the open medulla oblongata where longitudinal ridges can be seen. Immediately to either side of the midventral groove lies the somatic motor column; the next lateral ridge is the visceral sensory column; the groove between these two columns is the sunken visceral motor column. The lateral row of very tiny beadlike swellings is the somatic sensory column.

The paired optic lobes (tectum) lie immediately in front of the cerebellum (figure 10.7a). Cerebral hemispheres lie in front of the optic lobes and are the most anterior enlargements of the brain. Extending from each hemisphere, anterolaterally, is the olfactory tract. Before reaching the olfactory sac, the tract swells slightly to form the olfactory bulb. Olfactory nerves, impractical to see, run from the lining of the sac to the olfactory bulb. With good dissection, and a bit of luck, the very thin, threadlike nervus terminalis may be seen on the anterior side, next to and parallel with the olfactory tract (figure 10.7a, b).

The brain narrows between cerebral hemispheres and optic lobes. The small dorsal dark patch on this narrow region of the brain is a second tela choroidea. A delicate membrane, the primitive meninx (pl. meninges), covers the surface of the brain and carries blood vessels serving the brain. In life, cerebral spinal fluid fills the space between the brain and its cavity within the chondrocranium, and also fills ventricles, hollow chambers inside the brain. These ventricles will be examined later.

Cranial Nerves

Nervus Terminalis (CN 0)

→ This thin nerve runs from the olfactory sac caudally in parallel with the olfactory tract and passes into the median fissure (crease) between the two cerebral hemispheres (figure 10.7a). It has no olfactory fibers and its function remains somewhat in doubt. Most likely, however, it carries visceral sensory and visceral motor fibers. Thus, in part (visceral motor) it contributes to the autonomic system.

Olfactory Nerve (CN I)

→ It is impractical to see the short olfactory nerve fibers within the wall of the olfactory sac (figure 10.6a). They synapse in the olfactory bulb on the posterior wall of the olfactory sac. The olfactory bulb narrows into the olfactory tract, an elongate extension of the brain, which continues posteromedially to a cerebral hemisphere. Fibers carry olfactory sensations.

Optic Nerve (CN II)

→ The stout optic nerve fibers (figure 10.6a) depart the retina opposite the laterally positioned pupil on the medial wall of the eyeball. The optic nerve can be found somewhat between the superior and inferior oblique extrinsic eye muscles. The nerve runs directly medially to beneath the cerebral hemisphere, then turns posteriorly, coursing to and entering the optic lobe of the brain. Optic fibers carry only visual information (special somatic sensory).

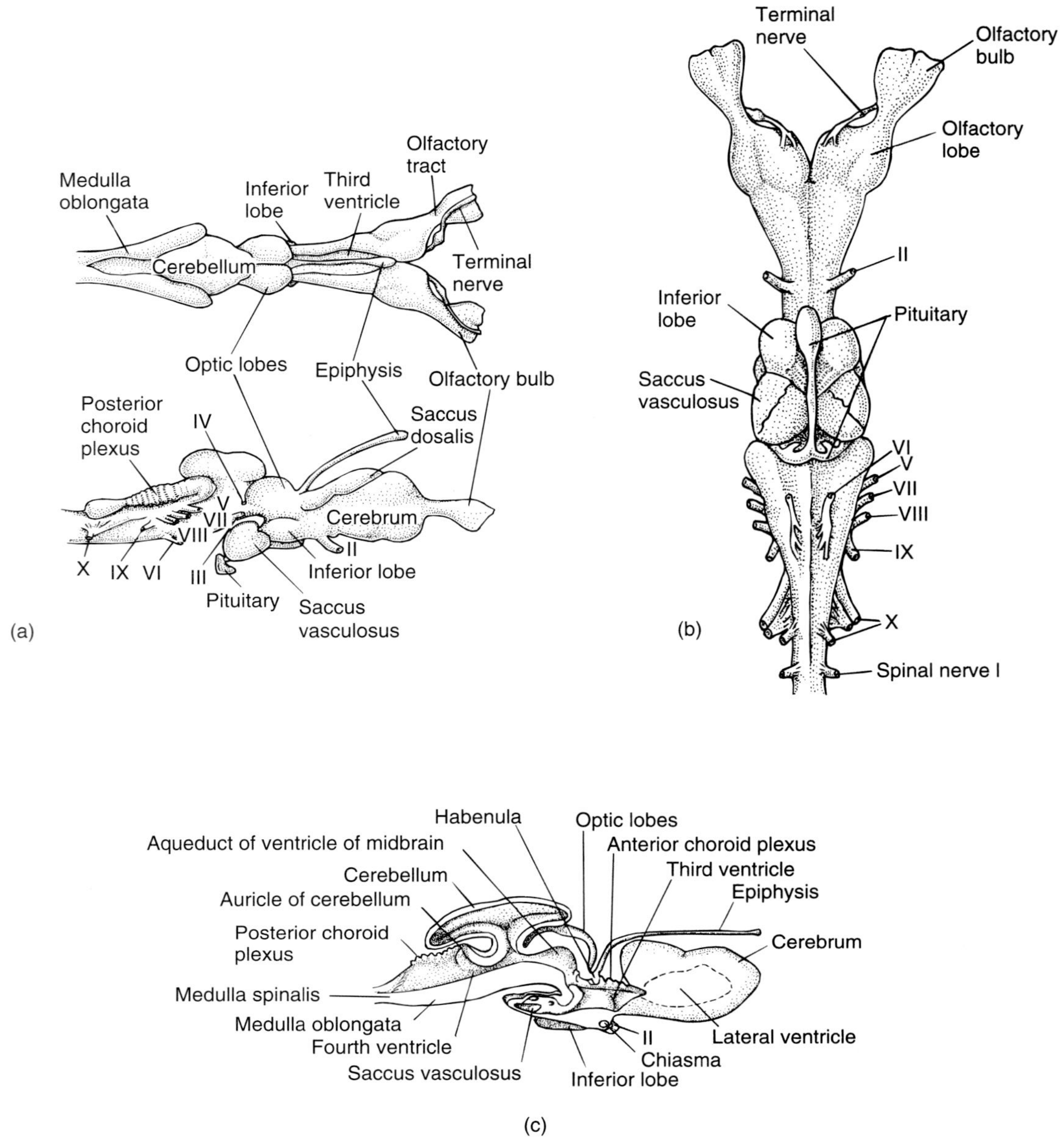

Figure 10.7
The brain in the shark *Scymnus*. (a) Dorsal (top) and lateral (bottom) views. (b) Ventral view. (c) Sagittal view.
(a) Redrawn from A.S. Romer and T.S. Parsons, 1985, The Vertebrate Body, Saunders College Publishing, after Butschli, Ahlborn; (b) Redrawn from A.S. Romer and T.S. Parsons, 1985, The Vertebrate Body, Saunders College Publishing, after Butschli, Weinstein, Sisson; (c) Redrawn from A.S. Romer and T.S. Parsons, 1985, The Vertebrate Body, Saunders College Publishing.

Oculomotor Nerve (CN III)

→ The oculomotor nerve (figure 10.6a) springs from the floor of the midbrain. It passes to the origins of the rectus extrinsic eye muscles, then branches to supply the inferior oblique, and the superior, inferior, and medial rectus muscles. Cut through the insertions of the superior oblique and superior rectus muscles and reflect them to aid in tracing out the distributions of this nerve. The branch to the inferior rectus muscle in turn sends off another branch that accompanies a blood vessel into the wall of the eyeball. This is the ciliary nerve, often impractical to see, part of the autonomic system. The oculomotor nerve carries mostly general somatic motor (and proprioceptive) fibers, but also contains a few general visceral motor fibers as well.

Trochlear Nerve (CN IV)

→ Externally, the trochlear nerve (figure 10.6a) springs from the roof of the midbrain between optic lobes and cerebellum (in fact, its motor neurons lie on the floor of the midbrain). Upon emergence from the chondrocranium, it passes under the superficial ophthalmic trunk and to the superior oblique muscle that it exclusively supplies. The trochlear nerve is a general somatic motor nerve (with a few proprioceptive fibers).

Abducens Nerve (CN VI)

→ To complete the identification of nerves supplying extrinsic eye muscles, skip to the abducens nerve. With the brain exposed only dorsally, it is not yet feasible to see the emergence of the abducens nerve from the floor of the medulla oblongata. However, it can be seen along the origin of the lateral rectus muscle that it exclusively supplies. The abducens is a general somatic motor (and proprioceptive) nerve.

Trigeminal Nerve (CN V)

→ Recall that the trigeminal nerve innervates derivatives of the mandibular arch and gathers sensory information from the skin of the head. In sharks, the origin of the trigeminal nerve (V) is merged with the facial (VII) and acoustic (VIII) nerves, giving rise to a large knot of nerves next to the auricle.

The trigeminal nerve of sharks has four main branches: superficial ophthalmic, deep ophthalmic, mandibular, and maxillary branches (figure 10.6b).

Superficial Ophthalmic Branch The superficial ophthalmic branch, together with a branch of the facial nerve of the same name, form the superficial ophthalmic trunk. This trunk was previously identified as it passed next to the rostrum of the chondrocranium on its way to the skin over the top and sides of the snout. The superficial ophthalmic branch carries only general somatic sensory fibers.

Deep Ophthalmic Branch Upon branching from the trigeminus, the deep ophthalmic branch (= profundus) passes through the chondrocranium, across the dorsomedial surface of the eyeball, and out the anterior wall of the orbit via the orbitonasal canal. Lift the deep ophthalmic branch slightly as it crosses the eyeball to find a few thin branches, ciliary nerves, that penetrate the wall of the eyeball. Trace the deep ophthalmic branch forward through the canal. It joins, then separates from, the superficial ophthalmic trunk to eventually reach the skin on the top and sides of the snout. The deep ophthalmic branch carries a few general visceral motor (ciliary nerves) and many general somatic sensory (from skin of snout) fibers.

Mandibular Branch The mandibular branch departs at about the same point as the superficial and deep ophthalmic branches, but lies beneath them, passing approximately under the lateral rectus muscle. The mandibular branch skirts the rear side of the orbit, supplying various gill arch muscles on the floor of the orbit, then arches ventrally into muscles of the lower jaw. Follow it for a distance through this muscle mass that it supplies. The mandibular branch carries general visceral motor (mandibular muscles) and a few general somatic sensory (skin over lower jaw) fibers.

Maxillary Branch The maxillary branch of the trigeminus, together with the buccal branch of the facial nerve, are merged into a prominent broad ribbonlike trunk running along the floor of the orbit, the infraorbital trunk (figure 10.6a) Remove the eyeball of one side by cutting further insertions of extrinsic muscles, nerves close to the wall of the eyeball, and by pulling out the optic pedicle. The infraorbital trunk crosses the medial wall of the orbit, then out the anterior wall. The maxillary branch returns fibers from the skin over the upper jaw and underside of the rostrum. The maxillary branch thus carries general somatic sensory fibers.

Facial Nerve (CN VII)

→ Recall that the facial nerve innervates derivatives of the hyoid arch and returns sensory information from the lateral line of the head, ampullae of Lorenzini, and areas of the mouth. The facial nerve also has four main branches: superficial ophthalmic, buccal, hyomandibular, and palatine branches (figure 10.6b). As just described above, these first two branches join with branches of the trigeminus to form composite trunks.

Superficial Ophthalmic Branch The superficial ophthalmic branch joins with the branch of the trigeminus of the same name to form the superficial ophthalmic trunk. The superficial ophthalmic branch of the facial nerve returns sensory fibers from the lateral line over the top and sides of the head and from the ampullae of Lorenzini. Thus, the superficial ophthalmic nerve carries special somatic sensory fibers.

Buccal Branch The buccal branch joins with the maxillary branch of the trigeminus to form the infraorbital trunk across the medial wall of the orbit. The buccal branch returns sensory fibers from the lateral line and ampullae of Lorenzini over the upper jaw and underside of the rostrum. The buccal branch carries special somatic sensory fibers.

Hyomandibular Branch The hyomandibular branch of the facial nerve skirts the rear edge of the spiracle. Near the base of the hyomandibular, where it first branches from other nerves, it bears a small swelling, the geniculate ganglion. Follow the nerve peripherally around the side of the head. The hyomandibular branch contains special visceral motor (hyoid muscles) fibers, special somatic sensory (lateral line) fibers, general visceral sensory (mouth lining) fibers, and special visceral sensory (taste buds) fibers.

Palatine Branch The palatine branch departs at the geniculate ganglion, then runs forward and ventrally to the lining of the mouth, from which it returns sensory fibers. The palatine branch carries general and special sensory fibers (mouth lining and taste buds, respectively).

Auditory Nerve (CN VIII)

→ The stout auditory nerve arises between the base of the hyomandibular nerve and medulla oblongata. It returns fibers from the inner ear. Pick away the otic capsule to better follow this nerve. The auditory nerve carries special somatic sensory fibers.

Glossopharyngeal Nerve (CN IX)

→ The glossopharyngeal nerve, slender at its base, is easily confused for part of the auditory nerve. The glossopharyngeal nerve supplies the first complete gill pouch. To find where it springs from the medulla oblongata, slightly lift this side of the brain. The glossopharyngeal nerve passes posterolaterally through the otic capsule, generally in parallel with the hyomandibular nerve. Pick away the otic capsule. Just before emerging from the capsule, the glossopharyngeal nerve swells, forming the petrosal ganglion (figure 10.6a,b). At or just distal to this ganglion, the nerve divides into pretrematic and posttrematic branches to the anterior and posterior walls of the gill pouch, respectively (figure 10.6b). A still smaller pharyngeal branch to the wall of the pharynx and a dorsal branch to the lateral line can occasionally be discovered. The glossopharyngeal nerve mostly carries visceral sensory fibers (general and special), but also a few somatic sensory (special) fibers from the lateral line.

Vagus Nerve (CN X)

→ Basically, the vagus nerve supplies the remaining gill pouches, lateral line, and viscera. Remove skin, muscles, and deeper cardinal sinus as needed to follow this nerve. It has a fan-shaped origin from the posterior end of the medulla oblongata. It sweeps posteriorly above the gill arches, giving off four branchial branches, each to separate gill pouches. Like the glossopharyngeal nerve previously examined, each of the four branchial branches of the vagus nerve has in turn pretrematic, posttrematic, and pharyngeal branches (figure 10.6b).

Formally, the four branches arise themselves from the visceral branch of the vagus. The visceral branch continues posteriorly, entering pericardial and pleuroperitoneal cavities to supply viscera within.

The medial coursing lateral branch of the vagus does not become anatomically separate until near the end of the row of gill pouches. The lateral branch carries somatic sensory fibers from the lateral line.

The vagus nerve carries all functional types of fibers, except somatic motor (and proprioceptive).

Occipital, Hypoglossal, and Spinal Nerves

→ At about the level of the last gill pouch, the hypoglossal nerve can be seen as it crosses the visceral branch of the vagus. Follow the hypoglossal nerve proximally toward the spinal cord. This nerve forms by the confluence of two occipital and several spinal nerves, but may be difficult to trace at this point in the dissection.

Expose the spinal cord behind the medulla oblongata by picking away the neural spines and neural arches of the first few vertebrae. Next to the spinal column, serially between vertebrae, are small swellings, the spinal ganglia of the dorsal root (figure 10.6a). From the side of the medulla oblongata, between the first such ganglion and the root of the vagus, spring two spinal roots without ganglia. These are the occipital nerves. The occipital nerves, like the next several spinal nerves that follow, can now be traced peripherally to where they contribute to the hypoglossal nerve.

The hypoglossal nerve swings around the gill region to supply hypobranchial muscles in the throat (figure 10.6b). The hypoglossal nerve carries general somatic motor (and proprioceptive) fibers.

Brain (Ventral Aspect)

Of all sharks dissected by members of the class, only the brains of one or two sharks should be removed. Check with your instructor before proceeding with the removal of the brain.

Removal of the Brain

→ To free the brain from the chondrocranium, first cut through the olfactory tracts and the spinal cord behind the medulla oblongata. Lift the anterior end of the brain, noting the optic nerves. Cut both optic nerves distally, leaving generous portions still attached to the ventral surface of the brain. Continue to lift the brain and cut remaining cranial nerves, leaving generous lengths of nerve attached. Continue until the brain is completely removed. Observation is sometimes aided by floating the brain in a shallow dissection pan of water.

Regions

→ Now in ventral view, refamiliarize yourself with the general brain regions already seen from the dorsal aspect (figure 10.7a,b).

Where the optic nerves converge and enter the brain is seen an unpaired medial swelling, the optic chiasma (figure 10.7b,c). Behind the optic chiasma are a pair of larger swellings, the inferior lobes of the infundibulum (figure 10.7a–c). Between the inferior lobes lies a median projection, sometimes torn off when the brain is removed, the hypophysis (= pituitary body) (figure 10.7a,b). To either side and slightly dorsally above the posterior tip of the hypophysis lie the heavily vascularized (and hence slightly discolored) vascular sacs, also part of the infundibulum (figure 10.7a–c).

Oculomotor nerves (CN III) spring from the wall of the midbrain just above the vascular sacs (figure 10.6a). Approximately between trigeminal nerves, on the floor of the medulla oblongata, lies the paired abducens nerve (CN VI; figure 10.7b). Lateral to the abducens is the glossopharyngeal nerve (IX).

Brain (Sagittal Aspect)

→ Use one of the brains removed previously for viewing ventral brain structures. With a razor blade or very sharp

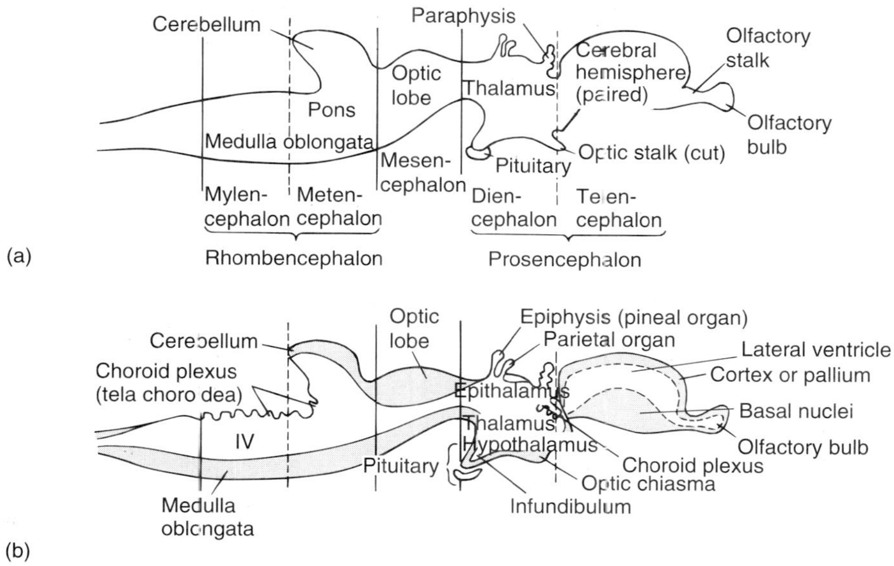

Figure 10.8
Regions of a shark brain. The bottom brain is further developed than the brain indicated at the top.

scalpel, make a midsagittal cut bisecting the brain into two equal, longitudinal halves.

The brain and spinal cord are hollow. This cavity in the spinal cord is the central canal. In the brain, the cavity is irregularly enlarged, forming four ventricles, continuous with each other and with the central canal (figure 10.7c). In life, the ventricles and central canal are filled with cerebral spinal fluid formed by the tela choroidea and probably reabsorbed in the primitive meninx. The first two ventricles (lateral ventricles) lie in the two cerebral hemispheres. The third ventricle resides in the optic lobes (here it is properly termed an optic ventricle) and extends into the hypophysis and forward up to the first two ventricles. The passage between the third and the first two ventricles is the foramen of Monro, not an obvious portal in the shark, but more so in mammals. The fourth ventricle occupies the medulla oblongata and extends up into the cerebellum, forming the cerebellar ventricle. The slender canal connecting the fourth and third ventricles is the cerebral aqueduct (of Sylvius).

Patterns & Connections

Based largely on early embryology, the brain is divided into three regions: forebrain (prosencephalon), midbrain (mesencephalon), and hindbrain (rhombencephalon). These in turn differentiate into five regions composed of specific structures indicated in figure 10.8. Reexamine the shark brain to identify roughly the boundaries of these five regions (telencephalon, diencephalon, mesencephalon, metencephalon, myelencephalon).

Below the anterior end of the optic lobes lies a small swelling, the habenula (figure 10.7c). The slender epiphysis (pineal body) arises from the habenula and projects dorsally, but is often lost when removing the brain (figure 10.8b). Directly across the ventricle from the habenula (sagittal view) is the optic chiasma (figure 10.8a,b). The walls of the brain between cerebral hemispheres and optic lobes represent the thalamus (figure 10.8b). The roof of the brain above the thalamus constitutes the epithalamus; the floor below and its derivatives constitute the hypothalamus (figure 10.8b).

Sheep Brain

Four preparations of the sheep brain will be studied: whole brains, sagittal sections, brain stems, and cross sections. Each will provide a different view of previously identified parts, and expose new structures for identification.

As with vertebrate brains generally, the mammalian brain early in ontogeny differentiates into three regions: prosencephalon, mesencephalon, rhombencephalon. In turn, these differentiate into specific regions of the adult nervous system (figure 10.9a,b).

Overview—Whole Brain

Meninges

The brain is wrapped by three layered meninges: from outside to inside, **dura**, **arachnoid**, and **pia mater** (figure 10.10). The outer two layers (dura mater and arachnoid) are usually left behind in the cranium when the skull is

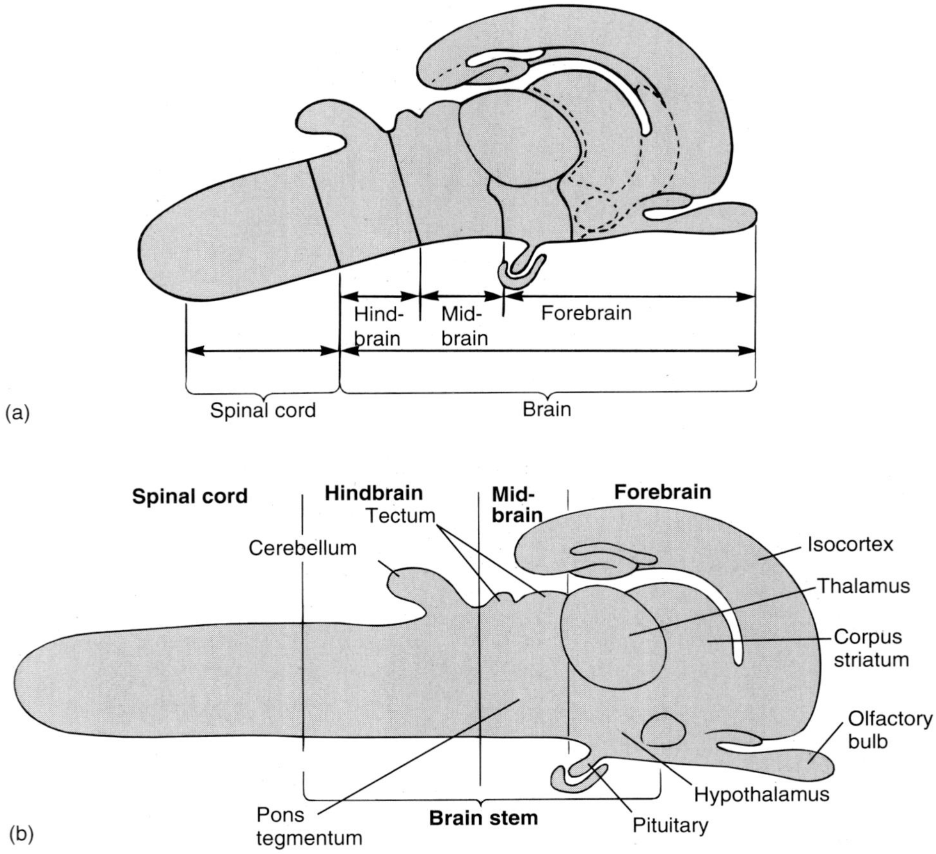

Figure 10.9
Anatomical regions of the adult vertebrate brain. (Top) General brain regions. (Bottom) Subdivisions of the main brain regions.
(a) After W.J.H. Nauta and M. Feirtag, 1979, "The organization of the brain," Scientific American, 1979(9):102.

Patterns & Connections

A bacterial or viral infection may cause an inflammation of the meninges of the brain, a condition called meningitis. If the infectious agents spread to the central nervous system, significant damage may result. Meningitis is typically diagnosed by sampling cerebrospinal fluid that surrounds the spinal cord to look for signs of immunological response to infection (lymphocytes) or for infectious agents (bacteria or viruses).

Patterns & Connections

In many mammals, the cerebral cortex is folded in a complicated fashion, increasing the volume of this region of the brain. But not all mammals show such folding. In the duckbill platypus, opossum, and many rodents, the cerebral cortex is smooth. In the echidna, kangaroo, and most primates, the degree of folding is variable. In all groups of mammals, the extent of folding seems to be more pronounced in large species.

removed. The inner most meningeal lining, the pia mater, carries blood vessels and tightly follows the contours of the brain.

Dorsal Aspect (Figure 10.11)

→ In dorsal view, the large, greatly expanded cerebral hemispheres constitute the most anterior part of the brain (figure 10.11). The convoluted ridges, termed gyri (sing. gyrus) are partially separated from one another by furrows, termed sulci (sing. sulcus). With a blunt probe, break through the pia mater over a sulcus and gently spread the adjacent gyri to see that the furrow continues down a few millimeters. Along the dorsal midline between hemispheres lies the superior sagittal sinus that receives blood exiting from the brain. In preserved specimens, it is usually not distinct. With a blunt probe, slit the superior sagittal sinus and gently spread the two hemispheres. This cleft between hemispheres is the median sulcus (figure 10.11). At the bottom of the sulcus lies the band joining the bases of the cerebral hemispheres, the corpus callosum.

The large, unpaired lobe of the brain in contact with the posterior ends of the cerebral hemispheres is the

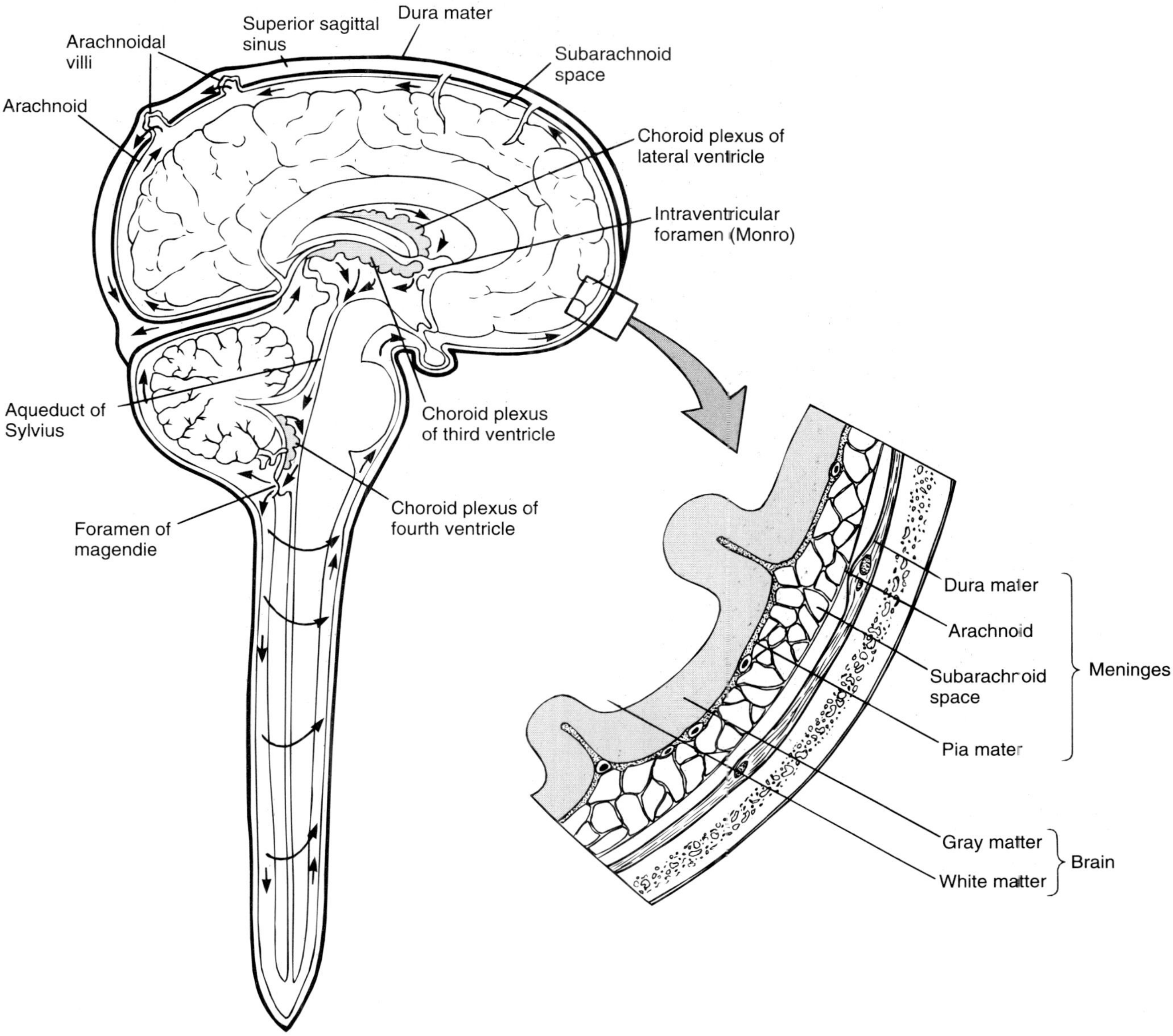

Figure 10.10
Cerebrospinal fluid and meninges. Arrows trace the circulation of cerebrospinal fluid through the brain and spinal cord of a mammal. The triple-layered meninges are enlarged to the right.
After H. Smith, 1960, Evolution of Chordate Structure, Holt, Rinehart and Winston, NY.

cerebellum (figure 10.11). Posterior to the cerebellum runs the spinal cord. The surface of the cerebellum is formed of many near regular folds termed folia, separated by sulci. Because it looks like a segmental worm tightly coiled into a circle, the median sagittal part of the cerebellum is termed the vermis; the lateral parts are termed the hemispheres. The most lateral superficial tip of each cerebellar hemisphere, in contact with the medulla oblongata, is the flocculus. It receives vestibular input and is thought to be homologous with the auricular lobes of lower vertebrates.

Gently spread the cerebral hemispheres and the cerebellum clearing connective tissue between these two parts of the brain (figure 10.14). Revealed are four swellings termed colliculi (= corpora quadrigemina), dorsal parts of the midbrain. The anterior pair, superior colliculi, look like two small embedded balls; the posterior pair form a transverse ridge with raised corners, the inferior colliculi.

Ventral Aspect (Figure 10.12)

→ In ventral view, the flattened oval-shaped olfactory bulbs lie at the anterior tip of the cerebral hemispheres. The broad olfactory tract begins on the posterior corner of the olfactory bulb, but almost immediately

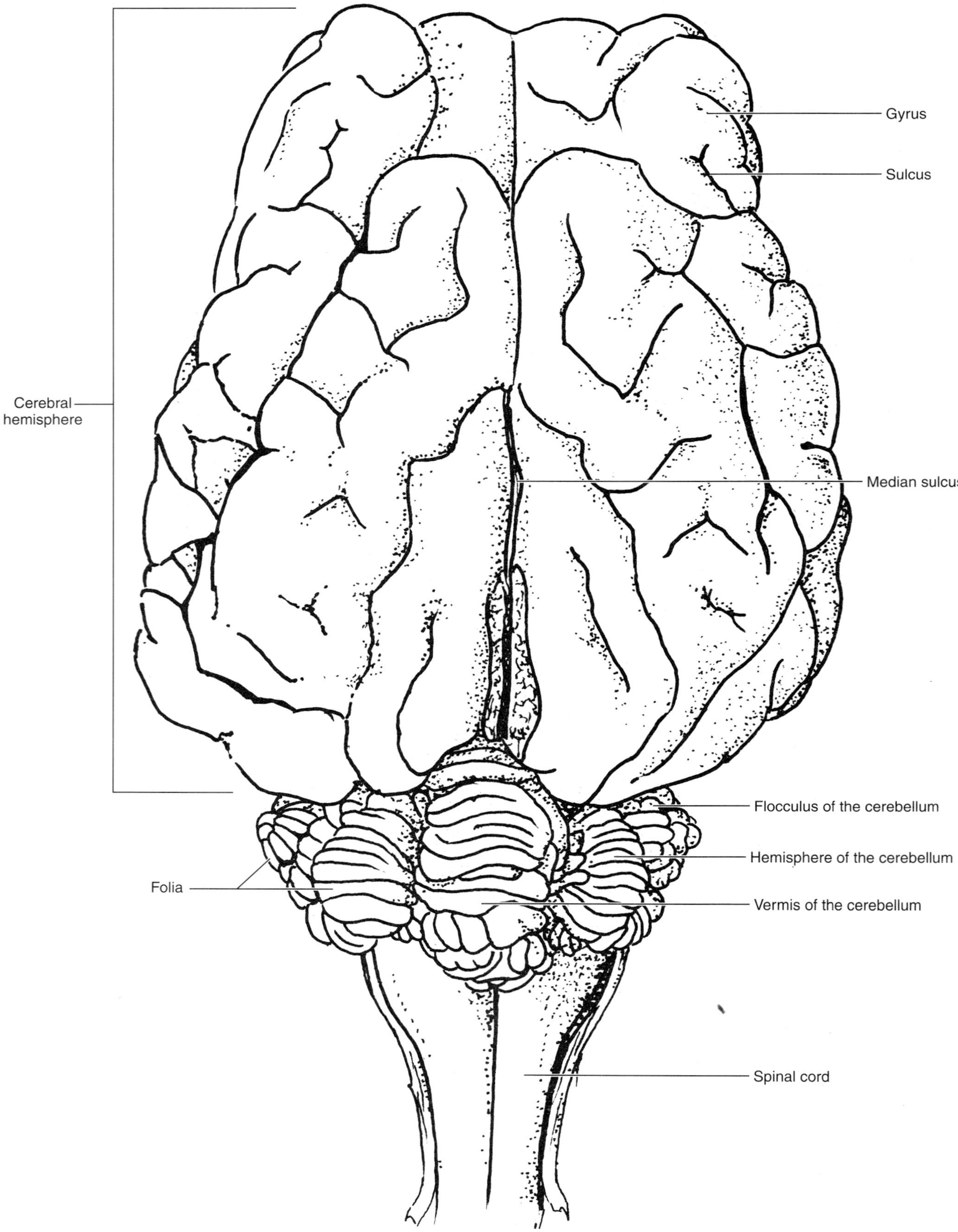

Figure 10.11
Dorsal view of the sheep brain.

all

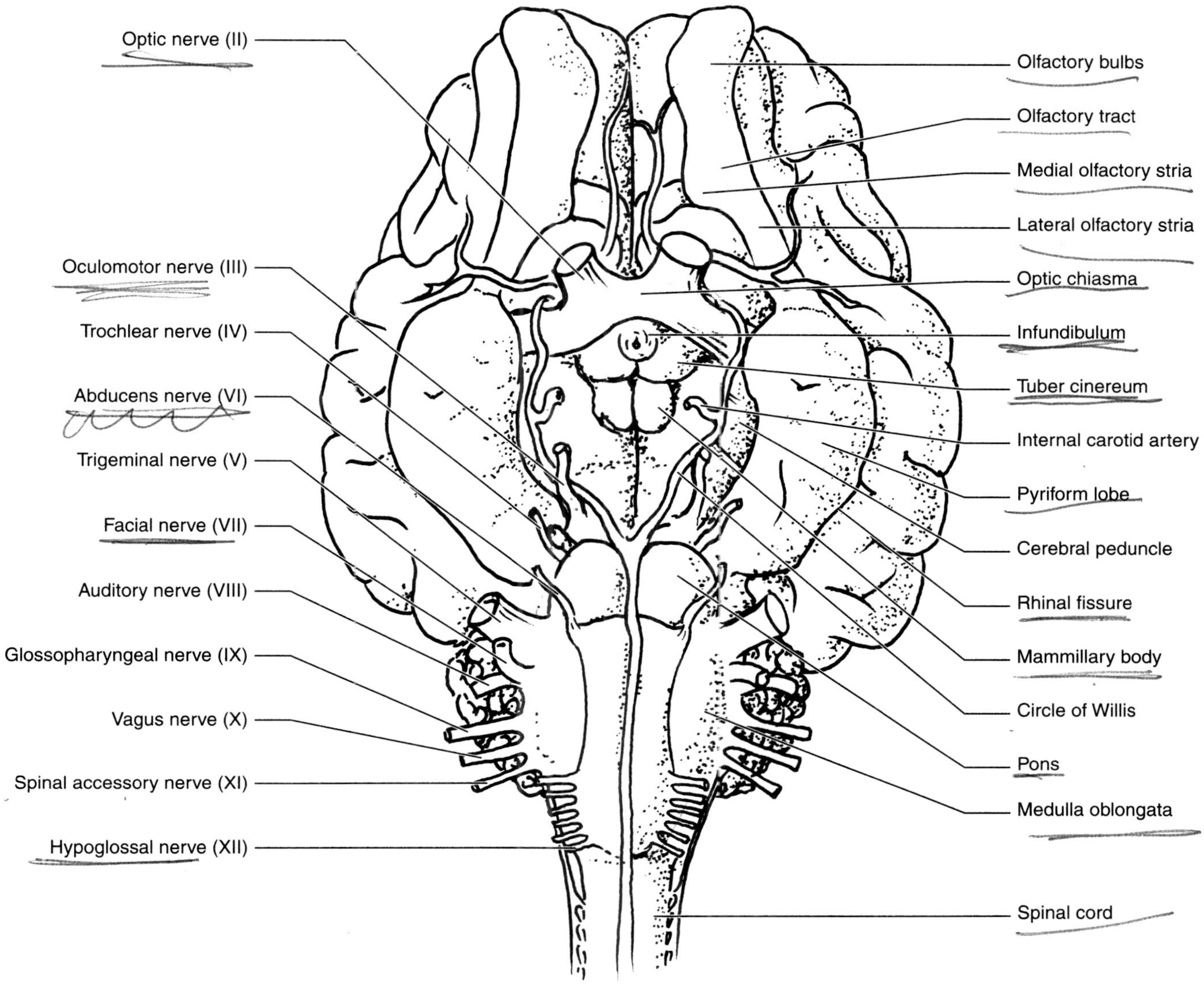

Figure 10.12
Ventral view of the sheep brain.

divides into medial and lateral bands. The medial division, the medial olfactory stria, is a broad, short band vaguely discernible. Clearing some pia mater may bring it into relief as it slants inward to the median sulcus. The lateral division of the olfactory tract is the lateral olfactory stria that passes posteriorly, eventually expanding into a prominent lobe on the base of the cerebral hemisphere, the pyriform (= piriform) lobe. The pyriform lobe lacks sulci within, but is set off along its dorsal edge from the rest of the cerebral hemisphere above by the rhinal fissure.

The most prominent nerves seen on the base of the brain are the X-shaped optic nerves. The optic chiasma is formed at their point of crossing. The prominent pituitary gland resides centrally on the base of the brain, but is almost invariably torn away. Instead, the remains of its hollow stalk of attachment to the brain, the infundibulum, can be found just behind the optic chiasma. Following the infundibulum is a small medial swelling, the mammillary body, continuous with two low ridges, tuber cinerei (sing. cinereum), that pass forward to either side of the infundibulum.

The circle of Willis is the blood vessel encircling the base of the optic nerves, mammillary body, and tuber cinerei. The internal carotid artery empties into the circle laterally, adjacent to each tuber cinereum. If the blood vessels are preserved, note how branches off the circle of Willis supply regions of the cerebrum.

The cerebral peduncle is the region lateral to and extending posteriorly from the mammillary body. The ribbonlike oculomotor nerve springs from this peduncle. The posterior part of this peduncle is crossed by a broad band, the pons, appearing as swellings on either side of the ventral midline. The pons continues laterally up into the cerebellum as a stout stalk, the cerebellar peduncle. The medulla oblongata follows posteriorly behind the pons, continuing until it constricts to give way to the spinal cord.

Cranial Nerves—Whole Brains (Figure 10.12)

The cranial nerves seem especially susceptible to destruction when the sheep brain is removed. You may thus need to inspect several whole brains to see them all.

Olfactory Nerve (I)

→ Fibers contributing to the olfactory nerve are short. They exit from the mucus membrane lining the nostrils, pass to and end in the olfactory bulb, and are impractical to see in gross dissection.

Optic Nerve (II)

The large optic nerves cross, forming the optic chiasma, before entering the base of the brain. In the chiasma, some fibers of one nerve turn and proceed with fibers of the other crossing nerve. Thus, visual information coming from one eye distributes to both sides of the brain.

Oculomotor Nerve (III)

The oculomotor nerve is broad and ribbonlike. It springs from the base of the cerebral peduncle.

Trochlear Nerve (IV)

The trochlear nerve springs from the dorsal surface of the brain stem next to the inferior colliculus. It passes around the sides of the brain stem where it can usually be found next to the cerebral peduncle.

Trigeminal Nerve (V)

The trigeminal nerve is the largest nerve of the hindbrain. It emerges from the posterior and lateral side of the pons. If the nerve has been cut, the remaining large stump can usually be located at this point.

Abducens Nerve (VI)

The slender abducens nerve arises from the medulla oblongata next to the ventral midline behind the pons. It runs forward, and in sheep brain preparations, its free cut end usually lies across the pons.

Facial Nerve (VII)

The cut stump of the facial nerve can be located in a low depression immediately posterior to the root of the trigeminus.

Auditory Nerve (VIII)

The root of the auditory nerve can be found just posterior to the facial nerve and ventral to the flocculus of the cerebellum. Note that the root of the auditory nerve rests on a low ridge, the trapezoid body, that runs downward and around the sides and ventral surface of the medulla oblongata. The facial nerve arises from the slight depression between trapezoid body and cerebellar peduncle. The abducens nerve arises from a more medial part of the trapezoid body.

Glossopharyngeal (IX), Vagus (X), Spinal Accessory (XI), and Hypoglossal (XII) Nerves

These four nerves arise from the medulla oblongata, but their close proximity to each other, and damaged condition resulting from the removal of the brain, can make their identification challenging.

The roots of the first three (IX, X, XI) are found in order (anterior to posterior) along the dorsolateral side of the medulla oblongata. The glossopharyngeal nerve, and vagus posterior to it, have small but defined stump-like roots. The spinal accessory, posterior to these two, arises as numerous tiny threadlike rootlets that combine to form the nerve. Rootlets of the last cranial nerve, hypoglossal (XII), are also tiny and threadlike. This final cranial nerve arises more medially and slightly caudal to the spinal accessory nerve, from the ventrolateral side of the medulla oblongata.

Brain Regions—Sagittal Sections (Figure 10.13)

This part of the dissection provides another view of structures already identified and reveals deep structures not yet described. However, it is equally intended to introduce the three brain regions, their subdivisions, and specific structures derived from each. Refer again to figure 10.9 to recall the five embryonic brain regions, discussed from posterior to anterior below.

Rhombencephalon (Hindbrain)

→ **Myelencephalon (= Medulla Oblongata)** The myelencephalon is composed of the medulla oblongata, the anatomical connection between the brain and spinal cord. It serves as a major reflex center and forms the base for cranial nerves (VII through XII).

Metencephalon Two parts compose the metencephalon, the pons and cerebellum. The outer surface of the cerebellum, the cerebellar cortex, is primarily highly folded gray matter. The finely branched white matter within is the arbor vitae, or "tree of life."

The fourth ventricle (figures 10.10, 10.13) lies beneath the cerebellum and is continuous posteriorly with the central canal of the spinal cord and anteriorly with

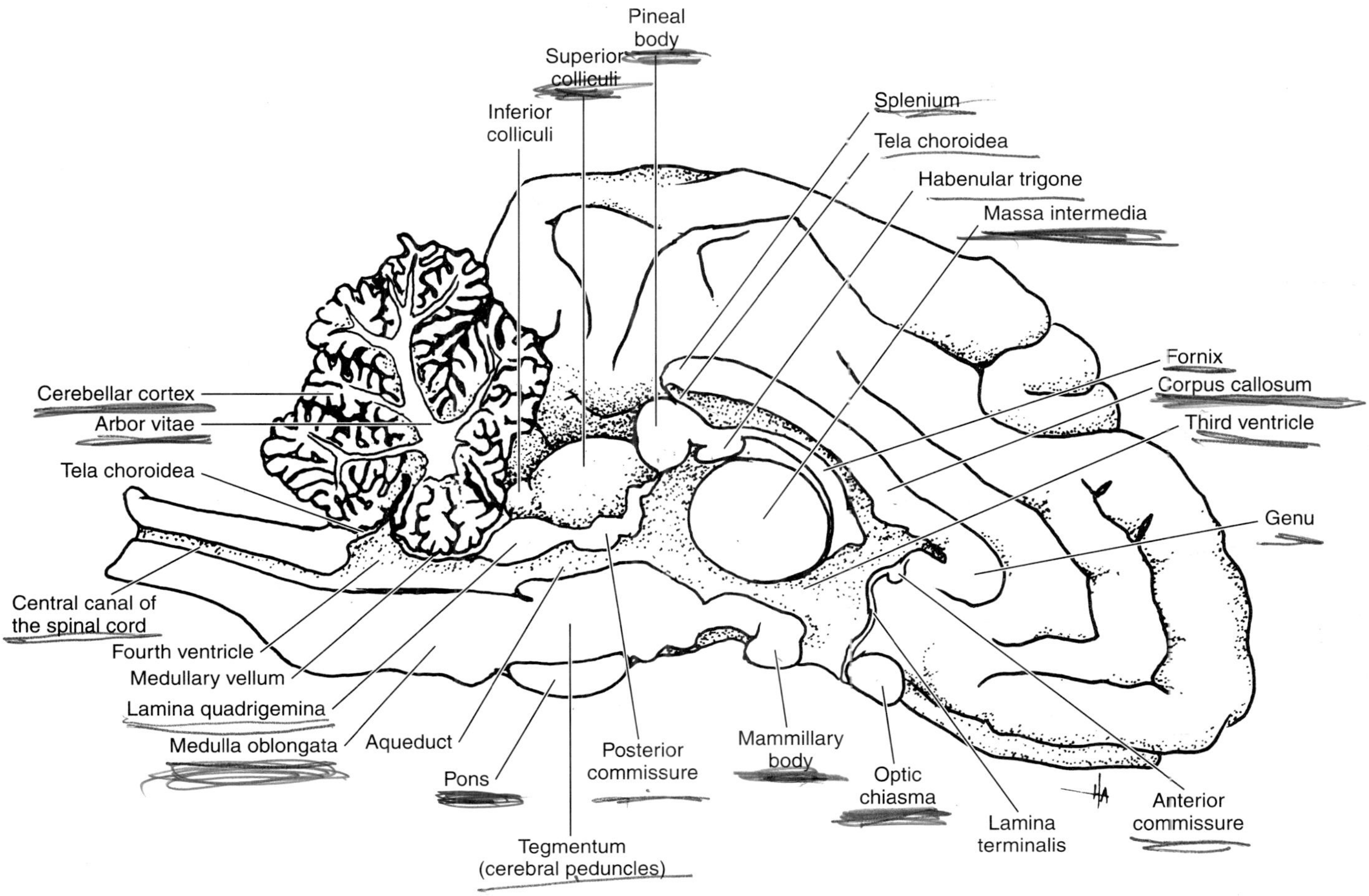

Figure 10.13
Sagittal view of the sheep brain.

the aqueduct of the midbrain. The thin membrane between the anterior base of the cerebellum and fourth ventricle is the medullary vellum. A similar membrane, the tela choroidea, occurs between the posterior base of the cerebellum and the fourth ventricle.

Mesencephalon (Midbrain)

→ **Corpora Quadrigemina** The roof, or tectum, of the midbrain over the aqueduct, differentiates into the lamina quadrigemina that supports two paired swellings, the superior and inferior colliculi (collectively the corpora quadrigemina).

→ **Cerebral Peduncle** The thick floor of the midbrain, just anterior to the pons, is the tegmentum or cerebral peduncles.

Prosencephalon (Forebrain)

→ **Diencephalon** The diencephalon encloses the third ventricle. The roof, sides, and floor compose the three basic regions of the diencephalon: epithalamus, thalamus, and hypothalamus, respectively. The most prominent part of the epithalamus is the small oval pineal body resting on the superior colliculus. The base of the pineal continues posteriorly into the posterior commissure, which in turn continues posteriorly into the lamina quadrigemina, a thicker strip just anterior to the cerebellum and below the superior colliculus. The base of the pineal continues anteriorly into the habenular trigone.

Form & Function

In fishes and amphibians, the midbrain is often the most prominent region of the brain. The tectum (roof) receives direct input from the eyes. In addition, information from the acousticolateralis system, the cerebellum, the olfactory epithelium, and the cutaneous sensors may be transmitted indirectly to the tectum. The tegmentum (floor) is also prominent in these vertebrates. In fact in some fishes, the tegmentum seems to be an important area associated with learned behaviors.

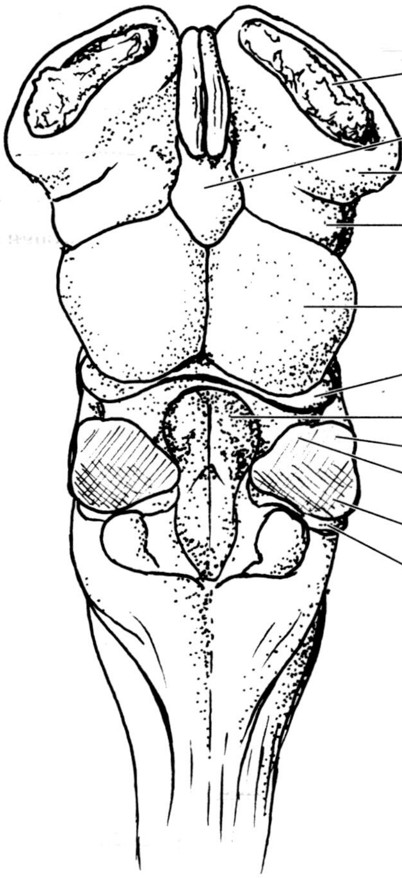

Figure 10.14
Dorsal view of the brain stem of a sheep brain.

Above the pineal lies another thin, vascular membrane, a tela choroidea. The largest part of the diencephalon, the thalamus, bears a prominent, circular, and medial enlargement, the massa intermedia (= intermediate mass), partially cut in making the sagittal section of brain.

Form & Function

The pineal organ participates in photoreception among lower vertebrates but tends to become an endocrine gland (thus, pineal "gland") in higher vertebrates. In lower vertebrates (including reptiles), it affects skin pigmentation and plays a role in regulating photoperiod. In birds and mammals, it helps to regulate biological rhythms.

→ The optic chiasma passes just ventral to a thin membrane, the lamina terminalis, the anterior boundary of the third ventricle. The tiny anterior commissure lies at the dorsal end of the lamina terminalis.

The ventral projection from the hypothalamus is the pituitary gland (= hypophysis), whose narrow stalk connecting it to the brain is the infundibulum. The mammillary body is a swelling found just posterior to the infundibulum.

Form & Function

The hypothalamus houses a collection of nuclei that regulate homeostasis to maintain the body's internal physiological balance. Homeostatic mechanisms adjusted by these nuclei pertain to temperature, water balance, appetite, metabolism, blood pressure, sexual behavior, alertness, and some aspects of emotional behavior. The hypothalamus stimulates the pituitary gland beneath it to regulate these homeostatic functions.

→ **Telencephalon (= Cerebrum)** The band of the corpus callosum connecting both hemispheres is seen in cross section. Both ends are curled and enlarged, forming a genu (anterior) and a splenium (posterior). The septum pellucidum is a thin membrane between the genu and splenium. It closes the entrance into the lateral ventricle of the cerebral hemisphere. Free the lower border of this septum from the fornix, a band of white matter seen in cross section and lying above the massa intermedia. The fornix begins anteriorly in the mammillary body (impractical to see this beginning in sagittal section alone). Pull the brain stem away from the hemisphere enough to allow you to follow the fornix posteriorly. The fornix sweeps around the inside of the cerebral hemisphere to end in the pyriform lobe. The entire free edge of this part of the fornix within the hemisphere is termed the fimbria. Fold out the edge of the fimbria to discover that its inner margin is greatly expanded. This expanded region is the hippocampus.

Brain Stem—Parts (Figure 10.14)

→ The brain regions, minus the cerebral hemispheres and cerebellum, collectively constitute the brain stem. Relocate structures already identified in whole and sagittal section, region by region.

Follow the optic tract as it sweeps upward along the side of the diencephalon. Most fibers reach the large rounded region on top of the thalamus. This swollen part of the thalamus is specifically termed the pulvinar nucleus. Optic fibers also synapse in the lateral geniculate body, a small lateral swelling at the base of the colliculi. The medial geniculate body lies immediately behind the lateral geniculate, but is not well defined here on the surface of the thalamus.

If still present, remove the cerebellum at its base. The pedicle by which the cerebellum attaches is formed by three peduncles: middle cerebellar peduncle (from pons along the lateral side of the myelencephalon), anterior cerebellar peduncle (parallel with and medial to the middle peduncle), and the posterior cerebellar peduncle or restiform (a short section completing the posterior side of the collective peduncle). The small, dorsal transverse swelling tucked in behind the posterior peduncle is the acoustic stria.

Brain—Optional Cross Sections

→ You may find it helpful to use cross sections to see structures from another view. Try following the caudate nucleus, fornix, hippocampus, corpus callosum, and other identified regions in the cerebral hemispheres through sequential cross sections. If you get lost, return to the earlier description to begin again.

In sections through the cerebral hemisphere, note the outer folded cortex.

Spinal Nerves

The dissection of the previous animal systems has made a thorough analysis of the spinal nerves difficult. However, in the cat, many cervical and thoracic spinal nerves should still be available for study. Spinal nerves often unite to form a complex network, or **plexus,** serving a broad region of the body. The last cervical and first thoracic spinal nerves form the **brachial plexus.** Similarly, the last lumbar and sacral spinal nerves form the **lumbosacral plexus.**

→ In the axillary region, identify the last cervical and first few thoracic spinal nerves that parallel the axillary artery to form the brachial plexus. These nerves will appear fibrous and have a tan or off-white color. In the lumbar region of the peritoneal cavity, identify several lumbar and sacral spinal nerves paralleling the femoral artery on the ventral side of the groin and thigh. These spinal nerves contribute to the lumbosacral plexus. The specific names of spinal nerves will not be followed.

Glossary

A

adaptation A phenotypic feature of an individual that contributes to that individual's survival; a feature's form/function and associated biological role with respect to a particular environment.

advanced *See* derived. *Compare* primitive.

afferent Bringing to, as sensory afferent fibers bringing impulses to the central nervous system. *Compare* efferent.

agnathan Vertebrate lacking jaws.

akinetic skull A skull lacking cranial kinesis.

amniote Vertebrate whose embryo is wrapped in an amnion.

amphystly Jaw suspension via two major attachments, the hyomandibula and the palatoquadrate.

analogy Features in two or more organisms performing a similar function; common function.

anamniote Vertebrate whose embryo lacks an amnion.

antagonist A muscle with an opposite action to other muscles. *Compare* synergist, fixator.

aponeurosis Broad, flat tendon.

archetype Fundamental type; basic underlying blueprint or model upon which the definitive animal or part of an animal is thought to be based.

archinephric duct General term for the urogenital duct that may go by alternative names (wolffian duct) at different embryonic stages (pronephric duct, mesonephric duct, opisthonephric duct) or in different functional roles (vas deferens).

artery Blood vessel carrying blood away from the heart; blood carried may be high or low in oxygen tension. *Compare* vein.

atrophy Decrease in size or density.

auditory Pertaining to the perception of sound.

autostyly Jaw suspension in which the jaws articulate directly with the braincase.

axon Nerve fiber of a neuron carrying an impulse away from the cell body.

B

baleen Keratinized straining plates arising from the integument in the mouth of some species of whales.

benthic Bottom dwelling.

bilateral symmetry A body in which left and right halves are mirror images of each other.

biological role How the form and function of a part performs in an environmental context contributing to survival. *Compare* function.

biomechanics The study of how physical forces affect and are incorporated into animal designs.

bipedal Walking or running by means of only the two hindlegs. *Compare* quadrupedal.

bolus Soft mass of food in the mouth or stomach. *Compare* chyme.

brachyodont Low crown teeth. *Compare* hypsodont.

braincase That part of the skull that contains the cranial cavities; part of the skull housing the brain.

brain stem Posterior part of brain composed of midbrain, pons, medulla.

branchial basket The expanded chordate pharynx serving suspension feeding.

bunodont Teeth with shape cusps. *Compare* lophodont, selenodont.

C

calcification The process of calcium deposition in tissue. *Compare* ossification.

carapace Dorsal, dome-shaped bony part of a turtle shell. *Compare* plastron.

carnassials Sectorial teeth of carnivores including upper premolar and lower molar. *See* sectorial teeth.

caudal Toward the tail or back end of the body; posterior.

cecum Blind-ended outpocketing from the intestines.

cementum Cellular and acellular layer on teeth, usually located on the roots but in some herbivores may contribute to the occlusal surface. *See* enamel, dentin.

central nervous system The brain and spinal cord.

centrum The body or base of a vertebra.

ceratotrichia Fan-shaped array of keratinized rods internally supporting the elasmobranch fin. *Compare* lepidotrichia.

chemoreceptor Sense organ responsive to chemical molecules. *Compare* radiation receptor, mechanoreceptor.

chiasma Crossing of fibers.

choana The internal naris; the openings of the nasal passage into the mouth.

chondrocranium That part of the skull formed by endochondral bone or cartilage that underlies and supports the brain, plus the fused or associated nasal capsules.

chromatophore General term for a pigment cell.

chyme The liquefied bolus of partially digested food after it leaves the stomach and enters the intestine; digesta. *Compare* bolus.

claws Sharp, curved laterally compressed nail at the end of a digit; talon.

coelom Fluid-filled body cavity formed in mesoderm.

copulation Coitus involving an intromittent organ.

coracoid Posterior coracoid; first evolved in early synapsids or their immediate ancestors. *Compare* procoracoid.

cornified Having a layer of keratin; keratinized.

cortex The outer portion or rim of an organ.

cosmine Older term for a derivative of dentin that covers some fish scales; cosmoid scale.

cranial Toward the head or front end of the body; anterior, rostral.

cranial kinesis Movement between the upper jaw and braincase about joints between them; in restricted sense, skulls with a movable joint across the roofing bones. *Compare* akinetic skull, prokinesis, mesokinesis, metakinesis.

cranial nerve Any nerve entering or departing from the brain.

crop Baglike expansion of the esophagus.

cutaneous respiration Gas exchange directly between blood and the environment through the skin.

D

deglutition The act of swallowing.

dendrite Nerve fiber of a neuron carrying impulses toward the cell body.

dendrograms A branching diagram representing relationships or history of a group of organisms.

dental formula Shorthand expression of the characteristic number of each type (incisor, canine, premolar, molar) of upper and lower teeth in a mammalian species.

dentin Forms the bulk of the tooth; similar in structure to bone, but harder; yellowish in color; composed of inorganic hydroxyapatite crystals and of collagen; secreted by odontoblasts of neural crest origin. *See* enamel, cementum.

dentition A set of teeth.

derived Denoting that an organism or species evolved late within its phylogenetic lineage; advanced; opposed to primitive.

dermal papilla The part of the tooth-forming primordium that is derived from neural crest cells, becomes associated with the enamel organ, and differentiates into odontoblasts that secrete dentin. *See* enamel organ.

dermatocranium That part of the skull formed from dermal bones.

dermis Skin layer beneath the epidermis derived from mesoderm.

design The structural and functional organization of a part related to its biological role.

diffusion The movement of molecules from an area of high to an area of low concentration; if the movement is random and unaided, it is passive diffusion.

digestion The mechanical and chemical breakdown of foods into their basic end products that can be absorbed, usually simple carbohydrates, proteins, fatty acids.

digitigrade Foot posture in which the balls of the feet, middle of the digits, support the weight (as in cats). *Compare* plantigrade, unguligrade.

dikinetic skull Kinetic skull with two joints passing transversely through the braincase. *Compare* monokinetic.

dioecious Female and male gonads in separate individuals.

diphyodont Pattern of tooth replacement involving only two sets of teeth, usually milk teeth and permanent teeth.

diplospondyly The condition in which a vertebral segment is composed of two centra. *Compare* monospondyly.

dissection The careful exposure of anatomical parts for the purpose of allowing students to discover and master for themselves the extraordinary morphological organization of an animal, in order to understand the processes these parts perform and the remarkable evolutionary history out of which they come. Pronounced *dis*-section, as opposed to *di*-section, which is chopping into two halves.

distal Toward the free end of an attached part such as the limb. *Compare* proximal.

dorsal Toward the back or upper surface of the body; opposite of ventral.

E

ecomorphology The study of the relationship between the form and function of a part and how it is actually used in a natural environmental setting; the basis for determining biological role.

effector An organ responding to nervous stimulation; muscle or gland.

efferent Carrying away, as motor neurons carry impulses away from the central nervous system.

electric organ Specialized block of muscle producing electrical fields and often high jolts of voltage.

emargination Large notches in the bony braincase. *Compare* fenestra.

emulsify To break up fats into smaller droplets. *Compare* digestion.

enamel Forms the occlusal cap on most teeth. Hardest substance in vertebrate body, consisting almost entirely of calcium salts as apatite crystals; secreted by ameloblasts of epidermal origin. *See* dentin, cementum.

enamel organ The part of the tooth-forming primordium that is derived from epidermis, becomes associated with the dermal papilla, and differentiates into the ameloblasts that secrete the enamel. *See* dermal papilla.

endochondral bone formation Embryonic formation of bone preceded by a cartilage precursor that is subsequently ossified; cartilage bone, replacement bone. *Compare* intramembranous bone formation.

endocrine Denoting a gland that releases its product directly into blood vessels. *Compare* exocrine.

endoskeleton Supportive or protective framework beneath the integument; within the body. *Compare* exoskeleton.

endothelium Single-celled lining to vascular channels.

ependyma The layer of cells lining the central canal of the chordate spinal cord.

epidermis Skin layer over the dermis derived from ectoderm.

epigenetics Above the genes; the study of developmental events above the level of the genes; the embryonic processes not directly arising from the genes that contribute to the developing phenotype.

epiphysis 1. Secondary center of ossification on the end of a bone; the end of a bone. 2. pineal gland.

evolutionary morphology The study of the relationship between change in anatomical design through time and the processes responsible.

excretion Removal from the body of wastes and substances in excess.

exocrine Denoting release of secretion into ducts. *Compare* endocrine.

exoskeleton Supportive or protective framework lying on the outside of the body. *Compare* endoskeleton.

extant Living.

extinct Dead.

extrinsic Originating outside the part on which it acts. *Compare* intrinsic.

F

fascicle Within a muscle organ, a bundle of muscle fibers defined by a connective tissue coat.

fenestra An opening within the bony braincase.

fermentation Anaerobic extraction of energy from food in vertebrates by microorganisms that release cellulases to enzymatically break down plant material.

fin An external, projecting plate or membrane from the body of an aquatic animal (as in fish).

fixator A muscle functioning to stabilize a joint. *Compare* synergist, antagonist.

foramen A perforation or hole through bone or cartilage.

foramen of ovale One-way connection between right and left atria in an embryonic mammal; closes at birth.

foramen of Panizza Connecting vessel between the bases of left and right aortic arches in crocodilians.

foregut Anterior embryonic gut that gives rise to pharynx, esophagus, stomach, and anterior intestine. *Compare* hindgut.

foregut fermentation *See* gastric fermentation.

frontal plane A plane passing from one side to the other of an organism so as to divide the body into dorsal and ventral parts. *Compare* transverse plane.

function How a part performs within the organism. *Compare* biological role.

functional morphology The study of the relationship between the anatomical design of a structure and the function or functions it performs.

G

gametogenesis The production of gametes.

ganglion Gathering of nerve cell bodies within the peripheral nervous system.

ganoin Older term for a derivative of enamel that covers some fish scales; ganoid scale.

gas bladder Gas-filled bag in fish derived from the gut. Because the composition of the gas may vary, the term "air bladder" is less appropriate. *Compare* swim bladder.

gastralia Rib-shaped dermal bones located in the abdominal region.

gastric fermentation Digestion involving microbial fermentation centered near specialized stomach; foregut fermentation. *See* intestinal fermentation.

gill Aquatic respiratory organ.

gill slit Pharyngeal slit associated with a gill.

gizzard An especially well muscularized region of the stomach used to grind hard foods.

gnathostome A vertebrate with jaws.

H

heterochrony Within an evolutionary lineage, the change in time of embryonic appearance of an organ relative to its time of embryonic appearance in a related member of the phylogeny. Usually heterochrony is concerned with the time of onset of sexual maturity relative to somatic development. *Compare* paedomorphosis.

heterodont Dentition in which the teeth are different in general appearance.

hindgut Posterior embryonic gut that gives rise to the posterior intestines. *Compare* foregut.

hindgut fermentation *See* intestinal fermentation.

holospondyly The condition in which vertebral elements, centra and spines, are anatomically fused into a single piece.

homeostasis The maintaining of a relatively stable internal environment within an organism.

homodont Dentition in which the teeth are similar in general appearance.

homology Features in two or more organisms resulting from the derivation of those features from common ancestors; common ancestry. *Compare* serial homology.

homoplasy Features in two or more organisms that look alike; similar appearance.

hoof Enlarged cornified plate on the end of an ungulate digit.

horn Unbranched keratinized sheath with a bony core located on the head; horns usually occur in both males and females and are retained year round.

hydrofoil Any object that produces lift when placed in a moving stream of water (as a shark pectoral fin).

hydrostatic organ A structure whose mechanical integrity depends upon a fluid-filled core enclosed by defining walls of connective tissue.

hydrostatic pressure Fluid force, as in blood resulting from heart contraction.

hyostyly Jaw suspension primarily through attachment with the hyomandibula.

hypsodont High crown teeth. *Compare* brachyodont.

I

incus Mammalian, middle ear bone, derived phylogenetically from the quadrate.

insertion For a muscle, its relatively movable point of attachment.

integument The skin covering the body

intervertebral body A pad of cartilage or sometime fibrous connective tissue between articular ends of successive vertebral centra.

intervertebral disk In the adult mammal, a pad of fibrocartilage with a gel-like core derived from the notochord and located between articular ends of successive vertebral centra. *Compare* intervertebral body.

intestinal fermentation Digestion involving microbial fermentation centered in the intestines; hindgut fermentation, cecal fermentation. *See* gastric fermentation.

intramembranous bone formation Embryonic formation of bone directly from mesenchyme without a cartilage precursor; dermal bone. *Compare* endochondral bone formation.

intratarsal joint An archosaur ankle type in which the line of flexion passes *between* calcaneus and astragalus. *Compare* mesotarsal joint.

intrinsic Belonging entirely to the part. *Compare* extrinsic.

intromittent organ Male reproductive organ inserted into the female reproductive tract to deliver sperm; penis or phallus.

invagination Indentation or infolding of the surface.

ischemia When blood flow to a region is insufficient to meet metabolic demands of the tissue.

isometry Geometric similarity in which proportions remain constant with a change in size.

J

jaws Skeletal elements of bone or cartilage that reinforce the lower borders of the mouth.

K

keratin Fibrous protein.

keratinization The process by which the proteins, especially keratin, are formed in the skin.

kinesis Denotes movement, usually in reference to relative movement of skull bones. *See* cranial kinesis.

L

lactation Release of milk from mammary glands to suckling young.

lamina Thin sheet, layer, or plate—e.g., gill lamella.

larva An immature (nonreproductive) stage morphologically different from the adult.

lateral Toward or on the side of the body.

lepidotrichia Fan-shaped array of ossified or chondrified dermal rods internally supporting the fin of bony fishes.

lepospondyly A holospondylous vertebra with a husk-shaped centrum usually pierced by a notochordal canal. *Compare* ceratotrichia.

lift The force produced by an airfoil perpendicular to its surface.

lingual feeding Prey are captured by use of the tongue.

load In mechanics, the forces to which a structure is subjected.

lophodont Teeth with broad, ridged cusps useful in grinding plant material. *Compare* bunodont, selenodont.

lymph Clear fluid carried in lymphatic vessels.

lymphoid tissue Blood-forming tissue outside of bone cavities—in the spleen, lymph nodes, etc.

M

malleus In mammals, one of the three middle ear bones phylogenetically derived from the articular.

mastication Chewing of food; mechanical breakdown of a large bolus of food into smaller pieces usually by use of teeth.

meatus A canal or channel.

mechanoreceptor Sense organ responsive to small changes in mechanical force. *Compare* chemoreceptor, photoreceptor.

medulla The inner portion or core of an organ.

merycism Remastication together with microbial fermentation in nonruminants. *Compare* rumination.

mesenchyme Loosely associated cells of mesodermal origin.

mesokinesis Skull with a transverse joint passing through the dermatocranium posterior to the orbit. *Compare* prokinesis, metakinesis.

mesonephros Kidney formed of nephric tubules arising in the middle of the nephric ridge; usually a transient embryonic stage replacing the pronephros, but itself replaced by the adult opisthonephros or metanephros. *Compare* pronephros, opisthonephros, metanephros.

mesotarsal joint An archosaur ankle type in which the calcaneous and astragalus fuse and the line of flexion passes between them and the distal tarsals. *Compare* intratarsal joint.

metakinesis Skull with a transverse hinge posteriorly between the deep neurocranium and outer dermatocranium. *Compare* prokinesis, mesokinesis.

metamorphosis Abrupt transformation from one anatomically distinct stage (juvenile) to another (adult).

metanephric duct Ureter; distinct from pronephric and mesonephric ducts.

metanephros Kidney formed of nephric tubules arising in the posterior region of the nephric ridge drained by a ureter; usually replaces embryonic pronephros and mesonephros. *Compare* pronephros, mesonephros, opisthonephros.

metapterygial fin Basic fin type in which the axis (metapterygial stem) is located posteriorly in the fin.

metapterygial stem The chain of endoskeletal elements within the fish fin that define the major internal supportive axis.

microcirculation The capillary beds plus the arterioles that supply them and the venules that drain them.

midsagittal plane Median sagittal plane passing through the central long axis of the body.

molariform Cheek teeth; general term for similar appearing premolar and molar teeth.

molt The shedding of parts or all of the cornified layer of the epidermis; loss of feathers or hair, usually on an annual rhythm; ecdysis.

monokinetic skull Kinetic skull with a single transverse joint passing through the braincase.

monospondyly The condition in which a vertebral segment is composed of one centrum. *Compare* diplospondyly.

morph A term referring to the general form or design of an animal—for example, juvenile morph (tadpole) and adult morph (sexually mature stage) of a frog.

morphological cross section The cross-sectional area of a structure perpendicular to its long axis, as in muscles. *Compare* physiological cross section.

morphology The study of anatomy and its significance.

motor neuron A nerve cell carrying impulses to an effector organ. *Compare* sensory neuron.

motor pattern A defined, local pattern of activity produced by muscles that shows little variation when repeated.

motor unit One motor neuron and the subset of muscle fibers it supplies; important in producing graded muscle force.

mucous gland An organ secreting a protein-rich mucin that is usually a thick fluid. *Compare* serous gland.

muscle fiber A muscle cell; a contractile part of a muscle organ.

muscle organ Muscle cells together with noncontractile tissues that support them (connective tissue, blood vessels, nerves).

myeloid tissue Blood-forming tissue housed inside bones.

myofibril Contractile unit in a muscle cell; a chain of repeating sarcomeres, composed of myofilaments.

myofilament Composed of myosin and actin.

myomeres Differentiated segments of adult body muscle.

myotomes Undifferentiated, embryonic blocks of presumptive muscle.

N

naris A nostril.

natural selection The process by which organisms with poorly suited features, on average, fare less well in a particular environment and tend to perish, thereby leaving (preserving) those individuals with more favorable adaptations; survival of the fittest.

neomorph Denoting in a derived species a new morphological structure that has no equivalent evolutionary antecedent.

neoteny Paedomorphosis produced by delaying the onset of somatic development.

nephridium Tubular excretory organ.

nephron The portion of the uriniferous tubule concerned with the formation of urine, composed of proximal, intermediate, and distal regions; nephric tubule.

nephrotome Segmental, early embryonic urinary structure that is the forerunner of a nephron.

network Anything reticulated or decussated at equal distances, with interstices between the intersections.

neural crest A paired strip of tissue that pinches off from the dorsal edges of the neural groove as it forms into the neural tube.

neurocranium That part of the braincase that contains cavities for the brain and associated sensory capsules (nasal, optic, otic).

neuroglia Nonnervous supportive cells of the nervous system.

neurotransmitter Chemical released into the synapse at the end of a nerve fiber, usually an axon.

nerve Collection of nerve fibers coursing together in the peripheral nervous system.

notochord Long, axial rod composed of a fibrous connective tissue wall around cells and/or a fluid-filled space.

nucleus 1. membrane-bound organelle within the body of a cell. 2. Gathering of nerve cell bodies within the central nervous system.

O

occlusion Closure; the meeting of opposite tooth rows.

olfaction The act of smelling.

ontogeny The course of embryonic development of an individual from egg to death of the adult.

operculum Lid or cover, as over the gills of fishes.

opisthonephros The adult kidney formed from the mesonephros together with additional tubules from the posterior region of the nephric ridge. *Compare* pronephros, mesonephros, metanephros.

origin For a muscle, its relatively fixed point of attachment. *Compare* insertion.

osmoregulation Active maintenance of water and solute levels.

ossification The process of bone formation; the appearance of bone cells and their surrounding matrix. *Compare* calcification.

osteoderm A dermal bone located under and in support of an epidermal scale.

osteon Highly ordered arrangement of bone cells into concentric rings and bone matrix around a central canal carrying blood vessels and nerves; haversian system.

otolith Single calcareous mass in the cupula of hair cells.

oviduct Urogenital duct transporting ova and often involved in protection and nourishment of the embryo; müllerian duct.

oviparity Egg laying.

oviposition The act of laying eggs.

ovulation Release of the ovum from the ovary.

P

paedomorphosis Juvenile-like. The general retention of juvenile features of ancestors in the late developmental stages of descendants. *See* neoteny.

parallel muscle A muscle organ in which all its muscle fibers lie in the same direction and are aligned with its long axis. *Compare* pinnate muscle.

parasagittal plane A sagittal plane parallel with the midsagittal plane.

parturition The act of giving birth via viviparity. *Compare* oviposition.

pentidactyl Five digits per limb; thought to be the basic pattern characteristic of tetrapods, but modified by functional demands.

peripheral nervous system The cranial and spinal nerves and their associated ganglia; the part of the nervous system outside the central nervous system.

peristalsis Muscular contractions in the walls of a tubular structure, as in the digestive tract.

pharyngeal slit An elongate opening in the lateral wall of the pharynx.

photoreceptor Radiation receptor responsive to visible light.

phylogeny The course of evolutionary change within a related group of organisms.

physiological cross section In muscles, the total cross-sectional area of all muscle fibers perpendicular to their long axes. *Compare* morphological cross section.

pinnate muscle A muscle organ in which all its muscle fibers are aligned oblique to its long axis. *Compare* parallel muscle.

plantigrade Foot posture in which the entire sole is in contact with the ground. *Compare* digitigrade, unguligrade.

plastron Ventral bony part of turtle shell. *Compare* carapace.

platysma Unspecialized muscle derived from hyoid arch musculature that spreads as a thin subcutaneous sheet into the neck and over the face.

plexus A network of intermingling parts, as in blood vessels or nerves.

polydactyly An increase in the number of digits over the basic pendidactylous number. *Compare* polyphalangy.

polyphalangy An increase in the number of phalanges in each digit. *Compare* polydactyly.

polyphyodont Pattern of continuous tooth replacement. *Compare* diphyodont.

polyspondyly The condition in which a vertebral segment is composed of two or more centra.

portal system Set of venous vessels beginning and ending in capillary beds.

preadaptation The concept that features possess the necessary properties of form and function to meet the demands of a particular environment before the organism experiences that particular environment.

prehension Prey capture characterized by rapid grasping of the prey, usually with the jaws or with claws.

primitive Denoting that an organism or species appeared early within its phylogenetic lineage; opposite of derived.

procoracoid Anterior coracoid (= precoracoid); first evolved in fishes. *Compare* coracoid.

proctodeum The embryonic invagination of surface ectoderm that contributes to the hindgut, usually giving rise to the cloaca.

project In the nervous system, to send neural impulses to.

prokinesis Skull with a transverse joint passing through the dermatocranium anterior to the orbit. *Compare* mesokinesis, metakinesis.

pronephros Kidney formed of nephric tubules arising in the anterior region of the nephric ridge; usually transient embryonic stage. *Compare* mesonephros, opisthonephros, metanephros.

proximal Toward the base of an attached part where it joins the body. *Compare* distal.

pterylae Feather tracks.

Q

quadrupedal Walking or running by means of four legs. *Compare* bipedal.

R

radial symmetry A regular arrangement of the body about a central axis.

radiation receptor Sense organ responsive to light and other forms of electromagnetic radiation.

reflex Involuntary action mediated by the nervous system.

reticulum The second of four chambers in the complex ruminant stomach; a specialized region of the esophagus. *Compare* rumen.

rhachitomous vertebra An aspidospondylous vertebra characteristic of some crossopterygians and some early amphibians.

rumen The first of four chambers in the complex ruminant stomach; an expanded specialization of the esophagus. *Compare* reticulum.

ruminant Placental mammals with a specialized expansion of the digestive tract, the rumen, for processing of plant material; Ruminatia.

rumination Remastication together with microbial fermentation in ruminants.

S

sagittal plane Any plane parallel with the long axis of the animal's body, oriented dorsoventrally.

secondary cartilage Cartilage that forms after initial bone ossification is complete; formed usually in response to mechanical stress, especially that cartilage that forms on the margins of intramembranous bone.

sectorial teeth Teeth with opposing sharp ridges specialized for cutting.

segmentation A body made up of repeating sections or parts; metamerism.

selection force or **pressure** The biological or physical demands arising from the environment that affect the survival of the individual living there.

selenodont Teeth with crescent-shaped cusps, as in artiodactyls. *Compare* bunodont, lophodont.

sensory neuron A nerve cell carrying responses from a sensory organ. *Compare* motor neuron.

serial homology Similarity between successively repeated features in the same individual.

serial theory The hypothesis that jaws evolved from one of the anterior branchial arches.

sesamoid bone A bone that develops directly in tendons.

sinusoids Tiny vascular channels slightly larger than capillaries lined or partially lined only by endothelium.

somatic Pertaining to the body, usually the skeleton, its muscles, and the skin, but not the viscera.

spermatophore A reproductive structure capped by sperm, used to deliver sperm to a female.

sphincter A band of muscle around a tube or opening, functioning to constrict or close it.

spiracle A reduced gill slit, first in series.

splanchnocranium That part of the skull arising first in support of the pharyngeal slits and later contributing to the jaws and other structures of the head; branchial arches and derivatives; visceral cranium.

stapes In mammals, one of the three middle ear bones phylogenetically derived from the columella (hyomandibula).

stereospondylous vertebra A monospondylous vertebra in which the single centrum (an intercentrum) is separate (aspidospondylous).

stigmata Extensively subdivided pharyngeal slit.

stolon Rootlike process of ascidians and other invertebrates that may fragment into pieces that asexually grow into more individuals.

stratified Formed of layers.

streptostyly The condition in which the quadrate is movable relative to the braincase.

suspension feeding Feeding based upon filtering suspended food from water, usually involving cilia and secreted mucus; filter feeding, cilia-mucus feeding.

swim bladder Gas bladder functioning primarily in buoyancy control.

synapse The region of contact between one nerve fiber and another, a nerve fiber and a neuron, or a nerve fiber and an effector.

synapticules Cross-linking connections between pharyngeal bars in amphioxus.

synergist Two or more muscles cooperating to produce motion in the same direction. *Compare* antagonist, fixator.

T

talons Specialized bird claws used in striking or catching live prey.

tendon Non-contractile, fibrous connective tissue band connecting a muscle organ to bone. *Compare* aponeurosis.

thermoreceptor Radiation receptor sensitive to infrared energy.

tract Collection of nerve fibers coursing together in the brain or spinal cord. *Compare* nerve.

transverse plane A plane passing from one side to the other of an organism so as to divide the body into anterior and posterior parts.

transverse process A general term for any bony or cartilaginous projection from the centrum or neural arch.

tusk Specialized, long teeth protruding from the mouth; elongate incisors (elephants), left upper incisor (narwhal), canines (walruses).

tympanum Ear drum; tympanic membrane.

U

ungulate Hoofed placental mammals belonging to the orders Perissodactyla (horses) and Artiodactlya (cattle, deer, pigs etc.).

unguligrade Foot posture in which the weight is carried on the tips of the toes (as in horses).

ureter Metanephric duct; urinary duct arising as a ureteric diverticulum and draining the metanephros.

V

vasoconstriction Narrowing a blood vessel; usually resulting from active smooth muscle contraction. *Compare* vasodilation.

vasodilation Widening a blood vessel; may be active or passive enlargement. *Compare* vasoconstriction.

vein Blood vessel carrying blood toward the heart; blood carried may be low or high in oxygen tension. *Compare* artery.

ventilation The active movement of water or air across respiratory exchange surfaces.

ventral Toward the belly or bottom of an animal; opposite of dorsal.

vertebra One of several bone or cartilage blocks firmly joined into a backbone, defining the major body axis of vertebrates.

villus A fingerlike projection of a tissue layer, as in the small intestine.

viviparity Live birth.

W

wolffian duct Mesonephric duct.

Z

zygapophysis Projection of a neural arch that establishes an articulation with the adjacent neural arch.

Credits

Chapter 2

Figure 2.4A: R.A. Cloney, "Ascidian larvae and the events of metamorphosis," *American Zoologist* 22:817–826, 1982. American Society of Zoologists. Reprinted by permission; Figure 2.4c: Copyright © Kenneth V. Kardong.

Chapter 3

Figures 3.6A-C, 3.8, and 3.10: © Kenneth V. Kardong and Edward J. Zalisko.

Chapter 4

Figure 4.3: Redrawn from M.H. Wake, ed., *Hyman's Comparative Vertebrate Anatomy,* 3d ed., © 1979 by The University of Chicago. All rights reserved. Reprinted by permission; Figure 4.11B: Copyright © Kenneth V. Kardong.

Chapter 5

Figures 5.9, 5.11A-C: © Kenneth V. Kardong and Edward J. Zalisko; Figure 5.11D: © Kenneth V. Kardong and Edward J. Zalisko; Redrawn from Edwin S. Goodrich, *Studies on the Structure & Development of Vertebrates,* 1930 London: Macmillan; and republished 1986, The University of Chicago Press, figure 80, p. 70; Figures 5.12, 5.13, 5.14, 5.16BC, 5.17, 5.18, 5.19, 5.20: © Kenneth V. Kardong and Edward J. Zalisko; Figure 5.21: © Kenneth V. Kardong and Edward J. Zalisko; Redrawn after Alexis L. Romanoff, *The Avian Embryo,* figure 37.1, p. 1010, The MacMillan Company; after Lebedinsky, 1913; Figures 5.22, 5.24, 5.29, 5.31, 5.32, 5.33, 5.35: © Kenneth V. Kardong and Edward J. Zalisko.

Chapter 6

Figures 6.6, 6.7: © Kenneth V. Kardong and Edward J. Zalisko; Figure 6.8: Copyright © Kenneth V. Kardong; Figures 6.9, 6.11, 6.12, 6.13, 6.14, 6.15, 6.16, 6.17, 6.18, 6.19AB, 6.20, 6.21: © Kenneth V. Kardong and Edward J. Zalisko.

Chapter 7

Figures 7.3, 7.5, 7.6, 7.7, 7.8, 7.9, 7.10: © Kenneth V. Kardong and Edward J. Zalisko.

Chapter 8

Figures 8.2, 8.3, 8.5, 8.6, 8.9, 8.10, 8.12, 8.13, 8.14: © Kenneth V. Kardong and Edward J. Zalisko.

Chapter 9

Figure 9.1: Redrawn from Ari van Tienhoven: *Reproductive Physiology of Vertebrates, Second Edition.* Copyright © 1983 by Cornell University Press. Used by permission of the publisher; Figures 9.2, 9.3, 9.4, 9.5, 9.6, 9.7, 9.9: © Kenneth V. Kardong and Edward J. Zalisko.

Chapter 10

Figures 10.11, 10.12, 10.13, 10.14: © Kenneth V. Kardong and Edward J. Zalisko.

Index

B

C

D

E

F

G

H

I

M

P

Q

R

S

T